Student Solutions Manual

for

Tussy/Gustafson's

Prealgebra

Third Edition

Brian F. Sucevic
Valencia Community College

THOMSON

BROOKS/COLE

Australia • Brazil • Canada • Mexico • Singapore • Spain • United Kingdom • United States

Printed in the United States of America
1 2 3 4 5 6 7   09 08 07 06 05

Printer: Thomson/West Group

ISBN: 0-534-40291-7
Cover Image: Eric Bean/Getty Images

Thomson Higher Education
10 Davis Drive
Belmont, CA 94002-3098
USA

For more information about our products,
contact us at:
Thomson Learning Academic Resource Center
1-800-423-0563

For permission to use material from this text or product, submit a request online at
http://www.thomsonrights.com.
Any additional questions about permissions can be submitted by email to **thomsonrights@thomson.com.**

TABLE OF CONTENTS

Chapter 1 Whole Numbers

1.1 An Introduction to Whole Numbers

Vocabulary

1. A <u>set</u> is a collection of objects.

3. When 297 is written as 2 hundreds + 9 tens + 7 ones, it is written in <u>**expanded**</u>   notation.

5. Using a process known as graphing, whole numbers can be represented as points on the <u>**number**</u> line.

Concepts

7. 3   9. 6   11. whole numbers

13.

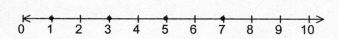

15.

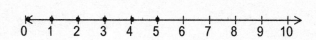

17. 47 > 41   19. 309 > 300

21. 2,052 < 2,502   23. Since 4 < 7, it is also true that 7 > 4.

Notation

25. The symbols { }, called <u>**braces**</u>, are used when writing a set.

Practice

27. 2 hundreds + 4 tens + 5 ones; two hundred forty-five

29. 3 thousands + 6 hundreds + 9 ones; three thousand six hundred nine

31. 3 ten thousands + 2 thousands + 5 hundreds; thirty-two thousand five hundred

33. 1 hundred thousand + 4 thousands + 4 hundreds + 1 one; one hundred four thousand four hundred one

35. 425   37. 2,736   39. 456

41. 27,598   43. 9,113   45. 10,700,506

47. 79,590   49. 80,000   51. 5,926,000

53. 5,900,000   55. $419,160   57. $419,000

Applications

59.   EATING HABITS

Country	Pounds
United States	261 lb
New Zealand	259 lb
Australia	239 lb
Cyprus	236 lb
Uruguay	230 lb
Austria	229 lb
Saint Lucia	222 lb
Denmark	219 lb
Canada	211 lb
Spain	211 lb

61a.   MISSIONS TO MARS

The 70's were most successful decade and had 7 successful missions.

61b.   The 60's were the most unsuccessful decade and had 9 unsuccessful missions.

63.   ENERGY RESERVES

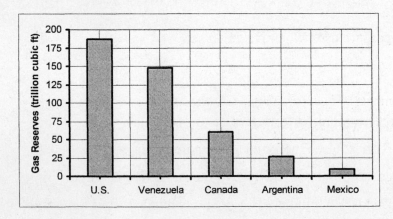

65.   COFFEE

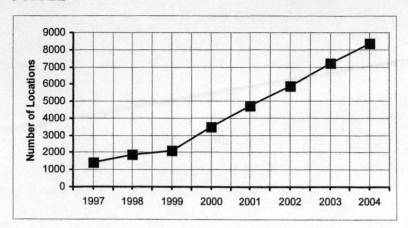

67a.   Fifteen thousand six hundred one and $\frac{00}{100}$, $15,601.00.

67b.   Three thousand four hundred thirty three and $\frac{00}{100}$, $3,433.00.

69.   EDITING

1,865,593; 482,880; 1,503; 269;

43,449

71a.   SPEED OF LIGHT   71b.   300,000,000 m/s

299,800,000 m/s

Writing

73.   They are the first numbers with which people learn to count.

75.   The low 130's would be $130,000-$134,999.

1.2 Adding and Subtracting Whole Numbers

Vocabulary

1. When two numbers are added, the result is called a **sum**. The numbers that are to be added are called **addends**.

3. When two numbers are added, the result is called a **sum**.

5. When two numbers are subtracted, the result is called a **difference**. In a subtraction problem, the **subtrahend** is subtracted from the **minuend**.

7. The property that allows us to group numbers in an addition in any way we want is called the **associative** property of addition.

Concepts

9. Commutative property of addition

11. Associative property of addition

13a. $x + y = y + x$

13b. $(x + y) + z = x + (y + z)$

15. Any number added to **zero** stays the same.

17. $4 + 3 = 7$

Notation

19. The grouping symbols ( ) are called **parenthesis**.

21. $(36 + 11) + 5 = 47 + 5$
$$= 52$$

23. $25 + 13 = 38$

25. $156 + 305 = 461$

27. $19 + 39 + 53 = 111$

29. $(95 + 16) + 39 = 111 + 39$
$$= 150$$

31. $25 + (321 + 17) = 25 + 338$
$$= 363$$

33.
$$\begin{array}{r} 632 \\ + \ 347 \\ \hline 979 \end{array}$$

35.
$$\begin{array}{r} 1,372 \\ + \ \ 613 \\ \hline 1,985 \end{array}$$

37.
$$\begin{array}{r} 6,427 \\ +\ 3,573 \\ \hline 10,000 \end{array}$$

39.
$$\begin{array}{r} 8,539 \\ +\ 7,368 \\ \hline 15,907 \end{array}$$

41.
$$\begin{array}{r} 1,246 \\ 578 \\ +\ \ \ 37 \\ \hline 1,861 \end{array}$$

43.
$$\begin{array}{r} 3,156 \\ 1,578 \\ +\ \ 578 \\ \hline 5,312 \end{array}$$

45. $P = \text{Perimeter}$
$P = 32 + 12 + 32 + 12$
$P = 88 \text{ ft}$

47. $P = 17 + 17 + 17 + 17$
$P = 68 \text{ in.}$

49. $17 - 14 = 3$

51. $39 - 14 = 25$

53. $174 - 71 = 103$

55. $633 - (598 - 30) = 633 - 568$
$\qquad\qquad\qquad\qquad = 65$

57. $160 - 15 - 4 = 145 - 4$
$\qquad\qquad\quad = 141$

59. $29 - 17 - 12 = 12 - 12$
$\qquad\qquad\quad = 0$

61.
$$\begin{array}{r} 367 \\ -343 \\ \hline 24 \end{array}$$

63.
$$\begin{array}{r} 423 \\ -305 \\ \hline 118 \end{array}$$

65.
$$\begin{array}{r} 1,537 \\ -\ \ 579 \\ \hline 958 \end{array}$$

67.
$$\begin{array}{r} 4,267 \\ -2,578 \\ \hline 1,689 \end{array}$$

69.
$$\begin{array}{r} 17,246 \\ -\ 6,789 \\ \hline 10,457 \end{array}$$

71.
$$\begin{array}{r} 15,700 \\ -15,397 \\ \hline 303 \end{array}$$

73. $43 - 12 + 9 = 31 + 9$
$\qquad\qquad\quad = 40$

75. $120 + 30 - 40 = 150 - 40$
$\qquad\qquad\qquad\quad = 110$

Applications

77. TAXIS

$\$23 - \$5 = \$18$

$\$18$ was the fare.

79. DOW JONES AVERAGE

$r =$ amount the Dow rose

$r = 11,305 - 11,272 = 33$

The Dow rose 33 points.

81. BANKING

$\$370 + \$40 - \$197 = \$410 - \$197$
$\qquad\qquad\qquad\qquad\quad = \$213$

There was $213 left in the account.

83. TAX DEDUCTIONS

$$
\begin{array}{r}
2,345 \\
1,712 \\
1,778 \\
445 \\
1,003 \\
+2,774 \\
\hline
10,057
\end{array}
$$

She drove 10,057 miles the first 6 months.

85a. INCOME

$$
\begin{array}{r}
26,785 \\
28,107 \\
29,429 \\
30,751 \\
+32,073 \\
\hline
147,145
\end{array}
$$

$\$147,145$

85b.

$$
\begin{array}{r}
29,701 \\
31,023 \\
32,345 \\
33,667 \\
+34,989 \\
\hline
161,725
\end{array}
$$

$\$161,725$

87.   BLUEPRINTS

L = Total length of house

$L = 24 + 35 + 16 + 16$

$L = 91$

Length of house is 91 ft.

89.   CANDY

$$\begin{array}{r} 1,040,000,000 \\ 1,810,000,000 \\ 1,993,000,000 \\ +1,390,000,000 \\ \hline 6,223,000,000 \end{array}$$

$6,223,000,000 is the total candy sales for the 4 holidays.

91.   CITY FLAGS

P = Perimeter of flag

$P = 64 + 34 + 64 + 34$

$P = 196$

196 inches must be purchased.

Writing

93.   It does not matter which order you add numbers, their sum will be the same.

Review

95.   3 thousands + 1 hundred + 2 tens + 5 ones

97.   6,354,780

99.   6,350,000

1.3 Multiplying and Dividing Whole Numbers

Vocabulary

1. **Multiplication** is repeated addition.

3. The statement $ab = ba$ expresses the **commutative** property of multiplication. The statement $(ab)c = a(bc)$ expresses the **associative** property of multiplication.

5. In a division, the dividend is divided by the **divisor**. The result of a division is called a **quotient.**

Concepts

7. $4 \bullet 8$

9. Multiply length times width

11a. $1 \bullet 25 = 25$ 11b. $62(1) = 62$

11c. $10 \bullet 0 = 0$ 11d. $0(4) = 0$

13. $5 \bullet 12$ using the fact that area is length times width.

Notation

15a. $\times, \bullet, (\ )$ 15b. $\overline{)\qquad}, \div, /\ or - (\text{fraction bar})$

17. ft^2 means square feet and is a unit of measure for area.

Practice

19. $12 \bullet 7 = 84$ 21. $27(12) = 324$

23. $9 \bullet (4 \bullet 5) = 9 \bullet 20$ 25. $5 \bullet 7 \bullet 3 = 35 \bullet 3$
$= 180$ $= 105$

27.
$$
\begin{array}{r}
99 \\
\times 77 \\
\hline
693 \\
+6930 \\
\hline
7,623
\end{array}
$$

29.
$$
\begin{array}{r}
20 \\
\times 53 \\
\hline
60 \\
+1000 \\
\hline
1,060
\end{array}
$$

31.
$$\begin{array}{r} 112 \\ \times\ 23 \\ \hline 336 \\ +2240 \\ \hline 2,576 \end{array}$$

33.
$$\begin{array}{r} 207 \\ \times\ 97 \\ \hline 1449 \\ +18630 \\ \hline 20,079 \end{array}$$

35.
$$\begin{array}{r} 13456 \\ \times\ 217 \\ \hline 94192 \\ 134560 \\ +2691200 \\ \hline 2,919,952 \end{array}$$

37.
$$\begin{array}{r} 3302 \\ \times\ 358 \\ \hline 26416 \\ 165100 \\ +990600 \\ \hline 1,182,116 \end{array}$$

39. $A = (14)(6)$
$A = 84\ \text{in}^2$ or 84 sq in

41. $A = (12)(12)$
$A = 144\ \text{in.}^2$ or 144 sq in.

43. $40 \div 5 = 8$

45. $42 \div 14 = 3$

47. $132 \div 11 = 12$

49. $\dfrac{221}{17} = 13$

51.
$$\begin{array}{r} 73 \\ 13\overline{)949} \\ \underline{91} \\ 39 \\ \underline{39} \\ 0 \end{array}$$

Answer is 73

53.
$$\begin{array}{r} 41 \\ 33\overline{)1353} \\ \underline{132} \\ 33 \\ \underline{33} \\ 0 \end{array}$$

Answer is 41

55.
$$
\begin{array}{r}
205 \\
39\overline{)7995} \\
\underline{78} \\
19 \\
\underline{0} \\
195 \\
\underline{195} \\
0
\end{array}
$$

Answer is 205

57.
$$
\begin{array}{r}
210 \\
29\overline{)6090} \\
\underline{58} \\
29 \\
\underline{29} \\
0
\end{array}
$$

Answer is 210

NOTE: You must add the zero at the end.

59.
$$
\begin{array}{r}
8 \\
31\overline{)273} \\
\underline{248} \\
25
\end{array}
$$

Answer is 8 R 25

61.
$$
\begin{array}{r}
20 \\
37\overline{)743} \\
\underline{74} \\
03 \\
\underline{0} \\
3
\end{array}
$$

Answer is 20 R 3

63.
$$
\begin{array}{r}
30 \\
42\overline{)1273} \\
\underline{126} \\
13 \\
\underline{0} \\
13
\end{array}
$$

Answer is 30 R 13

65.
$$
\begin{array}{r}
31 \\
57\overline{)1795} \\
\underline{171} \\
85 \\
\underline{57} \\
28
\end{array}
$$

Answer is 31 R 28

Applications

67. WAGES

Let P = pay earned

$P = (12)(11)$

$P = 132$

She earned $132.

69. FINDING DISTANCE

Let D = distance

$D = (14)(29)$

$D = 406$

The car can go 406 miles.

71. CONCERTS

Let A = total audience

$A = (2)(37)(1,700)$

$A = (74)(1,700)$

$A = 125,800$

125,800 people heard them.

73. ORANGE JUICE

Let J = total cans of juice

$J = (13)(24)$

$J = 312$

It takes 312 oranges.

75. ELEVATORS

Let W = total weight

$W = (14)(150)$

$W = 2100$

Yes, since the total is over 2000.

77. WORD PROCESSING

Let E = total entries

$E = (8)(9)$

$E = 72$

The table will hold 72 entries.

79. DISTRIBUTING MILK

$$\begin{array}{r} 3 \\ 23\overline{)73} \\ \underline{69} \\ 4 \end{array}$$

There were 4 left over.

81. MILAGE

$$\frac{700 \text{ miles}}{140 \text{ gallons}} = 5 \text{ miles per gallon}$$

5 miles per one gallon.

83. $(2)(5,280) = 10,560$

There are 10,560 ft in 2 miles.

$11,000 - 10,560 = 440$

Thus, 440 more feet.

85. PRICE OF TEXTBOOKS

$$\frac{954,193}{23,273} = 41$$

Each book costs $41.

87. VOLLEYBALL LEAGUES

$$\frac{216}{7} = 30 \, R \, 6$$

$$\frac{216}{8} = 27 \text{ but not an even number of teams}$$

$$\frac{216}{9} = 24$$

9 per team would satisfy all of the conditions.

89. COMPARING ROOMS

A_R = Area of rectangular room

$A_R = (14)(17)$

$A_R = 238$

Area of rectangular room is 238 ft^2

A_s = Area of square room

$A_s = (16)(16)$

$A_s = 256$

Area of square room is 256 ft^2

Therefore, the square room has a larger area.

91.   GARDENING

A_G = Area of garden

$A_G = (27)(19)$

$A_G = 513$

To find area used for planting, subtract the area of path from 513.

A_U = Area used for planting

$A_U = 513 - 125$

$A_U = 388$

There are 388 ft^2 of planting space in the garden

Writing

93. Because performing $10 \div 2$ gives 5, but attempting the reverse $2 \div 10$ would give a fraction for an answer and not 5, thus division is not commutative.

95. 1 foot is a measure of distance while 1 square foot is a measure of area where 1 square foot is a square that has sides of 1 foot.

Review

97.   8

99.   357
      39
   +476
   872

Estimation

1. 30000
    10000
    9000
    1000   No.
    10000
    +30000
    90000

3. 500
    × 70
    0
    35000
    35000

    No, since 451 was rounded to 500, the number shown is way too high.

5. 2000
    30)60000
    60
    0

    No, 200 is way too small.

7. CAMPAIGNING

    3500
    600
    1200
    300
    +2700
    8900

    Approx. 8,900 miles

9. GOLF COURSES

    30
    3000)90000
    9000
    0

    Approx. 30 bags of seed.

11. CURRENCY

    1,800,000,000
    5)9,000,000,000
    5
    40
    40
    0

    1,800,000,000 $5 bills.

1.4 Prime Factors and Exponents

Vocabulary

1. Numbers that are multiplied together are called **factors**.

3. To **factor** a whole number means to express it as the product of other whole numbers.

5. Whole numbers, greater than one, that are not prime numbers are called **composite** numbers.

7. To prime factor a number means to write it as a product of only **prime** numbers.

9. In the exponential expression 6^4, 6 is called the **base**, and 4 is called the **exponent**.

Concepts

11. $27 = 1 \bullet 27$
       $27 = 3 \bullet 9$

13a.   44   13b.   100

15a   1 and 11   15b.   1 and 23

15c.   1 and 37   15d.   They are prime numbers.

17.   Yes   19. $2 \bullet 3 \bullet 3 \bullet 5 = 6 \bullet 3 \bullet 5$
                                                 $= 18 \bullet 5$
                                                   $= 90$

21. $11^2 \bullet 5 = 121 \bullet 5$ 23.   No.
               $= 605$

25.   27.

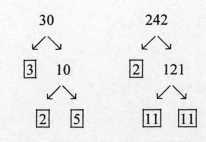

They have 2 and 5 in common.   They have 2 in common.

29.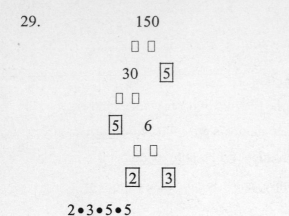

2•3•5•5

2•3•5•5

They are the same

31.

Product of the factors of 12	Sum of the factors of 12
1•12	13
2•6	8
3•4	7

33.   Start with 2

Notation

35. $7^3 = 7 \bullet 7 \bullet 7$

37. $3^5 = 3 \bullet 3 \bullet 3 \bullet 3 \bullet 3$

39. $5^2(11) = 5 \bullet 5 \bullet 11$

41. $10^1 = 10$

43. $2 \bullet 2 \bullet 2 \bullet 2 \bullet 2 = 2^5$

45. $5 \bullet 5 \bullet 5 \bullet 5 = 5^4$

47. $4(4)(5)(5) = 4^2 \bullet 5^2$

Practice

49. 1, 2, 5, 10

51. 1, 2, 4, 5, 8, 10, 20, 40

53. 1, 2, 3, 6, 9, 18

55. 1, 2, 4, 11, 22, 44

57. 1, 7, 11, 77

59. 1, 2, 4, 5, 10, 20, 25, 50, 100

61.
$$39$$

$$\boxed{3} \quad \boxed{13}$$

$$3 \bullet 13$$

63.
$$99$$

$$\boxed{3} \quad 33$$

$$\boxed{3} \quad \boxed{11}$$

$$3^2 \bullet 11$$

65.
$$162$$

$$\boxed{2} \quad 81$$

$$9 \qquad 9$$

$$\boxed{3} \quad \boxed{3}\boxed{3} \quad \boxed{3}$$

$$2 \bullet 3^4$$

67.
$$220$$

$$22 \quad 10$$

$$\boxed{2} \ \boxed{11} \ \boxed{2} \ \boxed{5}$$

$$2^2 \bullet 5 \bullet 11$$

69.

$$64$$

$$\boxed{2} \quad 32$$

$$\boxed{2} \quad 16$$

$$\boxed{2} \quad 8$$

$$\boxed{2} \quad 4$$

$$\boxed{2} \quad \boxed{2}$$

$$2^6$$

71.

$$147$$

$$\boxed{3} \quad 49$$

$$\boxed{7} \quad \boxed{7}$$

$$3 \bullet 7^2$$

73.
$$
\begin{aligned}
3^4 &= 3 \bullet 3 \bullet 3 \bullet 3 \\
&= 9 \bullet 3 \bullet 3 \\
&= 27 \bullet 3 \\
&= 81
\end{aligned}
$$

75.
$$
\begin{aligned}
2^5 &= 2 \bullet 2 \bullet 2 \bullet 2 \bullet 2 \\
&= 4 \bullet 2 \bullet 2 \bullet 2 \\
&= 8 \bullet 2 \bullet 2 \\
&= 16 \bullet 2 \\
&= 32
\end{aligned}
$$

77.
$$
\begin{aligned}
12^2 &= 12 \bullet 12 \\
&= 144
\end{aligned}
$$

79.
$$
\begin{aligned}
8^4 &= 8 \bullet 8 \bullet 8 \bullet 8 \\
&= 64 \bullet 8 \bullet 8 \\
&= 512 \bullet 8 \\
&= 4096
\end{aligned}
$$

81.
$$
\begin{aligned}
3^2\left(2^3\right) &= 3 \bullet 3 (2 \bullet 2 \bullet 2) \\
&= 3 \bullet 3 (4 \bullet 2) \\
&= 3 \bullet 3 \bullet 8 \\
&= 9 \bullet 8 \\
&= 72
\end{aligned}
$$

83.
$$
\begin{aligned}
2^3 \bullet 3^3 \bullet 4^2 &= 2 \bullet 2 \bullet 2 \bullet 3 \bullet 3 \bullet 3 \bullet 4 \bullet 4 \\
&= 4 \bullet 2 \bullet 3 \bullet 3 \bullet 3 \bullet 4 \bullet 4 \\
&= 8 \bullet 3 \bullet 3 \bullet 3 \bullet 4 \bullet 4 \\
&= 24 \bullet 3 \bullet 3 \bullet 4 \bullet 4 \\
&= 72 \bullet 3 \bullet 4 \bullet 4 \\
&= 216 \bullet 4 \bullet 4 \\
&= 864 \bullet 4 \\
&= 3,456
\end{aligned}
$$

85. $234^3 = 12,812,904$ 87. $23^2 \cdot 13^3 = 1,162,213$

Applications

89.   PERFECT NUMBERS

Factors are 1, 2, 4, 7, 14, 28, so to be a

perfect number

$$1 + 2 + 4 + 7 + 14 = 28$$
$$3 + 4 + 7 + 14 = 28$$
$$7 + 7 + 14 = 28$$
$$14 + 14 = 28$$

thus, it is a perfect number.

91.   LIGHT

2 yards = 4 square units = 2^2 square units

3 yards = 9 square units = 3^2 square units

4 yards = 16 square units = 4^2 square units

Writing

93.   Use a variety of divisibility tests to see if the number is divisible by other prime numbers 2, 3, 5, 7, etc…. If you get to the number, it is prime.

95.   Factors of a number are any numbers that can be divided into the number with no remainder while prime factorization of the number is the number reduced to the product of prime numbers.

Review

97.   231,000

99. $\dfrac{0}{15} = 0$ Zero divided by any number other than zero is zero.

101. $A = lw$, Area = (length) (width)

1.5 Order of Operations

Vocabulary

1. The grouping symbols ( ) are called **parenthesis**, and the symbols [ ] are called **brackets**.

3. The **evaluate** the expression $2 + 5 \bullet 4$ means to find its value.

Concepts

5. 3 operations. Square the 2, multiply the result by 5, subtract one from the result.

7. Multiply the 5 and 7 first in the numerator. Subtract the 4 from 8 in the parenthesis in the denominator first.

9. $\begin{aligned} 2 \bullet 3^2 &= 2 \bullet 9 \\ &= 18 \end{aligned}$ and $\begin{aligned} \left(2 \bullet 3\right)^2 &= \left(6\right)^2 \\ &= 36 \end{aligned}$

Notation

11. $\begin{aligned} 28 - 5\left(2\right)^2 &= 28 - 5\left(4\right) \\ &= 28 - 20 \\ &= 8 \end{aligned}$

13. $\begin{aligned} \left[4\left(2+7\right)\right] - 6 &= \left[4\left(9\right)\right] - 6 \\ &= 36 - 6 \\ &= 30 \end{aligned}$

Practice

15. $\begin{aligned} 7 + 4 \bullet 5 &= 7 + 20 \\ &= 27 \end{aligned}$

17. $\begin{aligned} 2 + 3\left(0\right) &= 2 + 0 \\ &= 2 \end{aligned}$

19. $\begin{aligned} 20 - 10 + 5 &= 10 + 5 \\ &= 15 \end{aligned}$

21. $\begin{aligned} 25 \div 5 \bullet 5 &= 5 \bullet 5 \\ &= 25 \end{aligned}$

23. $\begin{aligned} 7\left(5\right) - 5\left(6\right) &= 35 - 30 \\ &= 5 \end{aligned}$

25. $\begin{aligned} 4^2 + 3^2 &= 16 + 9 \\ &= 25 \end{aligned}$

27. $\begin{aligned} 2 \bullet 3^2 &= 2 \bullet 9 \\ &= 18 \end{aligned}$

29.
$$3 + 2 \bullet 3^4 \bullet 5 = 3 + 2 \bullet 81 \bullet 5$$
$$= 3 + 162 \bullet 5$$
$$= 3 + 810$$
$$= 813$$

31.
$$5 \bullet 10^3 + 2 \bullet 10^2 + 3 \bullet 10^1 + 9 = 5 \bullet 1,000 + 2 \bullet 100 + 3 \bullet 10 + 9$$
$$= 5,000 + 200 + 30 + 9$$
$$= 5,239$$

33.
$$3(2)^2 - 4(2) + 12 = 3(4) - 4(2) + 12$$
$$= 12 - 8 + 12$$
$$= 4 + 12$$
$$= 16$$

35.
$$(8-6)^2 + (4-3)^2 = (2)^2 + (1)^2$$
$$= 4 + 1$$
$$= 5$$

37.
$$60 - \left(6 + \frac{40}{8}\right) = 60 - (6 + 5)$$
$$= 60 - 11$$
$$= 49$$

39.
$$6 + 2(5+4) = 6 + 2(9)$$
$$= 6 + 18$$
$$= 24$$

41.
$$3 + 5(6-4) = 3 + 5(2)$$
$$= 3 + 10$$
$$= 13$$

43.
$$(7-4)^2 + 1 = (3)^2 + 1$$
$$= 9 + 1$$
$$= 10$$

45.
$$6^3 - (10+8) = 6^3 - 18$$
$$= 216 - 18$$
$$= 198$$

47.
$$50 - 2(4)^2 = 50 - 2(16)$$
$$= 50 - 32$$
$$= 18$$

49.
$$16^2 - 4(2)(5) = 256 - 4(2)(5)$$
$$= 256 - 8(5)$$
$$= 256 - 40$$
$$= 216$$

51. $39 - 5(6) + 9 - 1 = 39 - 30 + 9 - 1$
$$= 9 + 9 - 1$$
$$= 18 - 1$$
$$= 17$$

53. $(18 - 12)^3 - 5^2 = (6)^3 - 5^2$
$$= 216 - 5^2$$
$$= 216 - 25$$
$$= 191$$

55. $2(10 - 3^2) + 1 = 2(10 - 9) + 1$
$$= 2(1) + 1$$
$$= 2 + 1$$
$$= 3$$

57. $6 + \dfrac{25}{5} + 6(3) = 6 + 5 + 6(3)$
$$= 6 + 5 + 18$$
$$= 11 + 18$$
$$= 29$$

59. $3\left(\dfrac{18}{3}\right) - 2(2) = 3(6) - 2(2)$
$$= 18 - 2(2)$$
$$= 18 - 4$$
$$= 14$$

61. $(2 \bullet 6 - 4)^2 = (12 - 4)^2$
$$= (8)^2$$
$$= 64$$

63. $4\left[50 - (3^3 - 5^2)\right] = 4\left[50 - (27 - 25)\right]$
$$= 4[50 - 2]$$
$$= 4[48]$$
$$= 192$$

65. $80 - 2\left[12 - (5 + 4)\right] = 80 - 2[12 - 9]$
$$= 80 - 2[3]$$
$$= 80 - 6$$
$$= 74$$

67. $2\left[100 - (5 + 4)\right] - 45 = 2[100 - 9] - 45$
$$= 2[91] - 45$$
$$= 182 - 45$$
$$= 137$$

69.
$$\frac{10+5}{6-1} = \frac{15}{5}$$
$$= 3$$

71.
$$\frac{5^2+17}{6-2^2} = \frac{25+17}{6-4}$$
$$= \frac{42}{2}$$
$$= 21$$

73.
$$\frac{(3+5)^2+2}{2(8-5)} = \frac{(8)^2+2}{2(3)}$$
$$= \frac{64+2}{6}$$
$$= \frac{66}{6}$$
$$= 11$$

75.
$$\frac{(5-3)^2+2}{4^2-(8+2)} = \frac{(2)^2+2}{4^2-10}$$
$$= \frac{4+2}{16-10}$$
$$= \frac{6}{6}$$
$$= 1$$

77.
$$12,985-(1,800+689) = 12,985-2,489$$
$$= 10,496$$

79.
$$3,245-25(16-12)^2 = 3,245-25(4)^2$$
$$= 3,245-25(16)$$
$$= 3,245-400$$
$$= 2,845$$

Applications

81. BUYING GROCERIES

$$2(6)+4(2)+2(1)=12+8+2$$
$$=20+2$$
$$=22$$

The total cost is $22.

83. BANKING

$$24(1)+6(5)+10(10)+12(20)+2(50)+1(100)$$
$$=24+30+100+240+100+100$$
$$=54+100+240+100+100$$
$$=54+240+300 \qquad \text{Using the commutative property of addition}$$
$$=294+300$$
$$=594$$

The total amount is $594.

85. SCRABBLE

brick: $3(3)+1+1+3+3(5)$ aphid: $3\left[1+2(3)+4+1+2\right]$
$$=9+1+1+3+15$$
$$=29$$

$$=3\left[1+6+4+1+2\right]$$
$$=3\left[14\right]$$
$$=42$$

"Brick" was worth 29 points and "aphid" was worth 42 points.

87. CLIMATE

$$\frac{75+80+83+80+77+72+86}{7} = \frac{553}{7}$$
$$= 79$$

77° F is the average temperature.

89. NATURAL NUMBERS

$$\frac{1+2+3+4+5+6+7+8+9}{9} = \frac{45}{9}$$
$$= 5$$

The average of the first 9 natural numbers is 5.

91. FAST FOOD

$$\frac{237+289+295+302+303+312+348}{7} = \frac{2086}{7}$$
$$= 298$$

The average of calories per sandwich is 298..

Writing

93.   People could view the mathematical operations in different orders which would give different results.

95.   To find the mean you add the numbers then divide by the number of items you added. The average tells you a number around which the set of numbers are grouped.

Review

97.   4,029
        +3,271
         7,300

99.   417
        × 23
         1251
         8340
         9,591

1.6 Solving Equations by Addition and Subtraction

Vocabulary

1. An equation is a statement that two expressions are **equal**. An equation contains an $=$ symbol.

3. To **check** the solution of an equation, we substitute the value for the variable in the original equation and see whether the result is a true statement.

5. **Equivalent** equations have exactly the same solutions.

Concepts

7. If $x = y$ and c is any number, then

$$x + c = y + c$$

9. 6 is being added to x; subtract 6 from both sides

Notation

11. $x + 8 = 24$

$x + 8 - 8 = 24 - 8$

$x = 16$

Check:

$x + 8 = 24$

$16 + 8 = 24$

$24 = 24$

So 16 is a solution.

Practice

13. Yes, $x = 2$ is an equation.

15. No, $7x < 8$ is not an equation. It is an inequality.

17. Yes, $x + y = 0$ is an equation.

19. Yes, $1 + 1 = 3$ is an equation even though it is incorrect.

21. $x + 2 = 3$

$1 + 2 = 3$ Yes

$3 = 3$

23. $a - 7 = 0$

$7 - 7 = 0$ Yes

$0 = 0$

25. $8 - y = y$
     $8 - 5 = 5$ No
     $3 \neq 5$

27. $x + 32 = 0$
     $16 + 32 = 0$ No
     $48 \neq 0$

29. $z + 7 = z$
     $7 + 7 = 7$ No
     $14 \neq 7$

31. $x = x$
     $0 = 0$ Yes

33. $x - 7 = 3$
     $x - 7 + 7 = 3 + 7$
     $x = 10$

     Check:
     $x - 7 = 3$
     $10 - 7 = 3$
     $3 = 3$

35. $a - 2 = 5$
     $a - 2 + 2 = 5 + 2$
     $a = 7$

     Check:
     $a - 2 = 5$
     $7 - 2 = 5$
     $5 = 5$

37. $1 = b - 2$
     $1 + 2 = b - 2 + 2$
     $3 = b \ \ or \ \ b = 3$

     Check:
     $1 = b - 2$
     $1 = 3 - 2$
     $1 = 1$

39. $x - 4 = 0$
     $x - 4 + 4 = 0 + 4$
     $x = 4$

     Check:
     $x - 4 = 0$
     $4 - 4 = 0$
     $0 = 0$

41. $y - 7 = 6$
     $y - 7 + 7 = 6 + 7$
     $y = 13$

     Check:
     $y - 7 = 6$
     $13 - 7 = 6$
     $6 = 6$

43. $70 = x - 5$
     $70 + 5 = x - 5 + 5$
     $75 = x \ \ or \ \ x = 75$

     Check:
     $70 = x - 5$
     $70 = 75 - 5$
     $70 = 70$

45. $312 = x - 428$

$312 + 428 = x - 428 + 428$

$740 = x \ \text{ or } \ x = 740$

Check:

$312 = x - 428$

$312 = 740 - 428$

$312 = 312$

47. $x - 117 = 222$

$x - 117 + 117 = 222 + 117$

$x = 339$

Check:

$x - 117 = 222$

$339 - 117 = 222$

$222 = 222$

49. $x + 9 = 12$

$x + 9 - 9 = 12 - 9$

$x = 3$

Check:

$x + 9 = 12$

$3 + 9 = 12$

$12 = 12$

51. $y + 7 = 12$

$y + 7 - 7 = 12 - 7$

$y = 5$

Check:

$y + 7 = 12$

$5 + 7 = 12$

$12 = 12$

53. $t + 19 = 28$

$t + 19 - 19 = 28 - 19$

$t = 9$

Check:

$t + 19 = 28$

$9 + 19 = 28$

$28 = 28$

55. $23 + x = 33$

$23 - 23 + x = 33 - 23$

$x = 10$

Check:

$23 + x = 33$

$23 + 10 = 33$

$33 = 33$

57. $5 = 4 + c$

$5 - 4 = 4 - 4 + c$

$1 = c \ \text{ or } \ c = 1$

Check:

$5 = 4 + c$

$5 = 4 + 1$

$5 = 5$

59. $99 = r + 43$

$99 - 43 = r + 43 - 43$

$56 = r \ \text{ or } \ r = 56$

Check:

$99 = r + 43$

$99 = 56 + 43$

$99 = 99$

61.

$$512 = x + 428$$
$$512 - 428 = x + 428 - 428$$
$$84 = x \ \ or \ \ x = 84$$

Check:

$$512 = x + 428$$
$$512 = 84 + 428$$
$$512 = 512$$

63.

$$x + 117 = 222$$
$$x + 117 - 117 = 222 - 117$$
$$x = 105$$

Check:

$$x + 117 = 222$$
$$105 + 117 = 222$$
$$222 = 222$$

65.

$$3 + x = 7$$
$$3 - 3 + x = 7 - 3$$
$$x = 4$$

Check:

$$3 + x = 7$$
$$3 + 4 = 7$$
$$7 = 7$$

67.

$$y - 5 = 7$$
$$y - 5 + 5 = 7 + 5$$
$$y = 12$$

Check:

$$y - 5 = 7$$
$$12 - 5 = 7$$
$$7 = 7$$

69.

$$4 + a = 12$$
$$4 - 4 + a = 12 - 4$$
$$a = 8$$

Check:

$$4 + a = 12$$
$$4 + 8 = 12$$
$$12 = 12$$

71.

$$x - 13 = 34$$
$$x - 13 + 13 = 34 + 13$$
$$x = 47$$

Check:

$$x - 13 = 34$$
$$47 - 13 = 34$$
$$34 = 34$$

Applications

73.   ARCHAEOLOGY

Analyze the problem

- The manuscript is **1,700 years** old.

- The manuscript is **425 years** older than the jar.

- We are asked to find **the age of the jar**.

Form an equation

Since we want to find the age of the jar, we can let $x =$ **the age of the jar**. Now we look for a key word or phrase in the problem.

     Key Phrase: <u>older than</u>

     Translation: <u>add</u>

Now we translate the words of the problem into an equation.

The age of the manuscript is 425 plus the age of the jar.

$$1,700 \qquad\qquad = 425 \ + \quad x$$

Solve the equation

$$1,700 = 425 + x$$
$$1,700 - 425 = 425 + x - 425$$
$$1,275 = x$$

State the conclusion   <u>**The jar is 1,275 years old**</u>.

Check the result

If the jar is 1,275 years old, then the manuscript is 1,275+425=1,700 years old. The answer checks.

75. ELECTIONS

   x = total number of votes cast

   $x = 47,401,185 + 39,197,469 + 8,085,294$

   $x = 94,683,948$

   The total number of votes cast was 94,683,948.

77. PARTY INVITATIONS

   x = number of invitations sent

   $x = 59 + 3$

   $x = 62$

   62 invitations were sent.

79. FAST FOODS

   x = money to borrow

   $$x + 68,500 = 287,000$$
   $$x + 68,500 - 68,500 = 287,000 - 68,000$$
   $$x = 218,500$$

   She would need to borrow $218,500.

81. CELEBRITY EARNINGS

   x = Oprah's earnings in 2003

   $$x - 152 = 28$$
   $$x - 152 + 152 = 28 + 152$$
   $$x = 180$$

   Oprah's earnings in 2003 were $180 million.

83. POWER OUTAGES

   x = amount meter must increase

   $$x + 60 = 85$$
   $$x + 60 - 60 = 85 - 60$$
   $$x = 25$$

   The meter must increase by 25 units.

85. AUTO REPAIRS

x = amount paid at the muffler shop

$x = 219 - 29$

$x = 190$

She paid $190 at the muffler shop.

Writing

87. When a number can be substituted into an equation and it makes both sides equal, that number is said to satisfy the equation or be the solution to that equation.

89. It is trying to show that if you subtract the same amount from both sides of a scale, the scale is still in balance. It is a visual representation of how the same idea works for solving equations.

91. Answers will vary.

Review

93. 325,780

95. $2 \bullet 3^2 \bullet 5 = 2 \bullet 9 \bullet 5$
$= 18 \bullet 5$
$= 90$

97. $8 - 2(3) + 1^3 = 8 - 2(3) + 1$
$= 8 - 6 + 1$
$= 2 + 1$
$= 3$

1.7 Solving Equations by Division and Multiplication

Vocabulary

1. According to the **<u>division</u>** property of equality, "If equal quantities are divided by the same nonzero quantity, the results will be equal quantities."

Concepts

3. If we multiply x by 6 and then divide that product by 6, the result is <u>x</u>.

5. If $x = y$, then $\dfrac{x}{z} = \dfrac{y}{z}$ (where $z \neq 0$).

7. Variable is being multiplied by 4; Divide by 4.

9a. Subtract 5 from both sides.

9b. Add 5 to both sides.

9c. Divide both sides by 5.

9d. Multiply both sides by 5.

Notation

11. $3x = 12$

$\dfrac{3x}{3} = \dfrac{12}{3}$

$x = 4$

Check:

$3x = 12$

$3 \bullet 4 = 12$

$12 = 12$

So 4 is the solution.

Practice

13. $3x = 3$

$$\frac{3x}{3} = \frac{3}{3}$$

$$x = 1$$

Check:

$$3x = 3$$

$$3(1) = 3$$

$$3 = 3$$

15. $2x = 192$

$$\frac{2x}{2} = \frac{192}{2}$$

$$x = 96$$

Check:

$$2x = 192$$

$$2(96) = 192$$

$$192 = 192$$

17. $17x = 51$

$$\frac{17x}{17} = \frac{51}{17}$$

$$x = 3$$

Check:

$$17x = 51$$

$$17(3) = 51$$

$$51 = 51$$

19. $34y = 204$

$$\frac{34y}{34} = \frac{204}{34}$$

$$y = 6$$

Check:

$$34y = 204$$

$$34(6) = 204$$

$$204 = 204$$

21. $100 = 100x$

$$\frac{100}{100} = \frac{100x}{100}$$

$$1 = x \ \ or \ \ x = 1$$

Check:

$$100 = 100x$$

$$100 = 100(1)$$

$$100 = 100$$

23. $16 = 8r$

$$\frac{16}{8} = \frac{8r}{8}$$

$$2 = r \ \ or \ \ r = 2$$

Check:

$$16 = 8r$$

$$16 = 8(2)$$

$$16 = 16$$

25.
$$\frac{x}{7} = 2$$

$$7 \cdot \frac{x}{7} = 7 \cdot 2$$

$$x = 14$$

Check:

$$\frac{x}{7} = 2$$

$$\frac{14}{7} = 2$$

$$2 = 2$$

27.
$$\frac{y}{14} = 3$$

$$14 \cdot \frac{y}{14} = 14 \cdot 3$$

$$y = 42$$

Check:

$$\frac{y}{14} = 3$$

$$\frac{42}{14} = 3$$

$$3 = 3$$

29.
$$\frac{a}{15} = 5$$

$$15 \cdot \frac{a}{15} = 15 \cdot 5$$

$$a = 75$$

Check:

$$\frac{a}{15} = 5$$

$$\frac{75}{15} = 5$$

$$5 = 5$$

31.
$$\frac{c}{13} = 3$$

$$13 \cdot \frac{c}{13} = 13 \cdot 3$$

$$c = 39$$

Check:

$$\frac{c}{13} = 3$$

$$\frac{39}{13} = 3$$

$$3 = 3$$

33.
$$1 = \frac{x}{50}$$

$$50 \cdot 1 = 50 \cdot \frac{x}{50}$$

$$50 = x \ \ or \ \ x = 50$$

Check:

$$1 = \frac{x}{50}$$

$$1 = \frac{50}{50}$$

$$1 = 1$$

35.
$$7 = \frac{t}{7}$$

$$7 \cdot 7 = 7 \cdot \frac{t}{7}$$

$$49 = t \ \ or \ \ t = 49$$

Check:

$$7 = \frac{t}{7}$$

$$7 = \frac{49}{7}$$

$$7 = 7$$

37. $9z = 90$

$$\frac{9z}{9} = \frac{90}{9}$$

$$z = 10$$

Check:

$$9z = 90$$

$$9(10) = 90$$

$$90 = 90$$

39. $7x = 21$

$$\frac{7x}{7} = \frac{21}{7}$$

$$x = 3$$

Check:

$$7x = 21$$

$$7(3) = 21$$

$$21 = 21$$

41. $86 = 43t$

$$\frac{86}{43} = \frac{43t}{43}$$

$$2 = t \ \ or \ \ t = 2$$

Check:

$$86 = 43t$$

$$86 = 43(2)$$

$$86 = 86$$

43. $21s = 21$

$$\frac{21s}{21} = \frac{21}{21}$$

$$s = 1$$

Check:

$$21s = 21$$

$$21(1) = 21$$

$$21 = 21$$

45. $$\frac{d}{20} = 2$$

$$20 \bullet \frac{d}{20} = 20 \bullet 2$$

$$d = 40$$

Check:

$$\frac{d}{20} = 2$$

$$\frac{40}{20} = 2$$

$$2 = 2$$

47. $$400 = \frac{t}{3}$$

$$3 \bullet 400 = 3 \bullet \frac{t}{3}$$

$$1,200 = t \ \ or \ \ t = 1,200$$

Check: *Incorrect*

$$\frac{d}{20} = 2$$

$$\frac{40}{20} = 2$$

$$2 = 2$$

Applications

49. NOBEL PRIZE

Analyze the problem

- **3** people shared cash award.
- Each person received **$318,500**.
- We are asked to find **the Nobel Prize cash award**.

Form an equation

Since we want to find what the Nobel Prize cash award was,

we let $x = $ **the Nobel Prize cash award**. To form the equation, we look for a key word or phrase in the problem.

Key Phrase: shared the prize money

Translation: divide

Now we translate the words of the problem into an equation.

The Nobel Prize cash award	divided by	the number of recipients	was	$318,500
x	$\div$	3	=	$318,500

Solve the equation

$$\frac{x}{3} = 318,500$$

$$3 \bullet \frac{x}{3} = 3 \bullet 318,500$$

$$x = 955,500$$

State the conclusion **The Nobel Prize cash award was $955,500**.

Check the result

If we divide the Nobel Prize cash award by 3 we have

$\dfrac{\$955,500}{3} = \$318,500$. This was the amount each person received.

The answer checks.

51. SPEED READING

    x = speed after taking class

    $x = 3 \cdot 130$

    $x = 390$

    She would have a speed of 390 wpm after the class.

53. STAMPS

    x = number of rows of stamps per sheet

    $8x = 112$

    $\dfrac{8x}{8} = \dfrac{112}{8}$

    $x = 14$

    There would be 14 rows of stamps per sheet.

55. PHYSICAL EDUCATION

    x = students in PE class

    $x = 3 \cdot 32$

    $x = 96$

    There were 96 students in PE class.

57. ANIMAL SHELTERS

    x = calls after being featured on the news

    Quadrupled means multiplied by 4

    $x = 4 \cdot 8$

    $x = 32$

    The shelter received 32 calls per day after being featured on the news.

59. GRAVITY

x = weight of object on the moon

$$6x = 330$$

$$\frac{6x}{6} = \frac{330}{6}$$

$$x = 55$$

The scale on the moon would register 55 lbs.

Writing

61. It is trying to show that if you divide both sides of a scale by the same number, the scale is still in balance. It is a visual representation of how the same idea works for solving equations.

63. To solve an equation is to isolate the variable on one side of the equation and to isolate the solution on the other side of the equation.

Review

65. P = Perimeter of rectangle

$$P = 8 + 16 + 8 + 16$$

$$P = 48$$

The perimeter is 48 cm.

67.

$$2^3 \bullet 3 \bullet 5$$

69. $3^2 \bullet 2^3 = 9 \bullet 8$

$$= 72$$

71. FUEL ECONOMY

$$\frac{24 + 22 + 28 + 29 + 27}{5} = \frac{130}{5}$$

$$= 26$$

The average is 26 mpg.

Key Concept Variables

1. Let x = the monthly cost to lease the van

3. Let x = the width of the field

5. Let x = the distance traveled by the motorist

7. $a+b=b+a$

9. $\dfrac{b}{1}=b$

11. $n-1<n$

13. $(r+s)+t=r+(s+t)$

Chapter One Review

Section 1.1 An Introduction to Whole Numbers

1.

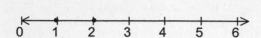

2.

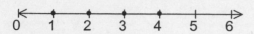

3.

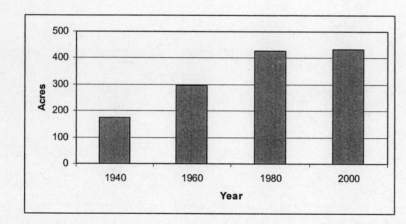

4.

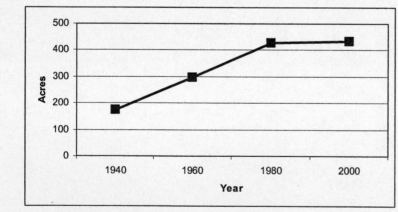

5.   6   6.   7

7.   5 hundred thousands + 7 ten thousands + 3 hundreds + 2 ones

8. 3 ten millions + 7 millions + 3 hundred thousands + 9 thousands +

   5 tens + 4 ones

9. 3,207

10. 23,253,412

11. 16,000,000,000

12. $9 > 7$

13. $3 < 5$

14. 2,507,300

15. 2,510,000

16. 2,507,350

17. 2,500,000

Section 1.2 Adding and Subtracting Whole Numbers

18. $56 + 22 = 78$

19. $137 + 0 = 137$

20. $\begin{aligned} 15 + (27 + 13) &= 15 + 40 \\ &= 55 \end{aligned}$

21. $\begin{aligned} 82 + 17 + 50 &= 99 + 50 \\ &= 149 \end{aligned}$

22. $\begin{aligned} (111 + 222) + 444 &= 333 + 444 \\ &= 777 \end{aligned}$

23. $0 + 2,332 = 2,332$

24. $\begin{array}{r} 236 \\ +282 \\ \hline 518 \end{array}$

25. $\begin{array}{r} 5,345 \\ +\ \ 655 \\ \hline 6,000 \end{array}$

26.
```
    135
    213
    615
  +  47
  ─────
  1,010
```

27.
```
    4,447
    7,478
 +13,061
 ───────
  24,986
```

28. Commutative property of addition

29. Associative property of addition

30. P = perimeter of square

$P = 24 + 24 + 24 + 24$

$P = 96$

Perimeter is 96 in.

31. $18 - 5 = 13$

32. $9 - (7 - 2) = 9 - 5$

$\qquad\qquad\quad = 4$

33. $22 - 5 - 6 = 17 - 6$

$\qquad\qquad\;\; = 11$

34. $5,231 - 5,177 = 54$

35.
```
   343
  −269
  ────
    74
```

36.
```
   17,800
  −15,725
  ───────
    2,075
```

37. $4 + 2 = 6$

38. $5 - 2 = 3$

39. TRAVEL

A = amount saved

$A = 237 - 192$

$A = 45$

$45 is the amount saved.

40. SAVINGS ACCOUNTS

$B = $ account balance

$B = 931 + 271 - 37 - 380$

$B = 1,202 - 37 - 380$

$B = 1,165 - 380$

$B = 785$

$785 is the final balance.

41. REBATE

$C = $ car's original sticker price

$C = 21,991 + 1,550$

$C = 23,541$

The car's original sticker price was $23,541.

Section 1.3 Multiplying and Dividing Whole Numbers

42. $8 \bullet 7 = 56$

43. $7(8) = 56$

44. $8 \bullet 0 = 0$

45. $7 \bullet 1 = 7$

46. $10 \bullet 8 \bullet 7 = 80 \bullet 7$

$\qquad = 560$

47. $5 \bullet (7 \bullet 6) = 5 \bullet 42$

$\qquad = 210$

48.
$$
\begin{array}{r}
157 \\
\times \ 21 \\
\hline
157 \\
3140 \\
\hline
3,297
\end{array}
$$

49.
$$
\begin{array}{r}
3,723 \\
\times \ 48 \\
\hline
29784 \\
148920 \\
\hline
178,704
\end{array}
$$

50.
$$
\begin{array}{r}
356 \\
\times \ 89 \\
\hline
3204 \\
28480 \\
\hline
31,684
\end{array}
$$

51.
$$
\begin{array}{r}
5,624 \\
\times \ 81 \\
\hline
5624 \\
449920 \\
\hline
455,544
\end{array}
$$

52. Associative property of multiplication

53. Commutative property of multiplication

54. WAGES

E = what she earned

$E = 38 \bullet 9$

$E = 342$

She earned \$342.

55. HORSESHOES

P = perimeter

$P = 6 + 48 + 6 + 48$

$P = 108$

The perimeter is 108 ft.

A = area

$A = 6 \bullet 48$

$A = 288$

The area is 288 ft^2.

56. PACKAGING

T = total eggs in 5 gross

$T = 12 \bullet 12 \bullet 5$

$T = 144 \bullet 5$

$T = 720$

There are 720 eggs in 5 gross.

57. $\dfrac{6}{3} = 2$

58. $\dfrac{15}{1} = 15$

59. $73 \div 0 = \text{undefined}$

60. $\dfrac{0}{8} = 0$

61. $357 \div 17 = 21$

$$
\begin{array}{r}
21 \\
17{\overline{)357}} \\
34 \\
\hline
17 \\
17 \\
\hline
0
\end{array}
$$

62. $1,443 \div 39 = 37$

$$
\begin{array}{r}
37 \\
39{\overline{)1443}} \\
117 \\
\hline
273 \\
273 \\
\hline
0
\end{array}
$$

63.
$$21\overline{)405} = 19\ R6$$
$$\underline{21}$$
$$195$$
$$\underline{189}$$
$$6$$

64.
$$54\overline{)1269} = 23\ R27$$
$$\underline{108}$$
$$189$$
$$\underline{162}$$
$$27$$

65. TREATS

T = treats each child gets

$T = \dfrac{745}{45}$ (division shown to the right)

Each child gets 16 candies.

There are 25 candies left.

$$45\overline{)745}$$
$$\underline{45}$$
$$295$$
$$\underline{270}$$
$$25$$

(quotient 16)

66. COPIES

C = number of copies of the test made

$C = \dfrac{84}{3}$

$C = 28$

There were 28 copies of the test made.

Section 1.4 Prime Factors and Exponents

67. 1, 2, 3, 6, 9, 18

68. 1, 5, 25

69. 31 is prime

70. 100 is composite

71. 1 is neither

72. 0 is neither

73. 125 is composite

74. 47 is prime

75. 171 is odd

76. 214 is even

77.   0 is even

78.   1 is odd

79.

$$42$$

$$\boxed{2} \quad 21$$

$$\boxed{3} \quad \boxed{7}$$

$$2 \bullet 3 \bullet 7$$

80.

$$375$$

$$\boxed{3} \quad 125$$

$$\boxed{5} \quad 25$$

$$\boxed{5} \quad \boxed{5}$$

$$3 \bullet 5^3$$

81. 6^4

82. $5^3 \bullet 13^2$

83. $\begin{aligned} 5^3 &= 5 \bullet 5 \bullet 5 \\ &= 25 \bullet 5 \\ &= 125 \end{aligned}$

84. $\begin{aligned} 11^2 &= 11 \bullet 11 \\ &= 121 \end{aligned}$

85. $\begin{aligned} 2^3 \bullet 5^2 &= 8 \bullet 25 \\ &= 200 \end{aligned}$

86. $\begin{aligned} 2^2 \bullet 3^3 \bullet 5^2 &= 4 \bullet 27 \bullet 25 \\ &= 108 \bullet 25 \\ &= 2,700 \end{aligned}$

Section 1.5 Order of Operations

87. $\begin{aligned} 13 + 12 \bullet 3 &= 13 + 36 \\ &= 49 \end{aligned}$

88. $\begin{aligned} 35 - 15 \div 5 &= 35 - 3 \\ &= 32 \end{aligned}$

89. $\begin{aligned} (13 + 12)3 &= (25)3 \\ &= 75 \end{aligned}$

90. $\begin{aligned} (8 - 2)^2 &= (6)^2 \\ &= 36 \end{aligned}$

91. $8 \bullet 5 - 4 \div 2 = 40 - 4 \div 2$
$$= 40 - 2$$
$$= 38$$

92. $8 \bullet (5 - 4 \div 2) = 8 \bullet (5 - 2)$
$$= 8 \bullet 3$$
$$= 24$$

93. $2 + 3(10 - 4 \bullet 2) = 2 + 3(10 - 8)$
$$= 2 + 3(2)$$
$$= 2 + 6$$
$$= 8$$

94. $4(20 - 5 \bullet 3 + 2) - 4$
$$= 4(20 - 15 + 2) - 4$$
$$= 4(5 + 2) - 4$$
$$= 4(7) - 4$$
$$= 28 - 4$$
$$= 24$$

95. $3^3 \left(\dfrac{12}{6} \right) - 1^4 = 3^3 (2) - 1^4$
$$= 27(2) - 1$$
$$= 54 - 1$$
$$= 53$$

96. $\dfrac{12 + 3 \bullet 7}{5^2 - 14} = \dfrac{12 + 21}{25 - 14}$
$$= \dfrac{33}{11}$$
$$= 3$$

97. $7 + 3\left[10 - 3(4 - 2) \right] = 7 + 3\left[10 - 3(2) \right]$
$$= 7 + 3[10 - 6]$$
$$= 7 + 3[4]$$
$$= 7 + 12$$
$$= 19$$

98. $5 + 2\left[(15 - 3 \bullet 4) - 2 \right] = 5 + 2\left[(15 - 12) - 2 \right]$
$$= 5 + 2[3 - 2]$$
$$= 5 + 2[1]$$
$$= 5 + 2$$
$$= 7$$

99. DICE GAMES

$$3(6)+2(5)=18+10$$
$$=28$$

100. YAHTZEE

$$\frac{159+244+184+240+166+213}{6}$$
$$=\frac{1206}{6}$$
$$=201$$

Section 1.6 Solving Equations by Addition and Subtraction

101. $x+2=13;\ x=5$

$5+2=13$

$7\neq13$

No

102. $x-3=1;\ x=4$

$4-3=1$

$1=1$

Yes

103. Variable is y

104. Variable is t

105. $x-7=2$

$x-7+7=2+7$

$x=9$

Check:

$x-7=2$

$9-7=2$

$2=2$

106. $x-11=20$

$x-11+11=20+11$

$x=31$

Check:

$x-11=20$

$31-11=20$

$20=20$

107. $225=y-115$

$225+115=y-115+115$

$340=y\ or\ y=340$

Check:

$225=y-115$

$225=340-115$

$225=225$

108. $101=p-32$

$101+32=p-32+32$

$133=p\ or\ p=133$

Check:

$101=p-32$

$101=133-32$

$101=101$

109.
$$x + 9 = 18$$
$$x + 9 - 9 = 18 - 9$$
$$x = 9$$

Check:
$$x + 9 = 18$$
$$9 + 9 = 18$$
$$18 = 18$$

110.
$$b + 12 = 26$$
$$b + 12 - 12 = 26 - 12$$
$$b = 14$$

Check:
$$b + 12 = 26$$
$$14 + 12 = 26$$
$$26 = 26$$

111.
$$175 = p + 55$$
$$175 - 55 = p + 55 - 55$$
$$120 = p \ \text{ or } \ p = 120$$

Check:
$$175 = p + 55$$
$$175 = 120 + 55$$
$$175 = 175$$

112.
$$212 = m + 207$$
$$212 - 207 = m + 207 - 207$$
$$5 = m \ \text{ or } \ m = 5$$

Check:
$$212 = m + 207$$
$$212 = 5 + 207$$
$$212 = 212$$

113.
$$x - 7 = 0$$
$$x - 7 + 7 = 0 + 7$$
$$x = 7$$

Check:
$$x - 7 = 0$$
$$7 - 7 = 0$$
$$0 = 0$$

114.
$$x + 15 = 1,000$$
$$x + 15 - 15 = 1,000 - 15$$
$$x = 985$$

Check:
$$x + 15 = 1,000$$
$$985 + 15 = 1,000$$
$$1,000 = 1,000$$

115. FINANCING

$x =$ amount of money needed to borrow

$$x = 122,750 - 25,500$$
$$x = 97,250$$

$97,250 will have to be borrowed.

116.   DOCTOR'S PATIENTS

x = number of original patients

$$x - 13 = 172$$
$$x - 13 + 13 = 172 + 13$$
$$x = 185$$

The doctor had 185 patients before moving.

Section 1.7 Solving Equations by Division and Multiplication

117. $3x = 12$

$$\frac{3x}{3} = \frac{12}{3}$$
$$x = 4$$

Check:

$$3x = 12$$
$$3(4) = 12$$
$$12 = 12$$

118. $15y = 45$

$$\frac{15y}{15} = \frac{45}{15}$$
$$y = 3$$

Check:

$$15y = 45$$
$$15(3) = 45$$
$$45 = 45$$

119. $105 = 5r$

$$\frac{105}{5} = \frac{5r}{5}$$
$$21 = r \ \ or \ \ r = 21$$

Check:

$$105 = 21r$$
$$105 = 21(5)$$
$$105 = 105$$

120. $224 = 16q$

$$\frac{224}{16} = \frac{16q}{16}$$
$$14 = q \ \ or \ \ q = 14$$

Check:

$$224 = 16q$$
$$224 = 16(14)$$
$$224 = 224$$

121.

$$\frac{x}{7} = 3$$

$$7 \bullet \frac{x}{7} = 7 \bullet 3$$

$$x = 21$$

Check:

$$\frac{x}{7} = 3$$

$$\frac{21}{7} = 3$$

$$3 = 3$$

122.

$$\frac{a}{3} = 12$$

$$3 \bullet \frac{a}{3} = 3 \bullet 12$$

$$a = 36$$

Check:

$$\frac{a}{3} = 12$$

$$\frac{36}{3} = 12$$

$$12 = 12$$

123.

$$15 = \frac{s}{21}$$

$$21 \bullet 15 = 21 \bullet \frac{s}{21}$$

$$315 = s \ \ or \ \ s = 315$$

Check:

$$15 = \frac{s}{21}$$

$$15 = \frac{315}{21}$$

$$15 = 15$$

124.

$$25 = \frac{d}{17}$$

$$17 \bullet 25 = 17 \bullet \frac{d}{17}$$

$$425 = d \ \ or \ \ d = 425$$

Check:

$$25 = \frac{d}{17}$$

$$25 = \frac{425}{17}$$

$$25 = 25$$

125.

$$12x = 12$$

$$\frac{12x}{12} = \frac{12}{12}$$

$$x = 1$$

Check:

$$12x = 12$$

$$12(1) = 12$$

$$12 = 12$$

126.

$$\frac{x}{12} = 12$$

$$12 \bullet \frac{x}{12} = 12 \bullet 12$$

$$x = 144$$

Check:

$$\frac{x}{12} = 12$$

$$\frac{144}{12} = 12$$

$$12 = 12$$

127. CARPENTRY

x = length of each piece

$3x = 72$

$$\frac{3x}{3} = \frac{72}{3}$$

$x = 24$

Each board is 24 in. long.

128. JEWELERY

x = cost of each chain

$$\frac{x}{4} = 32$$

$$4 \bullet \frac{x}{4} = 4 \bullet 32$$

$x = 128$

The chain cost $128 total.

Chapter One Test

1.

2.   5 thousands + 2 hundreds + 6 tens + 6 ones

3.   7,507

4.   35,000,000

5.

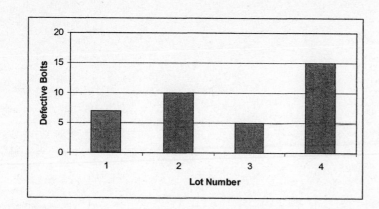

6.

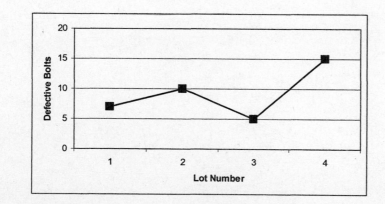

7. $15 > 10$ 8. $1,247 < 1,427$

9.
$$
\begin{array}{r}
327 \\
435 \\
123 \\
+\ 606 \\
\hline
1{,}491
\end{array}
$$

10.
$$
\begin{array}{r}
535 \\
-\ 287 \\
\hline
248
\end{array}
$$

11.
$$
\begin{array}{r}
44{,}526 \\
+\ 13{,}479 \\
\hline
58{,}105
\end{array}
$$

12.
$$
\begin{array}{r}
4{,}521 \\
-\ 3{,}579 \\
\hline
942
\end{array}
$$

13.   STOCKS

x = cost of stock on Thursday

$x = 73 + 12 - 9$

$x = 85 - 9$

$x = 76$

$76 was the cost of the stock on Thursday.

14.   1, 2, 4, 5, 10, 20

15.
$$
\begin{array}{r}
53 \\
\times\ 8 \\
\hline
424
\end{array}
$$

16.
$$
\begin{array}{r}
367 \\
\times\ 73 \\
\hline
1101 \\
25690 \\
\hline
26{,}791
\end{array}
$$

17.
$$
\begin{array}{r}
72 \\
63\overline{)4536} \\
\underline{441} \\
126 \\
\underline{126} \\
0
\end{array}
$$

18.
$$
73\overline{)8379} = 114 \text{ R}57
$$
$$
\begin{array}{r}
114 \\
73\overline{)8379} \\
\underline{73} \\
107 \\
\underline{73} \\
349 \\
\underline{292} \\
57
\end{array}
$$

19a. $15 \bullet 0 = 0$

19b. $\dfrac{0}{15} = 0$

20a.   associative property of multiplication

20b.   commutative property of addition

21.   FURNITURE SALES

P = perimeter
$P = 105 + 75 + 105 + 75$
$P = 360$

Perimeter is 360 ft.

A = area
$A = 105 \bullet 75$
$A = 7,875$

The area is $7,875$ ft^2.

22.

$$74\overline{)3451}$$

$$\begin{array}{r} 46 \\ 74\overline{)3451} \\ \underline{296} \\ 491 \\ \underline{444} \\ 47 \end{array}$$

47 would be left over.

23.   COLLECTABLES

T = total baseball cards in case
$T = 12 \bullet 24 \bullet 12$
$T = 288 \bullet 12$
$T = 3,456$

There are 3,456 baseball cards in a case.

24.

```
            252
           ↙  ↘
        [2]   126
             ↙  ↘
          [2]   63
               ↙  ↘
            [3]   21
                 ↙  ↘
              [3]   [7]
```

$2^2 \bullet 3^2 \bullet 7$

25. $9 + 4 \bullet 5 = 9 + 20$
$= 29$

26.
$$\frac{3 \bullet 4^2 - 2^2}{(2-1)^3} = \frac{3 \bullet 16 - 4}{(1)^3}$$

$$= \frac{48-4}{1}$$

$$= \frac{44}{1}$$

$$= 44$$

27.
$$10 + 2\left[12 - 2(6-4)\right]$$

$$= 10 + 2\left[12 - 2(2)\right]$$

$$= 10 + 2\left[12 - 4\right]$$

$$= 10 + 2\left[8\right]$$

$$= 10 + 16$$

$$= 26$$

28. GRADES

$$\frac{73 + 52 + 70 + 0 + 0}{5} = \frac{195}{5}$$

$$= 39$$

Average would be 39.

29.
$$x + 13 = 16$$

$$3 + 13 = 16$$

$$16 = 16$$

Yes, it makes a true statement when checked in the equation.

30.
$$100 = x + 1$$

$$100 - 1 = x + 1 - 1$$

$$99 = x \ \text{ or } \ x = 99$$

Check:

$$100 = x + 1$$

$$100 = 99 + 1$$

$$100 = 100$$

31.
$$y - 12 = 18$$

$$y - 12 + 12 = 18 + 12$$

$$y = 30$$

Check:

$$y - 12 = 18$$

$$30 - 12 = 18$$

$$18 = 18$$

32.
$$5t = 55$$

$$\frac{5t}{5} = \frac{55}{5}$$

$$t = 11$$

Check:

$$5t = 55$$

$$5(11) = 55$$

$$55 = 55$$

33.
$$\frac{q}{3} = 27$$

$$3 \bullet \frac{q}{3} = 3 \bullet 27$$

$$q = 81$$

Check:

$$\frac{q}{3} = 27$$

$$\frac{81}{3} = 27$$

$$27 = 27$$

34. PARKING

x = number of spaces the school currently has

Double means to multiply by 2

$$2x = 6,200$$

$$\frac{2x}{2} = \frac{6,200}{2}$$

$$x = 3,100$$

The school currently has 3,100 spaces.

35. LIBRARIES

x = current age of the building

$$x + 6 = 200$$

$$x + 6 - 6 = 200 - 6$$

$$x = 194$$

The building is 194 years old.

36. To solve an equation means to find all the values of the variable that, when substituted into the equation, make a true statement. This can be done algebraically by isolating the variable on one side of the equation.

Chapter 2 The Integers

2.1 An Introduction to the Integers

Vocabulary

1. Numbers can be represented by points equally spaced on a number **line**.

3. To **graph** a number means to locate it on a number line and highlight it with a **dot**.

5. The symbols > and < are called **inequality** symbols.

7. The **absolute** value of a number is the distance between the number and zero on a number line.

9. The set of all positive and negative whole numbers, along with 0, is called the set of **integers**.

Concepts

11a. $-2 < 2$ 11b. $0 > -1$ 11c. $-1 < 0$ and $0 > -1$

13. Yes   15. $15 - 8$

17. $15 > 12$

19a. -225 19b. -10 19c. -3

19d. $-12,000$ 19e. -2

21. It is negative   23. -4 25. -8 and 2

27. -7 29. $6 - 4, -6, -(-6)$
                         Answers may vary

Notation

31a. $-(-8)$ 31b. $|-8|$

31c. $8-8$ 31d. $-|-8|$

Practice

33. $|9| = 9$ 35. $|-8| = 8$ 37. $|-14| = 14$

39. $-|20| = -20$ 41. $-|-6| = -6$ 43. $|203| = 203$

45. $-0 = 0$ 47. $-(-11) = 11$ 49. $-(-4) = 4$

51. $-(-12) = 12$

53.

55.

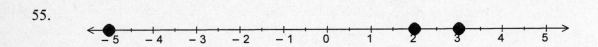

57. $-5 < 5$ 59. $-12 < -6$ 61. $-10 > -11$

63. $|-2| > 0$ 65. $-1,255 \le -(-1,254)$ 67. $-|-3| \le 4$

Applications

69. FLIGHT OF A BALL

Time	Position of ball
1 sec	2
2 sec	3
3 sec	2
4 sec	0
5 sec	−3
6 sec	−7

71. TECHNOLOGY

Peaks: 2, 4, 0; Valleys: −3, −5, −2

73a. GOLF

-1, one below par

73b.

-3, 3 below par

73c. Most of the scores are below par.

75a. WEATHER MAPS

−10° to −20°

75b. 10°

75c. 10°

77a. HISTORY   77b. A.D.   77c. B.C.   77d. The birth of Christ.

200 yr

79. LINE GRAPHS

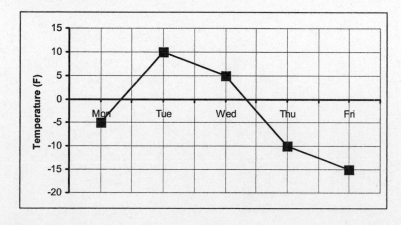

Writing

81.   Every number has a number that is it's opposite, for instance the opposite of 4 is −4, and the opposite of −10 is 10. It is the negative of the number.

83.   Since absolute value of a number is the distance away from the origin, distance can never be negative, thus absolute values cannot be negative.

85.   Positive buoyancy means you have a tendency to rise in the water, negative buoyancy means you have a tendency to sink in the water, neutral means that you do neither.

Review

87.   23,500   89.   Yes

91.   Associative property of multiplication

2.2 Adding Integers

Vocabulary

1. When 0 is added to a number, the number remains the same. We call 0 the additive **identity**.

Concepts

3.

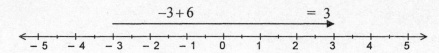

5.

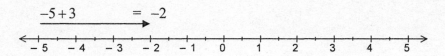

7a. yes   7b. Yes

9a. $|-7| = 7$ 9b. $|10| = 10$

11. To add two integers with unlike sign, **subtract** their absolute values, the smaller from the larger. Then attach to that result the sign of the number with the **larger** absolute value.

Notation

13. $-16 + (-2) + (-1) = -18 + (-1)$ 15. $(-3 + 8) + (-3) = 5 + (-3)$
$= -19$ $= 2$

17. -5 should be in parenthesis: $-6 + (-5)$.

Practice

19. 11

21. 23

23. 0

25. -99

27. $-6+(-3)=-9$

29. $-5+(-5)=-10$

31. $-6+7=1$

33. $-15+8=-7$

35. $20+(-40)=-20$

37. $30+(-15)=15$

39. $-1+9=8$

41. $-7+9=2$

43. $5+(-15)=-10$

45. $24+(-15)=9$

47. $35+(-27)=8$

49. $24+(-45)=-21$

51. $-2+6+(-1)=4+(-1)$
     $\qquad\qquad\qquad =3$

53. $-9+1+(-2)=-8+(-2)$
     $\qquad\qquad\qquad =-10$

55. $6+(-4)+(-13)+7=2+(-13)+7$
     $\qquad\qquad\qquad\qquad =-11+7$
     $\qquad\qquad\qquad\qquad =-4$

57. $9+(-3)+5+(-4)=6+5+(-4)$
     $\qquad\qquad\qquad\quad =11+(-4)$
     $\qquad\qquad\qquad\quad =7$

59. $-6+(-7)+(-8)=-13+(-8)$
     $\qquad\qquad\qquad\quad =-21$

61. $-7+0=-7$

63. $9+0=9$

65. $-4+4=0$

67. $2+(-2)=0$

69. 5

71. $2+(-10+8)=2+(-2)$
     $\qquad\qquad\qquad =0$

73. $(-4+8)+(-11+4)=4+(-7)$
     $\qquad\qquad\qquad\qquad =-3$

75. $\left[-3+(-4)\right]+(-5+2)=-7+(-3)$
$=-10$

77. $\left[6+(-4)\right]+\left[8+(-11)\right]=2+\left[-3\right]$
$=-1$

79. $-2+\left[-8+(-7)\right]=-2+\left[-15\right]$
$=-17$

81. $789+(-9,135)=-8,346$

83. $-675+(-456)+99=-1,131+99$
$=-1,032$

Applications

85.   **G FORCES**
3G, -3G

87.   **CASH FLOW**
$C=$ cashflow
$C=-900+450+380$
$C=-450+380$
$C=-70$

No, it produces -$70 per month.

89.   **MEDICAL QUESTIONAIRES**
$T=$ total points
$T=-4+3+(-3)+3+4+2$
$T=-1+(-3)+3+4+2$
$T=-4+3+4+2$
$T=-1+4+2$
$T=3+2$
$T=5$

Since T=5, he is at 4% risk.

91a.   ATOMS

$C = $ charge

$C = -4 + 3$

$C = -1$

The charge is -1.

91b.

$C = $ charge

$C = -4 + 4$

$C = 0$

The charge is 0, or it is neutral.

93.   FLOODING

$F = $ flood stage height

$F = -4 + 11$

$F = 7$

It will be 7 *ft* above flood stage.

95.   FILM PROFITS

$P = $ total profit

$P = 10 + (-5) + 15 + (-10)$

$P = 5 + 15 + (-10)$

$P = 20 + (-10)$

$P = 10$

The profit was $10 million dollars.

Writing

97.   No, it depends on which number has the larger absolute value.

99.   Adding 2 negatives can only produce a sum that is more negative than the 2 numbers added.

Review

101. $A = $ area of rectangle

$A = 5 \bullet 3$

$A = 15$

Area is 15 ft^2.

103. $D = $ distance the car can go

$D = 15 \bullet 25$

$D = 375$

The car can travel 375 miles.

105.   125

5   25

5   5

5^3

2.3 Subtracting Integers

Vocabulary

1. The answer to a subtraction problem is called the **difference**.

Concepts

3. **Subtraction** is the same as adding the opposite of the number to be subtracted.

5. Subtracting -6 is the same as adding 6.

7. For any numbers a and b, $a-b=a+\underline{(-b)}$.

9. After using parentheses as grouping symbols, if another set of grouping symbols is needed, we use **brackets**.

11. $-8-(-4)$

13. 7

15. No, $8-3=5$, but $3-8=-5$.

Notation

17. $1-3-(-2)=1+(-3)+2$
     $=-2+2$
     $=0$

19. $(-8-2)-(-6)=\left[-8+(-2)\right]-(-6)$
     $=-10-(-6)$
     $=-10+6$
     $=-4$

Practice

21. $8-(-1)=8+1$
     $=9$

23. $-4-9=-4+(-9)$
     $=-13$

25. $-5-5=-5+(-5)$
     $=-10$

27. $-5-(-4)=-5+4$
     $=-1$

29. $-1-(-1)=-1+1$
     $=0$

31. $-2-(-10)=-2+10$
     $=8$

33. $0-(-5)=0+5$
$\quad\quad\quad = 5$

35. $0-4=0+(-4)$
$\quad\quad = -4$

37. $-2-2=-2+(-2)$
$\quad\quad\quad = -4$

39. $-10-10=-10+(-10)$
$\quad\quad\quad\quad = -20$

41. $9-9=0$

43. $-3-(-3)=-3+3$
$\quad\quad\quad\quad = 0$

45. $-4-(-4)-15=-4+4-15$
$\quad\quad\quad\quad\quad = 0-15$
$\quad\quad\quad\quad\quad = -15$

47. $-3-3-3=-3+(-3)+(-3)$
$\quad\quad\quad\quad = -6+(-3)$
$\quad\quad\quad\quad = -9$

49. $5-9-(-7)=5+(-9)+7$
$\quad\quad\quad\quad = -4+7$
$\quad\quad\quad\quad = 3$

51. $10-9-(-8)=1+8$
$\quad\quad\quad\quad = 9$

53. $-1-(-3)-4=-1+3-4$
$\quad\quad\quad\quad = 2-4$
$\quad\quad\quad\quad = 2+(-4)$
$\quad\quad\quad\quad = -2$

55. $-5-8-(-3)=-5+(-8)+3$
$\quad\quad\quad\quad = -13+3$
$\quad\quad\quad\quad = -10$

57. $(-6-5)-3=\left[-6+(-5)\right]-3$
$\quad\quad\quad\quad = -11-3$
$\quad\quad\quad\quad = -11+(-3)$
$\quad\quad\quad\quad = -14$

59. $(6-4)-(1-2)=2-\left[1+(-2)\right]$
$\quad\quad\quad\quad = 2-[-1]$
$\quad\quad\quad\quad = 2+1$
$\quad\quad\quad\quad = 3$

61. $-9-(6-7)=-9-\left[6+(-7)\right]$
$\quad\quad\quad\quad = -9-[-1]$
$\quad\quad\quad\quad = -9+1$
$\quad\quad\quad\quad = -8$

63. $-8-\left[4-(-6)\right]=-8-[4+6]$
$\quad\quad\quad\quad = -8-10$
$\quad\quad\quad\quad = -8+(-10)$
$\quad\quad\quad\quad = -18$

65. $\left[-4+(-8)\right]-(-6)=-12+6$
$\quad\quad\quad\quad = -6$

67. $7-(-3)=7+3$
$\quad\quad\quad = 10$

69. $-10-(-6) = -10+6$
$\qquad\qquad\quad = -4$

71. $-1,557-890 = -2,447$

73. $20,007-(-496) = 20,503$

75. $-162-(-789)-2,303 = -1,676$

Applications

77. SCUBA DIVING

D = divers final depth

$D = -50+(-70)$

$D = -120$

The diver goes to -120 ft.

79. READING PROGRAMS

R = reading score improvement

$R = -7-(-23)$

$R = -7+23$

$R = 16$

The reading scores improved 16 points.

81. AMPERAGE

A = ammeter reading

$A = 5-7-6$

$A = 5+(-7)+(-6)$

$A = -2+(-6)$

$A = -8$

The ammeter reads -8.

83. GEOGRAPHY

D = difference in depths

$D = -283-(-1,290)$

$D = -283+1,290$

$D = 1,007$

The difference is 1,007 ft.

85. FOOTBALL

N = net gain or loss

$N = -1-6-5+8$

$N = -1+(-6)+(-5)+8$

$N = -7+(-5)+8$

$N = -12+8$

$N = -4$

The team gained -4 yd.

87a.   DIVING

Bottom   Water Line   Platform
-12   0   25

$$-12 \quad -8 \quad -4 \quad 0 \quad 4 \quad 8 \quad 12 \quad 16 \quad 20 \quad 24 \quad 28$$

87b. $T = $ total length
$$T = 25 - (-12)$$
$$T = 25 + 12$$
$$T = 37$$

The total length is 37ft.

89.   CHECKING ACCOUNTS

$B = $ balance in account
$$B = 1,303 - 676 - 121 - 750$$
$$B = 627 - 121 - 750$$
$$B = 506 - 750$$
$$B = 506 + (-750)$$
$$B = -244$$

No, he would have a $-\$244$ balance.

Writing

91. Subtraction can be shown as addition as follows: $10 - 2 = 10 + (-2)$.

93. You would change $-7 - 4$ to $-7 + (-4)$ which would be -11.

Review

95.   5,990

97.   1, 2, 4, 5, 10, 20

99.
$$12^2 - (5 - 4)^2 = 12^2 - (1)^2$$
$$= 144 - 1$$
$$= 143$$

101.
$$5x = 15$$
$$\frac{5x}{5} = \frac{15}{5}$$
$$x = 3$$

2.4 Multiplying Integers

Vocabulary

1. In the multiplication $-5(-4)$, the integers -5 and -4, which are being multiplied, are called **factors**. The answer, 20, is called the **product**.

3. In the expression -3^5, **3** is the base and 5 is the **exponent**.

Concepts

5. The product of two integers with **unlike** signs is negative.

7. The **commutative** property of multiplication implies that $-2(-3) = -3(-2)$.

9. -9; the opposite of that number.

11. (pos)(pos), (pos)(neg), (neg)(pos), (neg)(neg)

13a. $(-5)^{13}$ would be negative.   13b. $(-3)^{30}$ would be positive.

15a. $|-3| = 3$ 15b. $|12| = 12$ 15c. $|-5| = 5$

15d. $|9| = 9$ 15e. $|10| = 10$ 15f. $|-25| = 25$

17a

Problem	Number of negative factors	Answers
$-2(-2)$	2	4
$-2(-2)(-2)(-2)$	4	16
$-2(-2)(-2)(-2)(-2)(-2)$	6	64

17b. The answers entered in the table help to justify the following rule: The product of an **even** number of negative integers is positive.

Notation

19. $-3(-2)(-4) = 6(-4)$
     $= -24$

21. -5 should be in parentheses.

Practice

23. $-9(-6) = 54$

25. $-3 \bullet 5 = -15$

27. $12(-3) = -36$

29. $(-8)(-7) = 56$

31. $(-2)10 = -20$

33. $-40 \bullet 3 = -120$

35. $-8(0) = 0$

37. $-1(-6) = 6$

39. $-7(-1) = 7$

41. $1(-23) = -23$

43. $-6(-4)(-2) = 24(-2)$
     $= -48$

45. $5(-2)(-4) = -10(-4)$
     $= 40$

47. $2(3)(-5) = 6(-5)$
     $= -30$

49. $6(-5)(2) = -30(2)$
     $= -60$

51. $(-1)(-1)(-1) = 1(-1)$
     $= -1$

53. $-2(-3)(3)(-1) = 6(3)(-1)$
     $= 18(-1)$
     $= -18$

55. $3(-4)(0) = -12(0) = 0$

57. $-2(0)(10) = 0(10) = 0$

59. $-6(-10) = 60$

61. $(-4)^2 = (-4)(-4)$
     $= 16$

63. $(-5)^3 = (-5)(-5)(-5)$
     $= 25(-5)$
     $= -125$

65.
$$(-2)^3 = (-2)(-2)(-2)$$
$$= 4(-2)$$
$$= -8$$

67.
$$(-9)^2 = (-9)(-9)$$
$$= 81$$

69.
$$(-1)^5 = -1$$

71.
$$(-1)^8 = 1$$

73.
$$(-7)^2 = (-7)(-7) \quad -7^2 = -(7)(7)$$
$$= 49 \qquad\qquad = -49$$

75.
$$-12^2 = -(12)(12)$$
$$= -144$$
$$(-12)^2 = (-12)(-12)$$
$$= 144$$

77.
$$-76(787) = -59,812$$

79.
$$(-81)^4 = 43,046,721$$

81.
$$(-32)(-12)(-67) = -25,728$$

83.
$$(-25)^4 = 390,625$$

Applications

85a.   DIETING

Plan #1 = $(10)(-3) = -30 \text{ lbs}$

Plan #2 = $(14)(-2) = -28 \text{ lbs}$

85b.   Plan #1, the workout time is double that of plan #2.

87a.   MAGNIFICATION

High = 2; Low = -3

87b.   High = 4; Low = -6

89.   TEMPERATURE CHANGE

T = total temperature change
$T = 5(-4)$
$T = -20$

The total change would be $-20°$.

91.   EROSION

Decade = 10 years
E = total erosion
$E = 10(-2)$
$E = -20$

It will erode -20 ft in 10 years.

93.   WOMEN'S NATIONAL BASKETBALL ASSOCIATION

L = financial loss for giveaway

$L = -3(11,906)$

$L = -35,718$

The total loss would be $-\$35,718$.

Writing

95.   An even number of negatives will always multiply to give a positive answer.

97.   Since -1 is an odd negative, when multiplied by a positive number , the result would have to be negative, and when multiplied by a negative number, it would produce a positive result because there would be 2 negatives present in the multiplication.

Review

99. $3^2 \bullet 5 = 9 \bullet 5$

$ = 45$

101. x = increase in enrollment

$x = 12,300 - 10,200$

$x = 2,100$

The increase in enrollment is 2,100.

103. $<$ means less than

2.5 Dividing Integers

Vocabulary

1. In $\dfrac{-27}{3} = -9$, the number -9 is called the **quotient**, and the number 3 is the **divisor**.

3. The **absolute value** of a number is the distance between it and 0 on the number line.

5. The quotient of two negative integers is **positive**.

Concepts

7. $\quad 5(-5) = -25$

9. $\quad 0(?) = -6$

11. $\quad \dfrac{-20}{5} = -4$

13a. Always true

13b. Sometimes true

13c. Always true

Practice

15. $\dfrac{-14}{2} = -7$

17. $\dfrac{-8}{-4} = 2$

19. $\dfrac{-25}{-5} = 5$

21. $\dfrac{-45}{-15} = 3$

23. $\dfrac{40}{-2} = -20$

25. $\dfrac{50}{-25} = -2$

27. $\dfrac{0}{-16} = 0$

29. $\dfrac{-6}{0} = $ undefined

31. $\dfrac{-5}{1} = -5$

33. $-5 \div (-5) = 1$

35. $\dfrac{-9}{9} = -1$

37. $\dfrac{-10}{-1} = 10$

39. $\dfrac{-100}{25} = -4$

41. $\dfrac{75}{-25} = -3$

43. $\dfrac{-500}{-100} = 5$

45. $\dfrac{-200}{50} = -4$ 47. $\dfrac{-45}{9} = -5$ 49. $\dfrac{8}{-2} = -4$

51. $\dfrac{-13,550}{25} = -542$ 53. $\dfrac{272}{-17} = -16$

Applications

55.   TEMPERATURE DROP

Temperature drop was $-20°$ total.

$A = $ average temperature drop

$A = \dfrac{-20}{5}$

$A = -4$

The average temperature drop was $-4°$.

57.   SUBMARINE DIVES

$D = $ depth of each dive

$D = \dfrac{-3,000}{3}$

$D = -1,000$

The depth of each dive is $-1,000$ ft.

59.   BASEBALL TRADES

$B = $ games behind at end of season

$B = \dfrac{-12}{2}$

$B = -6$

They expect to be 6 games behind.

61.   MARKDOWNS

$M = $ markdown on the jeans

$M = \dfrac{-300}{20}$

$M = -15$

The markdown is $-\$15$.

63.   PAY CUTS

$P = $ employee's pay cut

$P = \dfrac{-9,135,000}{5,250}$

$P = -1,740$

Each employee would have their pay

cut by $\$-1,740$.

Writing

65. As with multiplication, an even number of negatives will always result in a positive answer.

67. $\dfrac{-10}{2} = -5$ because $2(-5) = -10$.

Review

69.
$$3\left(\dfrac{18}{3}\right)^2 - 2(2) = 3(6)^2 - 2(2)$$
$$= 3(36) - 2(2)$$
$$= 108 - 4$$
$$= 104$$

71.
$$210$$

$$21 \qquad 10$$

$$\boxed{3} \quad \boxed{7} \quad \boxed{2} \quad \boxed{5}$$

$$2 \bullet 3 \bullet 5 \bullet 7$$

73. Yes

75.
$$99 = r - 43$$
$$99 + 43 = r - 43 + 43$$
$$142 = r \ \text{ or } \ r = 142$$

2.6 Order of Operations and Estimation

Vocabulary

1. When asked to evaluate expressions containing more than one operation, we should apply the rules for the **order** of operations.

3. Absolute value symbols, parentheses, and brackets are types of **grouping** symbols.

Concepts

5.   3; exponent, multiplication, subtraction

7.   Numerator, multiplication
     Denominator, subtraction within the parentheses

9. $-3^2 = -(3)(3) = -9$

   $(-3)^2 = (-3)(-3) = 9$

   In -3^2 the negative is not part of the exponent and the base is 3, while in $(-3)^2$ the negative is located within the exponent and the base is -3.

Notation

11. $-8 - 5(-2)^2 = -8 - 5(4)$
   $= -8 - 20$
   $= -8 + (-20)$
   $= -28$

13. $\left[-4(2+7)\right] - 6 = \left[-4(9)\right] - 6$
   $= -36 - 6$
   $= -42$

Practice

15. $(-3)^2 - 4^2 = 9 - 16$
$$= 9 + (-16)$$
$$= -7$$

17. $3^2 - 4(-2)(-1) = 9 - 4(-2)(-1)$
$$= 9 - (-8)(-1)$$
$$= 9 - 8$$
$$= 1$$

19. $(2-5)(5+2) = [2 + (-5)](5+2)$
$$= (-3)(7)$$
$$= -21$$

21. $-10 - 2^2 = -10 - 4$
$$= -10 + (-4)$$
$$= -14$$

23. $\dfrac{-6-8}{2} = \dfrac{-6 + (-8)}{2}$
$$= \dfrac{-14}{2}$$
$$= -7$$

25. $\dfrac{-5-5}{2} = \dfrac{-5 + (-5)}{2}$
$$= \dfrac{-10}{2}$$
$$= -5$$

27. $-12 \div (-2)2 = (6)2$
$$= 12$$

29. $-16 - 4 \div (-2) = -16 - (-2)$
$$= -16 + 2$$
$$= -14$$

31. $|-5(-6)| = |30|$
$$= 30$$

33. $|-4 - (-6)| = |-4 + 6|$
$$= |2|$$
$$= 2$$

35. $5|3| = 5(3)$
$$= 15$$

37. $-6|-7| = -6(7)$
$$= -42$$

39.
$$(7-5)^2 - (1-4)^2 = (7-5)^2 - \left[1+(-4)\right]^2$$
$$= (2)^2 - (-3)^2$$
$$= 4 - 9$$
$$= 4 + (-9)$$
$$= -5$$

41.
$$-1(2^2 - 2 + 1^2) = -1(4 - 2 + 1)$$
$$= -1(2+1)$$
$$= -1(3)$$
$$= -3$$

43.
$$-50 - 2(-3)^3 = -50 - 2(-27)$$
$$= -50 - (-54)$$
$$= -50 + 54$$
$$= 4$$

45.
$$-6^2 + 6^2 = -36 + 36$$
$$= 0$$

47.
$$3\left(\frac{-18}{3}\right) - 2(-2) = 3(-6) - 2(-2)$$
$$= -18 - (-4)$$
$$= -18 + 4$$
$$= -14$$

49.
$$6 + \frac{25}{-5} + 6 \bullet 3 = 6 + (-5) + 6 \bullet 3$$
$$= 6 + (-5) + 18$$
$$= 1 + 18$$
$$= 19$$

51.
$$\frac{1-3^2}{-2} = \frac{1-9}{-2}$$
$$= \frac{1+(-9)}{-2}$$
$$= \frac{-8}{-2}$$
$$= 4$$

53.
$$\frac{-4(-5)-2}{-6} = \frac{20-2}{-6}$$
$$= \frac{18}{-6}$$
$$= -3$$

55.
$$-3\left(\frac{32}{-4}\right) - (-1)^5 = -3(-8) - (-1)^5$$
$$= -3(-8) - (-1)$$
$$= 24 - (-1)$$
$$= 24 + 1$$
$$= 25$$

57.
$$6(2^3)(-1) = 6(8)(-1)$$
$$= 48(-1)$$
$$= -48$$

59.
$$2+3\big[5-(1-10)\big]=2+3\Big[5-\big(1+[-10]\big)\Big]$$
$$=2+3\big[5-(-9)\big]$$
$$=2+3(5+9)$$
$$=2+3(14)$$
$$=2+42$$
$$=44$$

61.
$$-7(2-3\bullet5)=-7(2-15)$$
$$=-7\big[2+(-15)\big]$$
$$=-7(-13)$$
$$=91$$

63.
$$-\Big[6-(1-4)^2\Big]=-\Big[6-\big(1+[-4]\big)^2\Big]$$
$$=-\Big[6-(-3)^2\Big]$$
$$=-(6-9)$$
$$=-\big[6+(-9)\big]$$
$$=-(-3)$$
$$=3$$

65.
$$15+(-3\bullet4-8)=15+(-12-8)$$
$$=15+\big[-12+(-8)\big]$$
$$=15+(-20)$$
$$=-5$$

67.
$$\big|-3\bullet4+(-5)\big|=\big|-12+(-5)\big|$$
$$=\big|-17\big|$$
$$=17$$

69.
$$\big|(-5)^2-2\bullet7\big|=\big|25-2\bullet7\big|$$
$$=\big|25-14\big|$$
$$=\big|11\big|$$
$$=11$$

71.
$$-2+\big|6-4^2\big|=-2+\big|6-16\big|$$
$$=-2+\big|6+(-16)\big|$$
$$=-2+\big|-10\big|$$
$$=-2+10$$
$$=8$$

73.
$$2\big|1-8\big|\bullet\big|-8\big|=2\big|1+(-8)\big|\bullet\big|-8\big|$$
$$=2\big|-7\big|\bullet\big|-8\big|$$
$$=2(7)\bullet8$$
$$=14\bullet8$$
$$=112$$

75.
$$-2(-34)^2-(-605)=-2(1,156)-(-605)$$
$$=-2,312-(-605)$$
$$=-2,312+605$$
$$=-1,707$$

77.
$$-60-\frac{1,620}{-36}=-60-(-45)$$
$$=-60+45$$
$$=-15$$

79.
$$-30+(7-2)^2 = -30+(5)^2$$
$$= -30+25$$
$$= -5$$

81.
$$(3-4)^2-(3-9)^2 = (3+[-4])^2-(3+[-9])^2$$
$$= (-1)^2-(-6)^2$$
$$= 1-9 \quad 36$$
$$= 1+(-9) \quad (-36)$$
$$= -8 \quad -35$$

83. about -200

85. about -320

87. about $-9,000$

89. about $-1,200$

Applications

91. TESTING

G = grade received

$$G = 12(3)+3(-4)+5(-1)$$
$$G = 36+(-12)+(-5)$$
$$G = 24+(-5)$$
$$G = 19$$

The student received a grade of 19.

93. SCOUTING REPORTS

G = average gain for play

$$G = \frac{16+10+(-2)+0+4+(-4)+66+(-2)}{8}$$
$$G = \frac{88}{8}$$
$$G = 11$$

The team averages 11yd per play attempt.

95. OIL PRICES

E = estimation

$E = 70 + 90 + (-50) + (-20) + (-30)$

$E = 60$

There would be about a 60-cent gain.

Writing

97. Many people could read different ways to do the arithmetic without the order of operations.

99. Answers will vary.

Review

101. 5,000

103. Add the lengths of all sides.

105. T = total weight on the elevator

$T = 7(140)$

$T = 980$

No, they are under the limit.

2.7 Solving Equations Involving Integers

Vocabulary

1.   To **solve** an equation, we isolate the variable on one side of the = symbol.

Concepts

3. x

5a. $x+3-3=10-3$

5b. $x+3+(-3)=10+(-3)$

7a.   Multiplication by -2

7b.   Addition of -6

7c.   Multiplication by -4, subtraction of 8

7d.   Multiplication by -5, addition of -6

9.   When solving the equation $t-4=-8-2$, it is best to **simplify** the right-hand side of the equation first before undoing any operation performed on the variable.

11.   When solving an equation, we isolate the variable by undoing the operations performed on it in the **opposite** order.

13a.   Subtraction of 3

13b.   Addition of -6

Notation

15.
$$y+(-7)=-16+3$$
$$y+(-7)=-13$$
$$y+(-7)+7=-13+7$$
$$y=-6$$

17.
$$-13=-4y-1$$
$$-13+1=-4y-1+1$$
$$-12=-4y$$
$$\frac{-12}{-4}=\frac{-4y}{-4}$$
$$3=y$$
$$y=3$$

19. $-10 \bullet x$

Practice

21. $\quad -3x - 4 = 2; \ \ x = -2$

$\qquad -3(-2) - 4 = 2$

$\qquad\qquad 6 - 4 = 2$

$\qquad\qquad\qquad 2 = 2$

$\qquad\qquad\qquad$ yes

23. $\quad -x + 8 = -4; \ \ x = 4$

$\qquad -(-4) + 8 = -4$

$\qquad\qquad 4 + 8 = -4$

$\qquad\qquad 12 \neq -4$

$\qquad\qquad$ no

25. $\quad x + 6 = -12$

$\quad x + 6 - 6 = -12 - 6$

$\quad x = -12 + (-6)$

$\quad x = -18$

Check:

$\qquad x + 6 = -12$

$\quad -18 + 6 = -12$

$\qquad -12 = -12$

27. $\quad -6 + m = -20$

$\quad -6 + 6 + m = -20 + 6$

$\qquad\qquad m = -14$

Check:

$\qquad -6 + m = -20$

$\quad -6 + (-14) = -20$

$\qquad\qquad -20 = -20$

29. $\quad -5 + 3 = -7 + f$

$\qquad\quad -2 = -7 + f$

$\quad -2 + 7 = -7 + 7 + f$

$\qquad\quad 5 = f \ \ or \ \ f = 5$

Check:

$\quad -5 + 3 = -7 + f$

$\quad -5 + 3 = -7 + 5$

$\qquad -2 = -2$

31. $\quad h - 8 = -9$

$\quad h - 8 + 8 = -9 + 8$

$\qquad\quad h = -1$

Check:

$\qquad h - 8 = -9$

$\quad -1 - 8 = -9$

$\quad -1 + (-8) = -9$

$\qquad\quad -9 = -9$

33. $\quad 0 = y + 9$

$\quad 0 - 9 = y + 9 - 9$

$\quad -9 = y \ \ or \ \ y = -9$

Check:

$\quad 0 = y + 9$

$\quad 0 = -9 + 9$

$\quad 0 = 0$

35. $r-(-7)=-1-6$

    $r+7=-1+(-6)$

    $r+7=-7$

    $r+7-7=-7-7$

    $r=-7+(-7)$

    $r=-14$

    Check:

    $r-(-7)=-1-6$

    $-14+7=-1+(-6)$

    $-7=-7$

37. $t-4=-8-(-2)$

    $t-4=-8+2$

    $t-4=-6$

    $t-4+4=-6+4$

    $t=-2$

    Check:

    $t-4=-8-(-2)$

    $-2-4=-8+2$

    $-2+(-4)=-6$

    $-6=-6$

39. $x-5=-5$

    $x-5+5=-5+5$

    $x=0$

    Check:

    $x-5=-5$

    $0-5=-5$

    $-5=-5$

41. $-2s=16$

    $\dfrac{-2s}{-2}=\dfrac{16}{-2}$

    $s=-8$

    Check:

    $-2s=16$

    $-2(-8)=16$

    $16=16$

43. $-5t=-25$

    $\dfrac{-5t}{-5}=\dfrac{-25}{-5}$

    $t=5$

    Check:

    $-5t=-25$

    $-5(5)=-25$

    $-25=-25$

45. $-2+(-4)=-3n$

    $-6=-3n$

    $\dfrac{-6}{-3}=\dfrac{-3n}{-3}$

    $2=n \ \ or \ \ n=2$

    Check:

    $-2+(-4)=-3n$

    $-6=-3(2)$

    $-6=-6$

47. $-9h = -3(-3)$

$-9h = 9$

$\dfrac{-9h}{-9} = \dfrac{9}{-9}$

$h = -1$

Check:

$-9h = -3(-3)$

$-9(-1) = 9$

$9 = 9$

49. $\dfrac{t}{-3} = -2$

$-3\left(\dfrac{t}{-3}\right) = -2(-3)$

$t = 6$

Check:

$\dfrac{t}{-3} = -2$

$\dfrac{6}{-3} = -2$

$-2 = -2$

51. $0 = \dfrac{y}{8}$

$8(0) = 8\left(\dfrac{y}{8}\right)$

$0 = y \ \ or \ \ y = 0$

Check:

$0 = \dfrac{y}{8}$

$0 = \dfrac{0}{8}$

$0 = 0$

53. $\dfrac{x}{-2} = -6 + 3$

$\dfrac{x}{-2} = -3$

$-2\left(\dfrac{x}{-2}\right) = -2(-3)$

$x = 6$

Check:

$\dfrac{x}{-2} = -6 + 3$

$\dfrac{6}{-2} = -3$

$-3 = -3$

55.
$$\frac{x}{4} = -5 - 8$$
$$\frac{x}{4} = -5 + (-8)$$
$$\frac{x}{4} = -13$$
$$4\left(\frac{x}{4}\right) = 4(-13)$$
$$x = -52$$

Check:
$$\frac{x}{4} = -5 - 8$$
$$\frac{-52}{4} = -5 + (-8)$$
$$-11 = -11$$

57.
$$2y + 8 = -6$$
$$2y + 8 - 8 = -6 - 8$$
$$2y = -6 + (-8)$$
$$2y = -14$$
$$\frac{2y}{2} = \frac{-14}{2}$$
$$y = -7$$

Check:
$$2y + 8 = -6$$
$$2(-7) + 8 = -6$$
$$-14 + 8 = -6$$
$$-6 = -6$$

59.
$$-21 = 4h - 5$$
$$-21 + 5 = 4h - 5 + 5$$
$$-16 = 4h$$
$$\frac{-16}{4} = \frac{4h}{4}$$
$$-4 = h \ \ or \ \ h = -4$$

Check:
$$-21 = 4h - 5$$
$$-21 = 4(-4) - 5$$
$$-21 = -16 - 5$$
$$-21 = -16 + (-5)$$
$$-21 = -21$$

61.
$$-3v + 1 = 16$$
$$-3v + 1 - 1 = 16 - 1$$
$$-3v = 15$$
$$\frac{-3v}{-3} = \frac{15}{-3}$$
$$v = -5$$

Check:
$$-3v + 1 = 16$$
$$-3(-5) + 1 = 16$$
$$15 + 1 = 16$$
$$16 = 16$$

63.
$$8 = -3x + 2$$
$$8 - 2 = -3x + 2 - 2$$
$$6 = -3x$$
$$\frac{6}{-3} = \frac{-3x}{-3}$$
$$-2 = x \ \ or \ \ x = -2$$

Check:
$$8 = -3x + 2$$
$$8 = -3(-2) + 2$$
$$8 = 6 + 2$$
$$8 = 8$$

65.
$$-35 = 5 - 4x$$
$$-35 - 5 = 5 - 5 - 4x$$
$$-35 + (-5) = -4x$$
$$-40 = -4x$$
$$\frac{-40}{-4} = \frac{-4x}{-4}$$
$$10 = x \ \ or \ \ x = 10$$

Check:
$$-35 = 5 - 4x$$
$$-35 = 5 - 4(10)$$
$$-35 = 5 - 40$$
$$-35 = 5 + (-40)$$
$$-35 = -35$$

67.
$$4 - 5x = 34$$
$$4 - 4 - 5x = 34 - 4$$
$$-5x = 30$$
$$\frac{-5x}{-5} = \frac{30}{-5}$$
$$x = -6$$

Check:
$$4 - 5x = 34$$
$$4 - 5(-6) = 34$$
$$4 - (-30) = 34$$
$$4 + 30 = 34$$
$$34 = 34$$

69.
$$-5 - 6 - 5x = 4$$
$$-5 + (-6) - 5x = 4$$
$$-11 - 5x = 4$$
$$-11 + 11 - 5x = 4 + 11$$
$$-5x = 15$$
$$\frac{-5x}{-5} = \frac{15}{-5}$$
$$x = -3$$

Check:
$$-5 - 6 - 5x = 4$$
$$-5 + (-6) - 5(-3) = 4$$
$$-11 - (-15) = 4$$
$$-11 + 15 = 4$$
$$4 = 4$$

71.
$$4 - 6x = -5 - 9$$
$$4 - 6x = -5 + (-9)$$
$$4 - 6x = -14$$
$$4 - 4 - 6x = -14 - 4$$
$$-6x = -14 + (-4)$$
$$-6x = -18$$
$$\frac{-6x}{-6} = \frac{-18}{-6}$$
$$x = 3$$

Check:
$$4 - 6x = -5 - 9$$
$$4 - 6(3) = -5 + (-9)$$
$$4 - 18 = -14$$
$$4 + (-18) = -14$$
$$-14 = -14$$

73.
$$\frac{h}{-6} + 4 = 5$$
$$\frac{h}{-6} + 4 - 4 = 5 - 4$$
$$\frac{h}{-6} = 1$$
$$-6\left(\frac{h}{-6}\right) = (-6)1$$
$$h = -6$$

Check:
$$\frac{h}{-6} + 4 = 5$$
$$\frac{-6}{-6} + 4 = 5$$
$$1 + 4 = 5$$
$$5 = 5$$

75.

$$-2(4) = \frac{t}{-6} + 1$$

$$-8 = \frac{t}{-6} + 1$$

$$-8 - 1 = \frac{t}{-6} + 1 - 1$$

$$-8 + (-1) = \frac{t}{-6}$$

$$-9 = \frac{t}{-6}$$

$$-6(-9) = -6\left(\frac{t}{-6}\right)$$

$$54 = t \ \ or \ \ t = 54$$

Check:

$$-2(4) = \frac{t}{-6} + 1$$

$$-8 = \frac{54}{-6} + 1$$

$$-8 = -9 + 1$$

$$-8 = -8$$

77.

$$0 = 6 + \frac{c}{-5}$$

$$0 - 6 = 6 - 6 + \frac{c}{-5}$$

$$-6 = \frac{c}{-5}$$

$$-5(-6) = -5\left(\frac{c}{-5}\right)$$

$$30 = c \ \ or \ \ c = 30$$

Check:

$$0 = 6 + \frac{c}{-5}$$

$$0 = 6 + \frac{30}{-5}$$

$$0 = 6 + (-6)$$

$$0 = 0$$

79.

$$-1 = -8 + \frac{h}{-2}$$

$$-1 + 8 = -8 + 8 + \frac{h}{-2}$$

$$7 = \frac{h}{-2}$$

$$-2(7) = -2\left(\frac{h}{-2}\right)$$

$$-14 = h \ \ or \ \ h = -14$$

Check:

$$-1 = -8 + \frac{h}{-2}$$

$$-1 = -8 + \frac{-14}{-2}$$

$$-1 = -8 + 7$$

$$-1 = -1$$

81.

$$2x + 3(0) = -6$$

$$2x + 0 = -6$$

$$2x = -6$$

$$\frac{2x}{2} = \frac{-6}{2}$$

$$x = -3$$

Check:

$$2x + 3(0) = -6$$

$$2(-3) + 0 = -6$$

$$-6 = -6$$

83. $2(0) - 2y = 4$

$0 - 2y = 4$

$-2y = 4$

$$\frac{-2y}{-2} = \frac{4}{-2}$$

$y = -2$

Check:

$2(0) - 2y = 4$

$0 - 2(-2) = 4$

$4 = 4$

85. $-x = 8$

$$\frac{-x}{-1} = \frac{8}{-1}$$

$x = -8$

Check:

$-x = 8$

$-(-8) = 8$

$8 = 8$

87. $-15 = -k$

$$\frac{-15}{-1} = \frac{-k}{-1}$$

$15 = k \ \ or \ \ k = 15$

Check:

$-15 = -k$

$-15 = -(15)$

$-15 = -15$

Applications

89. SHARKS

    Analyze the problem

    - The first observations were at **–120** ft.

    - The next observations were at **–75** ft.

    - We must find **how many feet the cage was raised**.

    Form an equation

    Let x = **the number of feet the cage was raised**.

    Key Word: raised

    Translation: add

    Translate the words of the problem into an equation.

The first position of the cage	plus	the amount the cage was raised	is	the second position of the cage.
-120	$+$	x	$=$	-75

    Solve the equation

    $$-120 + x = -75$$
    $$-120 + x + 120 = -75 + 120$$
    $$x = 45$$

    State the conclusion

    The shark cage was raised 45 feet.

    Check the result

    If we add the number of feet the cage was raised to the first position, we get

    $-120 + 45 = -75$. The answer checks..

91.   MARKET SHARE

$x =$ gain in market share

$$-43 + x = -9$$
$$-43 + 43 + x = -9 + 43$$
$$x = 34$$

The company gained 34 points of market share.

93.   CHECKING ACCOUNTS

$x =$ balance before deposit

$$x + 220 = -215$$
$$x + 220 - 220 = -215 - 220$$
$$x = -215 + (-220)$$
$$x = -435$$

The balance was $-\$435$ before the deposit.

95.   POLLS

$x =$ points he gained in polls

$$-31 + x = -2$$
$$-31 + 31 + x = -2 + 31$$
$$x = 29$$

He gained 29 points in the polls.

97.   PRICE REDUCTIONS

$x =$ price drop each month

$$12x = -60$$
$$\frac{12x}{12} = \frac{-60}{12}$$
$$x = -5$$

The price dropped $5 per month.

99.   DREDGING A HARBOR

$x =$ feet needed to be dredged out

$$-47 - x = -65$$
$$-47 - x + 47 = -65 + 47$$
$$-x = -18$$
$$\frac{-x}{-1} = \frac{-18}{-1}$$
$$x = 18$$

18 feet must be dredged out.

101.   INTERNATIONAL TIME ZONES

$x =$ Seattle time zone

$$x = 9 - 17$$
$$x = 9 + (-17)$$
$$x = -8$$

Seattle is in zone -8.

Writing

103. There is still a negative sign on the variable. When a problem is solved, the variable should have a coefficient of positive one.

Review

105. $5^6 = 5 \bullet 5 \bullet 5 \bullet 5 \bullet 5 \bullet 5$

107.
$$7 + 3y = 43$$
$$7 - 7 + 3y = 43 - 7$$
$$3y = 36$$
$$\frac{3y}{3} = \frac{36}{3}$$
$$y = 12$$

109. $16 \div 8 = \dfrac{16}{8}$

Key Concept Signed Numbers

1. -5 3. -30

5. $+10$ or 10 7. -205

9.

negative   positive

11   $x < y$

13. **Addition**

Like signs: Add their absolute values and attach their common sign to the sum.

Unlike signs: Subtract their absolute values, the smaller from the larger, and attach the sign of the number with the larger absolute value to that result.

15. **Division**

Like signs: Divide their absolute values. The quotient is positive.

Unlike signs: Divide their absolute values. The quotient is negative.

Chapter Two Review

Section 2.1 An Introduction to the Integers

1.

2.

3. $-7 < 0$

4. $-20 < -19$

5. $|-16| \geq -16$

6. $56 \leq 56$
or $56 \geq 56$

7. WATER PRESSURE
-33 ft

8. $-\$1,200$

9. -10 sec

10. $|-4| = 4$

11. $|0| = 0$

12. $|-43| = 43$

13. $-|12| = -12$

14. negative

15. the opposite

16. negative

17. minus

18. $-(-12) = 12$

19. -8

20. 8

21. 0

Section 2.2 Adding Integers

22.

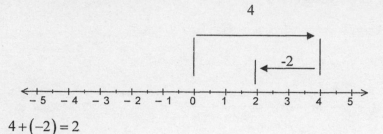

$$4 + (-2) = 2$$

23.

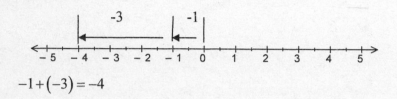

$$-1 + (-3) = -4$$

24. $-6 + (-4) = -10$ 25. $-23 + (-60) = -83$

26. $-1 + (-4) + (-3) = -5 + (-3)$ 27. $-4 + 3 = -1$
$$= -8$$

28. $-28 + 140 = 112$ 29. $9 + (-20) = -11$

30. $3 + (-2) + (-4) = 1 + (-4)$ 31. $(-2 + 1) + [(-5) + 4] = -1 + (-1)$
$$= -3$$ $$= -2$$

32. $-4 + 0 = -4$ 33. $0 + (-20) = -20$

34. $-8 + 8 = 0$ 35. $73 + (-73) = 0$

36. 11 37. -4

38.   DROUGHT

    $L =$ water level after rain

    $L = -100 + 35$

    $L = -65$

    The water level was 65 ft below normal.

Section 2.3 Subtracting Integers

39. $5 - 8 = 5 + (-8)$
          $= -3$

40. $-9 - 12 = -9 + (-12)$
             $= -21$

41. $-4 - (-8) = -4 + 8$
            $= 4$

42. $-6 - 106 = -6 + (-106)$
             $= -112$

43. $-8 - (-2) = -8 + 2$
            $= -6$

44. $7 - 1 = 6$

45. $0 - 37 = -37$

46. $0 - (-30) = 0 + 30$
              $= 30$

47.   Subtracting a number is the same as **adding** the **opposite** of that number.

48. $-9 - 7 + 12 = -9 + (-7) + 12$
                $= -16 + 12$
                $= -4$

49. $7 - [(-6) - 2] = 7 - [(-6) + (-2)]$
                $= 7 - (-8)$
                $= 7 + 8$
                $= 15$

50. $1 - (2 - 7) = 1 - [2 + (-7)]$
              $= 1 - (-5)$
              $= 1 + 5$
              $= 6$

51. $-12 - (6 - 10) = -12 - [6 + (-10)]$
                $= -12 - (-4)$
                $= -12 + 4$
                $= -8$

52. $-50-27 = -50+(-27)$
$$= -77$$

53. $2-\left[-(-3)\right] = 2-3$
$$= 2+(-3)$$
$$= -1$$

54. GOLD MINING

D = depth of second discovery

$D = -150-75$

$D = -150+(-75)$

$D = -225$

The second discovery was made at $-225\,\text{ft}$.

55. RECORD TEMPERATURES

Alaska $= 100-(-80)$ Virginia $= 110-(-30)$

Alaska $= 180$ Virginia $= 140$

The difference for Alaska was $180°$.   The difference for Virginia was $140°$.

Section 2.4 Multiplying Integers

56. $-9 \bullet 5 = -45$ 57. $-3(-6) = 18$

58. $7(-2) = -14$ 59. $(-8)(-47) = 376$

60. $-20 \bullet 5 = -100$ 61. $-1(-1) = 1$

62. $-1(25) = -25$ 63. $(5)(-30) = -150$

64. $(-6)(-2)(-3) = 12(-3)$
$= -36$

65. $4(-3)3 = (-12)3$
$= -36$

66. $0(-7) = 0$

67. $(-1)(-1)(-1)(-1) = 1$

68. TAX DEFICITS

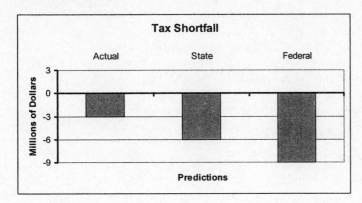

69. $(-5)^2 = 25$

70. $(-2)^5 = -32$

71. $(-8)^2 = 64$

72. $(-4)^3 = -64$

73. negative

74. $-2^2 = -4$; the base is 2
$(-2)^2 = 4$; the base is -2

Section 2.5 Dividing Integers

75. $\dfrac{-15}{5} = -3$ because $5(-3) = -15$

76. $\dfrac{-14}{7} = -2$

77. $\dfrac{25}{-5} = -5$

78. $-64 \div 8 = -8$

79. $\dfrac{-202}{-2} = 101$

80. $\dfrac{0}{-5} = 0$

81. $\dfrac{-4}{0}$ = undefined

82. $\dfrac{-673}{-673} = 1$

83. $\dfrac{-10}{-1} = 10$

84. PRODUCTION TIME

D = drop in production time per month

$D = \dfrac{-12}{6}$

$D = -2$

Production time dropped

2 min per month.

Section 2.6 Order of Operations and Estimation

85. $2 + 4(-6) = 2 + (-24)$
$= -22$

86. $7 - (-2)^2 + 1 = 7 - 4 + 1$
$= 3 + 1$
$= 4$

87. $2 - 5(4) + (-25) = 2 - 20 + (-25)$
$= 2 + (-20) + (-25)$
$= -18 + (-25)$
$= -43$

88. $-3(-2)^3 - 16 = -3(-8) - 16$
$= 24 - 16$
$= 8$

89. $-2(5)(-4) + \dfrac{|-9|}{3^2} = -2(5)(-4) + \dfrac{9}{9}$
$= -2(5)(-4) + 1$
$= -10(-4) + 1$
$= 40 + 1$
$= 41$

90. $-4^2 + (-4)^2 = -16 + 16$
$= 0$

91.
$$-12-(8-9)^2 = -12-(-1)^2$$
$$= -12-1$$
$$= -12+(-1)$$
$$= -13$$

92.
$$7|-8|-2(3)(4) = 7(8)-2(3)(4)$$
$$= 56-2(3)(4)$$
$$= 56-6(4)$$
$$= 56-24$$
$$= 32$$

93.
$$-4\left(\frac{15}{-3}\right)-2^3 = -4(-5)-2^3$$
$$= -4(-5)-8$$
$$= 20-8$$
$$= 12$$

94.
$$-20+2(12-5\bullet 2) = -20+2(12-10)$$
$$= -20+2(2)$$
$$= -20+4$$
$$= -16$$

95.
$$-20+2\left[12-(-7+5)^2\right] = -20+2\left[12-(-2)^2\right]$$
$$= -20+2(12-4)$$
$$= -20+2(8)$$
$$= -20+16$$
$$= -4$$

96.
$$8-|-3\bullet 4+5| = 8-|-12+5|$$
$$= 8-|-7|$$
$$= 8-7$$
$$= 1$$

97.
$$\frac{10+(-6)}{-3-1} = \frac{4}{-3+(-1)}$$
$$= \frac{4}{-4}$$
$$= -1$$

98.
$$\frac{3(-6)-11+1}{4^2-3^2} = \frac{-18-11+1}{16-9}$$
$$= \frac{-18+(-11)+1}{7}$$
$$= \frac{-29+1}{7}$$
$$= \frac{-28}{7}$$
$$= -4$$

99.   about -70

100.   about 20

101.   about $-7{,}000$

102.   about 1,100

Section 2.7 Solving Equations Involving Integers

103.
$$2x + 6 = -2$$
$$2(-4) + 6 = -2$$
$$-8 + 6 = -2$$
$$-2 = -2$$

Yes, it gives a true statement when checked in the equation.

104.
$$6 + \frac{x}{2} = -4$$
$$6 + \frac{-4}{2} = -4$$
$$6 + (-2) = -4$$
$$4 \neq -4$$

No, it gives a false statement when checked in the equation.

105.
$$t + (-8) = -18$$
$$t + (-8) + 8 = -18 + 8$$
$$t = -10$$

Check:
$$t + (-8) = -18$$
$$-10 + (-8) = -18$$
$$-18 = -18$$

106.
$$\frac{x}{-3} = -4$$
$$-3\left(\frac{x}{-3}\right) = -3(-4)$$
$$x = 12$$

Check:
$$\frac{x}{-3} = -4$$
$$\frac{12}{-3} = -4$$
$$-4 = -4$$

107.
$$y + 8 = 0$$
$$y + 8 - 8 = 0 - 8$$
$$y = -8$$

Check:

$$y + 8 = 0$$
$$-8 + 8 = 0$$
$$0 = 0$$

108.
$$-7m = -28$$
$$\frac{-7m}{-7} = \frac{-28}{-7}$$
$$m = 4$$

Check:

$$-7m = -28$$
$$-7(4) = -28$$
$$-28 = -28$$

109.
$$-x = -15$$
$$\frac{-x}{-1} = \frac{-15}{-1}$$
$$x = 15$$

Check:

$$-x = -15$$
$$-(15) = -15$$
$$-15 = -15$$

110.
$$4 = -y$$
$$\frac{4}{-1} = \frac{-y}{-1}$$
$$-4 = y \ or \ y = -4$$

Check:

$$4 = -y$$
$$4 = -(-4)$$
$$4 = 4$$

111.
$$-5t + 1 = -14$$
$$-5t + 1 - 1 = -14 - 1$$
$$-5t = -14 + (-1)$$
$$-5t = -15$$
$$\frac{-5t}{-5} = \frac{-15}{-5}$$
$$t = 3$$

Check:

$$-5t + 1 = -14$$
$$-5(3) + 1 = -14$$
$$-15 + 1 = -14$$
$$-14 = -14$$

112.
$$3(2) = 2 - 2x$$
$$6 = 2 - 2x$$
$$6 - 2 = 2 - 2x - 2$$
$$4 = -2x$$
$$\frac{4}{-2} = \frac{-2x}{-2}$$
$$-2 = x \ or \ x = -2$$

Check:

$$3(2) = 2 - 2x$$
$$6 = 2 - 2(-2)$$
$$6 = 2 - (-4)$$
$$6 = 2 + 4$$
$$6 = 6$$

113.

$$\frac{x}{-4} - 5 = -1 - 1$$

$$\frac{x}{-4} - 5 = -1 + (-1)$$

$$\frac{x}{-4} - 5 = -2$$

$$\frac{x}{-4} - 5 + 5 = -2 + 5$$

$$\frac{x}{-4} = 3$$

$$-4\left(\frac{x}{-4}\right) = -4(3)$$

$$x = -12$$

Check:

$$\frac{x}{-4} - 5 = -1 - 1$$

$$\frac{(-12)}{-4} - 5 = -1 + (-1)$$

$$3 - 5 = -2$$

$$3 + (-5) = -2$$

$$-2 = -2$$

114.

$$c - (-5) = 5$$

$$c + 5 = 5$$

$$c + 5 - 5 = 5 - 5$$

$$c = 0$$

Check:

$$c - (-5) = 5$$

$$0 + 5 = 5$$

$$5 = 5$$

115.   CHECKING ACCOUNTS

x = account balance before deposit

$$x + 85 = -47$$

$$x + 85 - 85 = -47 - 85$$

$$x = -47 + (-85)$$

$$x = -132$$

His balance was −$132 before the
deposit.

116.   WIND-CHILL FACTOR

x = perceived temperature change

$$-5 + x = -51$$

$$-5 + x + 5 = -51 + 5$$

$$x = -46$$

The perceived change is $-46°$.

117.   CREDIT CARD PROMOTION

x = customers that applied for credit

$$-8x = -968$$

$$\frac{-8x}{-8} = \frac{-968}{-8}$$

$$x = 121$$

121 people applied for credit.

118.   BANK FAILURE

x = each members debt

$$7x = 57,400$$

$$\frac{7x}{7} = \frac{57,400}{7}$$

$$x = 8,200$$

Each person assumed $8,200 of debt.

Chapter Two Test

1a. $-8 > -9$ 1b. $-8 < |-8|$ 1c.   the opposite of $5 < 0$

2. $\{..., -3, -2, -1, 0, 1, 2, 3, ...\}$

3.   SCHOOL ENROLLMENTS

Monroe high school

4.

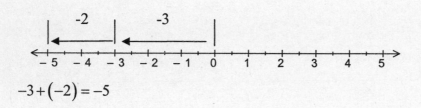

$-3 + (-2) = -5$

5a. $-65 + 31 = -34$ 5b. $-17 + (-17) = -34$

5c. $\left[6 + (-4)\right] + \left[-6 + (-4)\right] = 2 + (-10)$

$= -8$

6a. $-7 - 6 = -7 + (-6)$ 6b. $-7 - (-6) = -7 + 6$

$= -13$ $= -1$

6c. $0 - 15 = 0 + (-15)$ 6d. $-60 - 50 - 40 = -60 + (-50) + (-40)$

$= -15$ $= -150$

7a. $-10 \cdot 7 = -70$

7b. $-4(-2)(-6) = 8(-6)$
$$= -48$$

7c. $(-2)(-2)(-2)(-2) = 4(-2)(-2)$
$$= -8(-2)$$
$$= 16$$

7d. $-55(0) = 0$

8. $(-4)(5) = -20$

9a. $\dfrac{-32}{4} = -8$

9b. $\dfrac{8}{6-6} = \dfrac{8}{0} = \text{undefined}$

9c. $\dfrac{-5}{1} = -5$

9d. $\dfrac{0}{-6} = 0$

10.   BUSINESS TAKEOVERS

Total debt = $-6 + (-10) + (-2) = -18$

x = each businessman's debt

$6x = -18$

$\dfrac{6x}{6} = \dfrac{-18}{6}$

$x = -3$

Each businessman will incur a debt of $3 million.

11.   GEOGRAPHY

x = difference in elevations

$x = -282 - (-436)$

$x = -282 + 436$

$x = 154$

The difference is 154 ft.

12a.　$-(-6) = 6$

12b　$|-7| = 7$

12c.　$|-9+3| = |-6|$
　　　　$= 6$

12d.　$2|-66| = 2(66)$
　　　　$= 132$

13a.　$(-4)^2 = (-4)(-4)$
　　　　$= 16$

13b.　$-4^2 = -(4)(4)$
　　　　$= -16$

13a.　$(-4)^2 = (-4)(-4)$
　　　　$= 16$

14.　$-18 \div 2 \bullet 3 = -9 \bullet 3$
　　　　　　$= -27$

15.　$4 - (-3)^2 + 6 = 4 - 9 + 6$
　　　　　　　　$= 4 + (-9) + 6$
　　　　　　　　$= -5 + 6$
　　　　　　　　$= 1$

16.　$-3 + \left(\dfrac{-16}{4} \right) - 3^3 = -3 + (-4) - 3^3$
　　　　　　　　　　$= -3 + (-4) - 27$
　　　　　　　　　　$= -7 - 27$
　　　　　　　　　　$= -7 + (-27)$
　　　　　　　　　　$= -34$

17.　$-10 + 2 \left[6 - (-2)^2 (-5) \right]$
　　　$= -10 + 2 \left[6 - (4)(-5) \right]$
　　　$= -10 + 2 \left[6 - (-20) \right]$
　　　$= -10 + 2 [6 + 20]$
　　　$= -10 + 2 [26]$
　　　$= -10 + 52$
　　　$= 42$

18.　$\dfrac{4(-6) - 2^2}{-3 - 4} = \dfrac{4(-6) - 4}{-3 + (-4)}$
　　　　　　　$= \dfrac{-24 - 4}{-7}$
　　　　　　　$= \dfrac{-24 + (-4)}{-7}$
　　　　　　　$= \dfrac{-28}{-7}$
　　　　　　　$= 4$

19.　**TEMPERATURE CHANGE**

$t =$ temperature change

$t = -6(12)$

$t = -72° \text{ F}$

20. GAUGES

The arrow will move to the left.

$n = $ new reading

$n = 5 - 8 - 4$

$n = 5 + (-8) + (-4)$

$n = -7$

The new reading will be -7.

21. CARD GAMES

$s = $ score after 2nd round

$s = 8 - (3 + 10 + 9 + 1)$

$s = 8 - 23$

$s = 8 + (-23)$

$s = -15$

His score is -15 after 2 rounds.

22.
$$\frac{-10}{5} - 16 = -14$$
$$-2 - 16 = -14$$
$$-2 + (-16) = -14$$
$$-18 \neq -14$$
no

23.
$$c - (-7) = -8$$
$$c + 7 = -8$$
$$c + 7 - 7 = -8 - 7$$
$$c = -8 + (-7)$$
$$c = -15$$

24.
$$6 - x = -10$$
$$6 - x - 6 = -10 - 6$$
$$-x = -10 + (-6)$$
$$-x = -16$$
$$\frac{-x}{-1} = \frac{-16}{-1}$$
$$x = 16$$

25.
$$\frac{x}{-4} = 10$$
$$-4\left(\frac{x}{-4}\right) = -4(10)$$
$$x = -40$$

26.
$$-6x = 0$$
$$\frac{-6x}{-6} = \frac{0}{-6}$$
$$x = 0$$

27.
$$3x + (-7) = -11 + (-11)$$
$$3x + (-7) = -22$$
$$3x + (-7) + 7 = -22 + 7$$
$$3x = -15$$
$$\frac{3x}{3} = \frac{-15}{3}$$
$$x = -5$$

28.
$$-5 = -6a + 7$$
$$-5 - 7 = -6a + 7 - 7$$
$$-5 + (-7) = -6a$$
$$-12 = -6a$$
$$\frac{-12}{-6} = \frac{-6a}{-6}$$
$$2 = a \;\; or \;\; a = 2$$

29.
$$\frac{x}{-2} + 3 = (-2)(-6)$$
$$\frac{x}{-2} + 3 = 12$$
$$\frac{x}{-2} + 3 - 3 = 12 - 3$$
$$\frac{x}{-2} = 9$$
$$-2\left(\frac{x}{-2}\right) = -2(9)$$
$$x = -18$$

30.   CHECKING ACCOUNTS

x = balance before deposit

$$x + 225 = -19$$
$$x + 225 - 225 = -19 - 225$$
$$x = -19 + (-225)$$
$$x = -244$$

The balance was –$244.

31.   HOSPITAL CAPACITY

x = patients released

$$-3 - x = -21$$
$$-3 - x + 3 = -21 + 3$$
$$-x = -18$$
$$\frac{-x}{-1} = \frac{-18}{-1}$$
$$x = 18$$

18 patients were released.

32.
$$5(-4) = -4 + (-4) + (-4) + (-4) + (-4)$$
$$= -20$$

33.   The absolute value of a number is the distance that number is from zero, and distance can never be negative.

34.   It is true because $12 = 12$.

Chapter 1-2 Cumulative Review Exercises

1. 1, 2, 5, 9

2. 0, 1, 2, 5, 9

3. $-2, -1$

4. $-2, -1, 0, 1, 2, 5, 9$

5. 6

6. 3

7. 7,326,500

8. 7,330,000

9. BIDS

CRF cable should get the contract.

10. NUCLEAR POWER

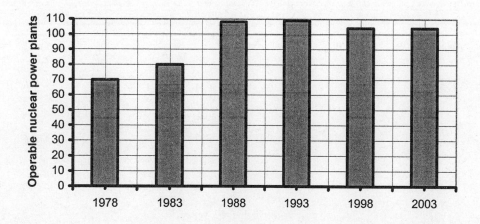

11. $237 + 549 = 786$

12. $6,375 - 2,569 = 3,806$

13.
$$
\begin{array}{r}
5,369 \\
-\ \ 685 \\
\hline
4,684
\end{array}
$$

14.
$$
\begin{array}{r}
7,899 \\
+\ 5,237 \\
\hline
13,136
\end{array}
$$

15. P=perimeter of garden

$P = 17 + 35 + 17 + 35$

$P = 104$

The perimeter is 104 ft.

A=area of garden

$A = (17)(35)$

$A = 595$

The area is 595 ft^2.

16. x = number of wooden chairs

$x = 147 - 27 - 55$

$x = 65$

There are 65 wooden chairs.

17.
```
    435
 ×   27
   3045
   8700
 11,745
```

18.
```
        13
 97)1261 = 13
     97
    291
    291
      0
```

19.
```
   4,587
 ×    67
   32109
  275220
 307,329
```

20.
```
       467
 38)17746 = 467
    152
    254
    228
    266
    266
      0
```

21. SHIPPING

T = number of tennis balls

$T = 12(12)(12)$

$T = 144(12)$

$T = 1,728$

There are 1,728 tennis balls in 12 gross.

22. 1, 2, 3, 6, 9, 18

23. Prime, odd

24. Composite, even

25. Even (neither prime nor composite)

26. Odd, (neither prime nor composite)

27.

504

2 252

2 126

2 63

3 21

3 7

$2^3 \cdot 3^2 \cdot 7$

28. 11^4

29. $5^2 \cdot 7 = 25 \cdot 7$

$\quad = 175$

30.

$16 + 2\left[14 - 3(5-4)^2\right]$

$= 16 + 2\left[14 - 3(1)^2\right]$

$= 16 + 2\left[14 - 3(1)\right]$

$= 16 + 2\left[14 - 3\right]$

$= 16 + 2\left[11\right]$

$= 16 + 22$

$= 38$

31. $25 + 5 \cdot 5 = 25 + 25$

$\quad = 50$

32.

$\dfrac{16 - 2 \cdot 3}{2 + (9-6)} = \dfrac{16 - 6}{2 + 3}$

$\qquad\qquad = \dfrac{10}{5}$

$\qquad\qquad = 2$

33.   SPEED CHECK

A=average speed

$$A = \frac{38 + 42 + 36 + 38 + 48 + 44}{6}$$

$$A = \frac{246}{6}$$

$$A = 41$$

No, they are not obeying the speed limit.

34. $3x - 2 = 16$

$3(6) - 2 = 16$

$18 - 2 = 16$

$16 = 16$

Yes, when substituted into the equation, you obtain a true statement.

35      $50 = x + 37$

$50 - 37 = x + 37 - 37$

$13 = x \;\; or \;\; x = 13$

Check:

$50 = x + 37$

$50 = 13 + 37$

$50 = 50$

36      $a - 12 = 41$

$a - 12 + 12 = 41 + 12$

$a = 53$

Check:

$a - 12 = 41$

$53 - 12 = 41$

$41 = 41$

37. $5p = 135$

$$\frac{5p}{5} = \frac{135}{5}$$

$$p = 27$$

Check:

$5p = 135$

$5(27) = 135$

$135 = 135$

38.

$$\frac{y}{8} = 3$$

$$8\left(\frac{y}{8}\right) = 8(3)$$

$$y = 24$$

Check:

$$\frac{y}{8} = 3$$

$$\frac{24}{8} = 3$$

$$3 = 3$$

39.

40.

41. True

42. $$3^2 = (3)(3) = 9$$
$$-3^2 = -(3)(3) = -9$$

43. $$-2 + (-3) = -5$$

44. $$-15 + 10 + (-9) = -5 + (-9)$$
$$= -14$$

45. $$-3 - 5 = -3 + (-5)$$
$$= -8$$

46. $$-15^2 - 2|-3| = -15^2 - 2(3)$$
$$= -225 - 2(3)$$
$$= -225 - 6$$
$$= -225 + (-6)$$
$$= -231$$

47. $$(-8)(-3) = 24$$

48. $$5(-7)^3 = 5(-343)$$
$$= -1,715$$

49. $$\frac{-14}{-7} = 2$$

50. $$\frac{450}{-9} = -50$$

51. $$5 + (-3)(-7) = 5 + 21$$
$$= 26$$

52. $$-20 + 2\big[12 - 5(-2)(-1)\big]$$
$$= -20 + 2\big[12 - (-10)(-1)\big]$$
$$= -20 + 2\big[12 - 10\big]$$
$$= -20 + 2\big[2\big]$$
$$= -20 + 4$$
$$= -16$$

53. $$\frac{10 - (-5)}{1 - 2 \bullet 3} = \frac{10 + 5}{1 - 6}$$
$$= \frac{15}{1 + (-6)}$$
$$= \frac{15}{-5}$$
$$= -3$$

54. $$\frac{3(-6) - 10}{3^2 - 4^2} = \frac{-18 - 10}{9 - 16}$$
$$= \frac{-18 + (-10)}{9 + (-16)}$$
$$= \frac{-28}{-7}$$
$$= 4$$

55.

$$-5t + 1 = -14$$
$$-5t + 1 - 1 = -14 - 1$$
$$-5t = -14 + (-1)$$
$$-5t = -15$$
$$\frac{-5t}{-5} = \frac{-15}{-5}$$
$$t = 3$$

Check:

$$-5t + 1 = -14$$
$$-5(3) + 1 = -14$$
$$-15 + 1 = -14$$
$$-14 = -14$$

56.

$$\frac{x}{-3} - 2 = -2(-2)$$
$$\frac{x}{-3} - 2 = 4$$
$$\frac{x}{-3} - 2 + 2 = 4 + 2$$
$$\frac{x}{-3} = 6$$
$$-3\left(\frac{x}{-3}\right) = -3(6)$$
$$x = -18$$

Check:

$$\frac{x}{-3} - 2 = -2(-2)$$
$$\frac{-18}{-3} - 2 = 4$$
$$6 - 2 = 4$$
$$4 = 4$$

57. BUYING A BUSINESS

x = each person's share of the debt

$$12x = 1,512,444$$
$$\frac{12x}{12} = \frac{1,512,444}{12}$$
$$x = 126,037$$

Each investor assumes $126,037.

58. THE MOON

x = minimum temperature

$$261 - x = 540$$
$$261 - x - 261 = 540 - 261$$
$$-x = 279$$
$$\frac{-x}{-1} = \frac{279}{-1}$$
$$x = -279$$

$-279°$ F is the minimum temperature.

Chapter 3 The Language of Algebra

3.1 Variables and Algebraic Expressions

Vocabulary

1. An algebraic **expression** is a combination of variables, numbers, and the operation symbols for addition, subtraction, multiplication, and division.

3. A **variable** is a letter that is used to stand for a number.

Concepts

5. $10 + 3x$, $3x - 10$, answers will vary   7.   Mr. Lamb, 15 miles farther

9.

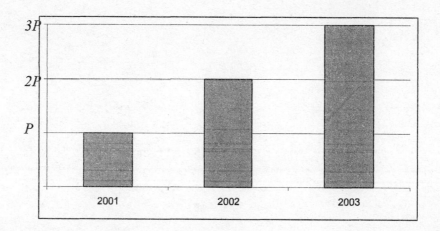

11.

Wind conditions	Speed of the jet (mph)
In still air	500
With the tail wind	$500 + x$
Against the head wind	$500 - x$

13. $\dfrac{h}{4}$ 15. $450 - x$ inches of tape left on the roll.

Notation

17. $8x$

19. $\dfrac{10}{g}$

Practice

21. $x-9$

23. $\dfrac{2}{3}p$

25. $6+r$

27. $d-15$

29. $1-s$

31. $2p$

33. $s+14$

35. $\dfrac{35}{b}$

37. $x-2$

39. c increased by 7

41.   7 less than c

43a. $60m$

43b. $3{,}600h$

45a. $\dfrac{s}{12}$

45b. $\dfrac{s}{52}$

47a. $12f$ in.

47b. $\dfrac{f}{3}$ yd

49. $j-5$

51. $6s$

53. $\dfrac{p}{15}$ days

55. $t+2$

57. $w=$ width
$w+6=$ length

59. $g=$ gallons drained out
$6-g=$ gallons remaining

61. $3x+5$

63. $10a+12$

Applications

65. ELECTIONS

x = votes received by Nixon

$x+118,550$ = votes received by Kennedy

67. THE BEATLES

c = number of copies of I Want to Hold Your Hand sold

$c-2,000,000$ = number of copies of Hey Jude sold

Writing

69. Variables are used to express unknown quantities.

71. A better approach would be to say Let x = the height of the building. How can x be a building?

Review

73. $-5+(-6)+1=-11+1$
$$=-10$$

75. $-x=4$
$$\frac{-x}{-1}=\frac{4}{-1}$$
$$x=-4$$

77. $\{...,-3,-2,-1,0,1,2,3,...\}$

79. $-3+(-2)+7=-5+7$
$$=2$$

3.2 Evaluating Algebraic Expressions and Formulas

Vocabulary

1. A **formula** is an equation that states a relationship between two or more variables.

3. To evaluate an algebraic expression, we **substitute** specific numbers for the variables in the expression and apply the rules for the order of operations.

Concepts

5. $2-8+10$ looks like subtraction if you do not put parentheses around the -8.

7a. $x =$ length of part 1
$x - 40 =$ length of part 2
$x + 16 =$ length of part 3

(answers may vary)

7b.   part 1 = 60 in.
part 2 = 60 − 40 = 20 in.
part 3 = 60 + 16 = 76 in.

9a.

Ticket price	Service charge	Total cost
20	2	22
25	2	27
p	2	$p + 2$

9b. $T = p + 2$

11.

	Rate (mph)	• Time (hr) =	Distance (miles)
Bike	12	4	48
Walking	3	t	$3t$
Car	x	3	$3x$

13a.   personal trainer

13b.   mechanic

13c.   paleontologist

13d.   realtor

13e.   doctor

13f.   economist

Notation

15a. $\quad d = rt$

15b. $\quad C = \dfrac{5}{9}(F - 32)$

15c. $\quad d = 16t^2$

Practice

17. $\quad 3x + 5 = 3(4) + 5$
$$= 12 + 5$$
$$= 17$$

19. $\quad -p = -(-4)$
$$= 4$$

21. $\quad -4t = -4(-10)$
$$= 40$$

23. $\quad \dfrac{x - 8}{2} = \dfrac{-4 - 8}{2}$
$$= \dfrac{-4 + (-8)}{2}$$
$$= \dfrac{-12}{2}$$
$$= -6$$

25. $\quad 2(p + 9) = 2(-12 + 9)$
$$= 2(-3)$$
$$= -6$$

27. $\quad x^2 - x - 7 = (-5)^2 - (-5) - 7$
$$= 25 + 5 - 7$$
$$= 30 - 7$$
$$= 23$$

29. $\quad -s^3 + 8s = -(-2)^3 + 8(-2)$
$$= -(-8) + 8(-2)$$
$$= 8 - 16$$
$$= 8 + (-16)$$
$$= -8$$

31. $\quad 4x^2 = 4(5)^2$
$$= 4(25)$$
$$= 100$$

33.

$$-b^2 + 3b = -(-4)^2 + 3(-4)$$
$$= -16 + 3(-4)$$
$$= -16 + (-12)$$
$$= -28$$

35.

$$\frac{24+k}{3k} = \frac{24+3}{3(3)}$$
$$= \frac{27}{9}$$
$$= 3$$

37.

$$|6-x| = |6-50|$$
$$= |6+(-50)|$$
$$= |-44|$$
$$= 44$$

39.

$$-2|x| - 7 = -2|-7| - 7$$
$$= -2(7) - 7$$
$$= -14 - 7$$
$$= -14 + (-7)$$
$$= -21$$

41.

$$\frac{x}{y} = \frac{30}{-10}$$
$$= -3$$

43.

$$-x - y = -(-1) - 8$$
$$= 1 - 8$$
$$= 1 + (-8)$$
$$= -7$$

45.

$$x(5h-1) = -2(5[2]-1)$$
$$= -2(10-1)$$
$$= -2(9)$$
$$= -18$$

47.

$$b^2 - 4ac = (-3)^2 - 4(4)(-1)$$
$$= 9 - 4(4)(-1)$$
$$= 9 - 16(-1)$$
$$= 9 - (-16)$$
$$= 9 + 16$$
$$= 25$$

49.

$$x^2 - y^2 = (5)^2 - (-2)^2$$
$$= 25 - 4$$
$$= 21$$

51.

$$\frac{50-6s}{-t} = \frac{50-6(5)}{-(4)}$$
$$= \frac{50-30}{-4}$$
$$= \frac{20}{-4}$$
$$= -5$$

53.
$$-5abc+1=-5(-2)(-1)(3)+1$$
$$=10(-1)(3)+1$$
$$=-10(3)+1$$
$$=-30+1$$
$$=-29$$

55.
$$5s^2t=5(-3)^2(-1)$$
$$=5(9)(-1)$$
$$=45(-1)$$
$$=-45$$

57.
$$\left|a^2-b^2\right|=\left|(-2)^2-(-5)^2\right|$$
$$=\left|4-25\right|$$
$$=\left|4+(-25)\right|$$
$$=\left|-21\right|$$
$$=21$$

59.
$$r=c+m$$
$$r=20+50$$
$$r=70$$
70 cents.

61.
$$p=r-c$$
$$p=13,500-5,300$$
$$p=8,200$$
$8,200

63.
$$r=c+m$$
$$r=18+5$$
$$r=23$$
$23

65.
$$d=rt$$
$$d=(60)(5)$$
$$d=300$$
300 miles

67.
$$C=\frac{5(F-32)}{9}$$
$$C=\frac{5(14-32)}{9}$$
$$C=\frac{5(14+[-32])}{9}$$
$$C=\frac{5(-18)}{9}$$
$$C=\frac{-90}{9}$$
$$C=-10$$
$-10°C$

69.

$$A = \frac{s}{n}$$

$$A = \frac{254 + 225 + 238}{3}$$

$$A = \frac{717}{3}$$

$$A = 239$$

239 average

71.

$$d = 16t^2$$

$$d = 16(2)^2$$

$$d = 16(4)$$

$$d = 64$$

64 ft

Applications

73.   FINANCIAL STATEMENTS

Annual Financials: Income Statement

All dollar amounts in millions

	Dec '98	Dec '97	Dec '96
Revenue	5,213	5,079	4,814
Cost of goods sold	2,053	2,051	1,921
Gross profit	3,160	3,028	2,893

75.   THERMOMETERS

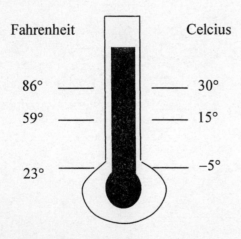

77. FALLING OBJECT

Time falling	Distance traveled (ft)	Time intervals
1 sec	16	Distance traveled from 0 sec to 1 sec 16 ft
2 sec	64	Distance traveled from 1 sec to 2 sec 48 ft
3 sec	144	Distance traveled from 2 sec to 3 sec 80 ft
4 sec	256	Distance traveled from 3 sec to 4 sec 112 ft

79 CUSTOMER SATISFACTION

$$A = \frac{s}{n}$$

$$A = \frac{53(5) + 26(3) + 9(1)}{53 + 26 + 9}$$

$$A = \frac{265 + 78 + 9}{88}$$

$$A = \frac{352}{88}$$

$$A = 4$$

Writing

81. If $a = -6$, then the opposite of a is positive 6.

83. Revenue is how much money a business brings in; markup is how much the price of an item is increased before it is sold; profit is revenue minus the cost of doing business.

85. Answers may vary.

Review

87. 17, 37, 41

89. $\left| -2 + (-5) \right| = \left| -7 \right|$
$$= 7$$

91. Division by 3

93. $-3 - (-6) = -3 + 6$
$$= 3$$

3.3 Simplifying Algebraic Expressions and the Distributive Property

Vocabulary

1. The **distributive** property tells us how to multiply $5(x+7)$. After doing the multiplication to obtain $5x+35$, we say that the parentheses have been **removed**.

3. When an algebraic expression is simplified, the result is an **equivalent** expression.

Concepts

5. $x(y+z)=xy+xz$

7. $(w+7)5$

9. $6 \bullet 2 + 6 \bullet 3 = 6(2+3)$

11. $-(y+9)=-y-9$

Notation

13. $-5(7n)=(-5 \bullet 7)n$
    $$= -35n$$

15. $-9(-4-5y)=(-9)(-4)-(-9)(5y)$
    $$= 36-(-45y)$$
    $$= 36+45y$$

17a. $-(-x)=x$

17b. $x-(-5)=x+5$

17c. $5x-10y+(-15)=5x-10y-15$

17d. $5 \bullet x = 5x$

Practice

19. $2(6x)=12x$

21. $-5(6y)=-30y$

23. $-10(-10t)=100t$

25. $(4s)3=12s$

27. $\quad 2c \bullet 7 = 14c$

29. $\quad -5 \bullet 8h = -40h$

31. $\quad -7x(6y) = -42xy$

33. $\quad 4r \bullet 4s = 16rs$

35. $\quad 2x(5y)(3) = 30xy$

37. $\quad 5r(2)(-3b) = -30br$

39. $\quad 5 \bullet 8c \bullet 2 = 80c$

41. $\quad (-1)(-2e)(-4) = -8e$

43. $\quad 4(x+1) = 4x+4$

45. $\quad 4(4-x) = 16-4x$

47. $\quad -2(3e+3) = -6e-6$

49. $\quad -8(2q-6) = -16q+48$

51. $\quad -4(-3-5s) = 12+20s$

53. $\quad (7+4d)6 = 42+24d$

55. $\quad (5r-6)(-5) = -25r+30$

57. $\quad (-4-3d)6 = -24-18d$

59. $\quad 3(3x-7y+2) = 9x-21y+6$

61. $\quad -3(-3z-3x-5y) = 9z+9x+15y$

63. $\quad -(x+3) = -x-3$

65. $\quad -(4t+5) = -4t-5$

67. $\quad -(-3w-4) = 3w+4$

69. $\quad -(5x-4y+1) = -5x+4y-1$

71. $\quad 2(4x)+2(5) = 2(4x+5)$

73. $\quad -4(5)-3x(5) = (-4-3x)5$

75. $\quad -3(4y)-(-3)(2) = -3(4y-2)$

77. $\quad 3(4)-3(7t)-3(5s) = 3(4-7t-5s)$

Writing

79. Simplifying an algebraic expression is to clear the expression of any arithmetic that can be performed. $7(5x) + 2(5y - 7) = 35x + 10y - 14$. (examples will vary)

81. The distributive property is applied by multiplying the expression on the outside of the parentheses by each term on the inside of the parentheses.

Review

83. $|-6 + 1| = |-5|$
     $\qquad\quad = 5$

85. Product implies multiplication; quotient implies division; difference implies subtraction; and sum implies addition.

87. $-6 > -7$

89. carpeting a room, painting a wall

3.4 Combining Like Terms

Vocabulary

1. A **term** is a number or a product of a number and one or more variables.

3. When we write $9x + x$ as $10x$, we say that we have **combined** like terms.

5. $2(x+3) = 2x + 2(3)$ is an example of the **distributive** property.

7. Simplifying the **sum** (or difference) of like terms is called combining like terms.

Concepts

9a.   term

9b.   factor

9c.   factor

9d.   factor

11a.   11

11b.   8

11c.   −4

11d.   1

11e.   −1

11f.   102

13.

Term	Coefficient	Variable part
$6m$	6	m
$-75t$	-75	t
w	1	w
$4bh$	4	bh

15.   Underlining like terms can help students identify like terms.

17. $d + 15 + d = (2d + 15)\,\text{mi}$

19a.   To add like terms, add 9 and 5 and keep the variable.

19b.   To subtract like terms, subtract 4 from 12 and keep the variable.

Notation

21. $5x + 7x = (5 + 7)x$
      $= 12x$

23. $2(x - 1) + 3x = 2x - 2 + 3x$
      $= 5x - 2$

25a.   The perimeter of a rectangle

25b.   2 times the length

25c.   2 times the width

Practice

27. $3x^2, 5x, 4$

29. $5, 5t, -8t, 4$

31.   2

33.   5

35. $6t + 9t = 15t$

37. $5s - s = 4s$

39. $-5x + 6x = 1x = x$

41. $-5d + 9d = 4d$

43. $3e - 7e = 3e + (-7e)$
      $= -4e$

45. $h - 7$ cannot be simplified

47. $4z - 10z = 4z + (-10z)$
      $= -6z$

49. $-3x - 4x = -3x + (-4x)$
      $= -7x$

51. $2t - 2t = 0$

53. $-6s + 6s = 0$

55. $x + x + x + x = 4x$

57. $2x + 2y$ cannot be simplified

59. $0 - 2y = -2y$

61. $3a - 0 = 3a$

63. $6t + 9 + 5t + 3 = 11t + 12$

65. $3w - 4 - w - 1 = 2w - 4 + (-1)$
$\qquad\qquad\qquad\quad = 2w - 5$

67. $-4r + 8R + 2R - 3r + R = -4r + (-3r) + 11R$
$\qquad\qquad\qquad\qquad\qquad = -7r + 11R$

69. $-45d - 12a - 5d + 12a = -45d + (-5d) + 0$
$\qquad\qquad\qquad\qquad\qquad = -50d$

71. $4x - 3y - 7 + 4x - 2 - y = 8x - 3y + (-y) - 7 + (-2)$
$\qquad\qquad\qquad\qquad\qquad\qquad = 8x - 4y - 9$

73. $4(x + 1) + 5(6 + x) = 4x + 4 + 30 + 5x$
$\qquad\qquad\qquad\qquad = 9x + 34$

75. $5(3 - 2s) + 4(2 - 3s) = 15 - 10s + 8 - 12s$
$\qquad\qquad\qquad\qquad\quad = 23 - 10s + (-12s)$
$\qquad\qquad\qquad\qquad\quad = -22s + 23$

77. $-4(6 - 4e) + 3(e + 1) = -24 + 16e + 3e + 3$
$\qquad\qquad\qquad\qquad\quad = 19e - 21$

79. $3t - (t - 8) = 3t - t + 8$
$\qquad\qquad\quad = 2t + 8$

81. $-2(2 - 3x) - 3(x - 4) = -4 + 6x - 3x + 12$
$\qquad\qquad\qquad\qquad\quad = 3x + 8$

83. $-4(-4y+5)-6(y+2)=16y-20-6y-12$
$$=10y-20+(-12)$$
$$=10y-32$$

Applications

85.   MOBILE HOMES

There are 4 exterior walls.

The front and back walls are the same.   The right and left walls are the same.

P_F = Perimeter of front wall P_R = Perimeter of right wall

$P_F = 2(10)+2(60)$ $P_R = 4(10)$

$P_F = 20+120$ $P_R = 40$ ft

$P_F = 140$ ft

The total perimeter for all walls is the sum of the perimeters of each wall.

P = Perimeter of all walls combined

P = front + back + right + left

$P = 140+140+40+40$

$P = 360$ ft

C = Cost of wood needed.

$C = (360)(80)$

$C = 28,800$ cents, which would be $288

87.   PARTY PREPARATIONS

Slow dancers	Fast dancers	Size of floor	Perimeter of floor
8	5	9×9	$P = 4(9) = 36$ ft
14	9	12×12	$P = 4(12) = 48$ ft
22	15	15×15	$P = 4(15) = 60$ ft
32	20	18×18	$P = 4(18) = 72$ ft
50	30	21×21	$P = 4(21) = 84$ ft

Writing

89. Like terms have the same variables and the same exponents on the variables.

91. A term is a number or product of a number and one or more variables, for example,

$5x, 7x^2, -3xy$. A factor is a number or variable involved in multiplication, for example, in

$3x,$ x is a factor of the term $3x$.

Review

93.

$$-4t - 3 = -11$$
$$-4t - 3 + 3 = -11 + 3$$
$$-4t = -8$$
$$\frac{-4t}{-4} = \frac{-8}{-4}$$
$$t = 2$$

95.

$$2^2 \cdot 5^2$$

97. The **<u>absolute</u> <u>value</u>** of a number is the distance between it and 0 on the number line.

3.5 Simplifying Expressions to Solve Equations

Vocabulary

1. To **solve** an equation means to find all values of the variable that make the equation a true statement.

3. In $2(x+4)$, to remove parentheses means to apply the **distributive** property.

5. The phrase "**combine** like terms" refers to the operations of addition and subtraction.

Concepts

7. $5x-3x=-9$ for $x=-5$

$$5(-5)-3(-5)=-9$$
$$-25+15=-9$$
$$-10 \neq -9$$

Therefore -5 is not a solution.

9. $5k$

11a. $4x$

11b. $2x$

13a. $3t-t-8=2t-8$

13b.
$$3t-t=-8$$
$$2t=-8$$
$$\frac{2t}{2}=\frac{-8}{2}$$
$$t=-4$$

13c. $3t-t-8$ for $t=-4$
$$=3(-4)-(-4)-8$$
$$=-12+4-8$$
$$=-16$$

Notation

15.
$$4x-2x=-20$$
$$2x=-20$$
$$\frac{2x}{2}=\frac{-20}{2}$$
$$x=-10$$

17.
$$5(x-9)=5$$
$$5x-5(9)=5$$
$$5x-45=5$$
$$5x-45+45=5+45$$
$$5x=50$$
$$\frac{5x}{5}=\frac{50}{5}$$
$$x=10$$

Practice

19. $5f + 8 = 4f + 11$ for $f = 3$

    $5(3) + 8 = 4(3) + 11$

      $15 + 8 = 12 + 11$

        $23 = 23$

  yes

21. $2(x - 1) = 33$ for $x = 12$

    $2(12 - 1) = 33$

      $2(11) = 33$

        $22 \neq 33$

  no

23. $3x + 6x = 54$

    $9x = 54$

    $\dfrac{9x}{9} = \dfrac{54}{9}$

    $x = 6$

25. $6x - 3x = 9$

    $3x = 9$

    $\dfrac{3x}{3} = \dfrac{9}{3}$

    $x = 3$

27. $60 = 3v - 5v$

    $60 = 3v + (-5v)$

    $60 = -2v$

    $\dfrac{60}{-2} = \dfrac{-2v}{-2}$

    $-30 = v$ or $v = -30$

29. $-28 = -m + 2m$

    $-28 = m$ or $m = -28$

31. $x + x + 6 = 90$

    $2x + 6 = 90$

    $2x + 6 - 6 = 90 - 6$

    $2x = 84$

    $\dfrac{2x}{2} = \dfrac{84}{2}$

    $x = 42$

33. $T + T - 17 = 57$

    $2T - 17 = 57$

    $2T - 17 + 17 = 57 + 17$

    $2T = 74$

    $\dfrac{2T}{2} = \dfrac{74}{2}$

    $T = 37$

35.
$$600 = m - 12 + m$$
$$600 = 2m - 12$$
$$600 + 12 = 2m - 12 + 12$$
$$612 = 2m$$
$$\frac{612}{2} = \frac{2m}{2}$$
$$306 = m \ \ or \ \ m = 306$$

37.
$$1,500 = b + 30 + b$$
$$1,500 = 2b + 30$$
$$1,500 - 30 = 2b + 30 - 30$$
$$1,470 = 2b$$
$$\frac{1,470}{2} = \frac{2b}{2}$$
$$735 = b \ \ or \ \ b = 735$$

39.
$$7x = 3x + 8$$
$$7x - 3x = 3x + 8 - 3x$$
$$4x = 8$$
$$\frac{4x}{4} = \frac{8}{4}$$
$$x = 2$$

41.
$$x - 14 = 2x$$
$$x - 14 - x = 2x - x$$
$$-14 = x \ \ or \ \ x = -14$$

43.
$$9t - 40 = 14t$$
$$9t - 40 - 9t = 14t - 9t$$
$$-40 = 5t$$
$$\frac{-40}{5} = \frac{5t}{5}$$
$$-8 = t \ \ or \ \ t = -8$$

45.
$$25 + 4j = 9j$$
$$25 + 4j - 4j = 9j - 4j$$
$$25 = 5j$$
$$\frac{25}{5} = \frac{5j}{5}$$
$$5 = j \ \ or \ \ j = 5$$

47.
$$-48 + 12t = 16t$$
$$-48 + 12t - 12t = 16t - 12t$$
$$-48 = 4t$$
$$\frac{-48}{4} = \frac{4t}{4}$$
$$-12 = t \ \ or \ \ t = -12$$

49.
$$-5g - 40 = -15g$$
$$-5g - 40 + 15g = -15g + 15g$$
$$10g - 40 = 0$$
$$10g - 40 + 40 = 0 + 40$$
$$10g = 40$$
$$\frac{10g}{10} = \frac{40}{10}$$
$$g = 4$$

51.
$$3s + 1 = 4s - 7$$
$$3s + 1 - 3s = 4s - 7 - 3s$$
$$1 = s - 7$$
$$1 + 7 = s - 7 + 7$$
$$8 = s \ \ or \ \ s = 8$$

53.
$$50a - 1 = 60a - 101$$
$$50a - 1 - 50a = 60a - 101 - 50a$$
$$-1 = 10a - 101$$
$$-1 + 101 = 10a - 101 + 101$$
$$100 = 10a$$
$$\frac{100}{10} = \frac{10a}{10}$$
$$10 = a \ \ or \ \ a = 10$$

55.
$$-7 + 5r = 83 - 10r$$
$$-7 + 5r + 10r = 83 - 10r + 10r$$
$$-7 + 15r = 83$$
$$-7 + 15r + 7 = 83 + 7$$
$$15r = 90$$
$$\frac{15r}{15} = \frac{90}{15}$$
$$r = 6$$

57.
$$100 - y = 100 + y$$
$$100 - y + y = 100 + y + y$$
$$100 = 2y + 100$$
$$100 - 100 = 2y + 100 - 100$$
$$0 = 2y$$
$$\frac{0}{2} = \frac{2y}{2}$$
$$0 = y \ \ or \ \ y = 0$$

59.
$$2(x + 6) = 4$$
$$2x + 12 = 4$$
$$2x + 12 - 12 = 4 - 12$$
$$2x = 4 + (-12)$$
$$2x = -8$$
$$\frac{2x}{2} = \frac{-8}{2}$$
$$x = -4$$

61.
$$-16 = 2(t + 2)$$
$$-16 = 2t + 4$$
$$-16 - 4 = 2t + 4 - 4$$
$$-16 + (-4) = 2t$$
$$-20 = 2t$$
$$\frac{-20}{2} = \frac{2t}{2}$$
$$-10 = t \ \ or \ \ t = -10$$

63.
$$-3(2w - 3) = 9$$
$$-6w + 9 = 9$$
$$-6w + 9 - 9 = 9 - 9$$
$$-6w = 0$$
$$\frac{-6w}{-6} = \frac{0}{-6}$$
$$w = 0$$

65.
$$-(c-4)=3$$
$$-c+4=3$$
$$-c+4-4=3-4$$
$$-c=3+(-4)$$
$$-c=-1$$
$$\frac{-c}{-1}=\frac{-1}{-1}$$
$$c=1$$

67.
$$4(p-2)=0$$
$$4p-8=0$$
$$4p-8+8=0+8$$
$$4p=8$$
$$\frac{4p}{4}=\frac{8}{4}$$
$$p=2$$

69.
$$2(4y+8)=3(2y-2)$$
$$8y+16=6y-6$$
$$8y+16-6y=6y-6-6y$$
$$2y+16=-6$$
$$2y+16-16=-6-16$$
$$2y=-6+(-16)$$
$$2y=-22$$
$$\frac{2y}{2}=\frac{-22}{2}$$
$$y=-11$$

71.
$$8+4(x-2)=-16$$
$$8+4x-8=-16$$
$$4x=-16$$
$$\frac{4x}{4}=\frac{-16}{4}$$
$$x=-4$$

73.
$$-15=5+5(2x+10)$$
$$-15=5+10x+50$$
$$-15=10x+55$$
$$-15-55=10x+55-55$$
$$-15+(-55)=10x$$
$$-70=10x$$
$$\frac{-70}{10}=\frac{10x}{10}$$
$$-7=x \ \text{ or } \ x=-7$$

75.
$$2x+3(x-4)=23$$
$$2x+3x-12=23$$
$$5x-12=23$$
$$5x-12+12=23+12$$
$$5x=35$$
$$\frac{5x}{5}=\frac{35}{5}$$
$$x=7$$

77.
$$10q + 3(q - 7) = 18$$
$$10q + 3q - 21 = 18$$
$$13q - 21 = 18$$
$$13q - 21 + 21 = 18 + 21$$
$$13q = 39$$
$$\frac{13q}{13} = \frac{39}{13}$$
$$q = 3$$

79.
$$16 - (x + 3) = -13$$
$$16 - x - 3 = -13$$
$$13 - x = -13$$
$$13 - x - 13 = -13 - 13$$
$$-x = -13 + (-13)$$
$$-x = -26$$
$$\frac{-x}{-1} = \frac{-26}{-1}$$
$$x = 26$$

81.
$$5 - (7 - y) = -5$$
$$5 - 7 + y = -5$$
$$5 + (-7) + y = -5$$
$$-2 + y = -5$$
$$-2 + y + 2 = -5 + 2$$
$$y = -3$$

Writing

83. $4x - x = 3x$, not 4.

85. To solve an equation means to find all values of the variable that make the equation a true statement

Review

87.
$$-7 - 9 = -7 + (-9)$$
$$= -16$$

89.
$$\frac{-8 + 2}{-2 + 4} = \frac{-6}{2}$$
$$= -3$$

91. $-(-5) = 5$

93. always positive

3.6 Problem Solving

Vocabulary

1. The words *increased by, longer, taller, higher, total*, and, *more than* indicate that the operation of **addition** should be used.

Concepts

3.   BUSINESS ACCOUNTS

  $5x$

5.   SERVICE STATIONS

  $g - 100$

7.   OCEAN TRAVEL

  $3m$

9.

  $2w$

11.   COMMISIONS

Type of shoe	Number sold	Commission per shoe ($)	Total commission ($)
Dress	10	3	30
Athletic	12	2	24
Child's	x	5	$5x$
Sandal	$9 - x$	4	$4(9 - x)$

13a.   9 pairs

13b. $9 - d$

15. AIRLINE SEATING

Analyze the problem

- There are **88** seats on the plane.

- There are **10** times as many economy as first-class seats.

- Find **the number of first-class seats**.

Form an equation

Let $x =$ **the number of first-class seats**.

Key Phrase: ten times as many

Translation: multiply by 10

So, $10x =$ **the number of economy** seats

The number of first-class seats	plus	the number of economy seats	is	88
x	$+$	$10x$	$=$	88

Solve the equation

$$x + 10x = 88$$
$$11x = 88$$
$$x = 8$$

State the conclusion **There are 8 first-class seats**.

Check the result

If there are 8 first-class seats, there are $10 \bullet 8 = 80$ economy seats. Adding 8 and 80, we get 88. The answer checks.

17.   BUSINESS

m = number of months to reach 100

$$5m + 15 = 100$$
$$5m + 15 - 15 = 100 - 15$$
$$5m = 85$$
$$\frac{5m}{5} = \frac{85}{5}$$
$$m = 17$$

It would take 17 months.

19.   ANTIQUES

y = years it takes to reach 100

$$4y + 56 = 100$$
$$4y + 56 - 56 - 15 = 100 - 56$$
$$4y = 44$$
$$\frac{4y}{4} = \frac{44}{4}$$
$$y = 11$$

It takes 11 years.

21.   RENTALS

m = monthly rent for apartment

$$\frac{m}{3} + 100 = 425$$
$$\frac{m}{3} + 100 - 100 = 425 - 100$$
$$\frac{m}{3} = 325$$
$$3\left(\frac{m}{3}\right) = 3(325)$$
$$m = 975$$

The monthly rent is $975.

23.   BOTTLED WATER DELIVERY

n = number of office buildings

$$300 - 3n = 117$$
$$300 - 3n - 300 = 117 - 300$$
$$-3n = 117 + (-300)$$
$$-3n = -183$$
$$\frac{-3n}{-3} = \frac{-183}{-3}$$
$$n = 61$$

He delivered to 61 office buildings.

25.   NUMBER PROBLEMS

$$5x - 10 = x + 6$$
$$5x - 10 - x = x + 6 - x$$
$$4x - 10 = 6$$
$$4x - 10 + 10 = 6 + 10$$
$$4x = 16$$
$$\frac{4x}{4} = \frac{16}{4}$$
$$x = 4$$

The number is 4.

27.   SERVICE STATIONS

g = gallons the premium tank holds

$g - 100$ = gallons the regular tank holds

$$g + g - 100 = 700$$
$$2g - 100 = 700$$
$$2g - 100 + 100 = 700 + 100$$
$$2g = 800$$
$$\frac{2g}{2} = \frac{800}{2}$$
$$g = 400$$

The premium tank holds 400 gallons.

29. OCEAN TRAVEL

d = distance freighter travelled

$3d$ = distance passenger ship travelled

$$d + 3d = 84$$
$$4d = 84$$
$$\frac{4d}{4} = \frac{84}{4}$$
$$d = 21$$

The freighter was 21 miles from port.

31. INTERIOR DECORATING

w = width of room

$2w$ = length of room

$$w + 2w + w + 2w = 60$$
$$6w = 60$$
$$\frac{6w}{6} = \frac{60}{6}$$
$$w = 10$$

The width of the room is 10 feet.

33. COMMERCIALS

t = time devoted to commericals

$t + 18$ = program time

$$t + t + 18 = 30$$
$$2t + 18 = 30$$
$$2t + 18 - 18 = 30 - 18$$
$$2t = 12$$
$$\frac{2t}{2} = \frac{12}{2}$$
$$t = 6$$

There were 6 minutes of commercials.

35.   COMMISSIONS

	Number sold	Commission per pair ($)	Total commission ($)
Dress	x	3	$3x$
Athletic	$9-x$	2	$2(9-x)$
Total	9		24

$$3x + 2(9-x) = 24$$
$$3x + 18 - 2x = 24$$
$$x + 18 = 24$$
$$x + 18 - 18 = 24 - 18$$
$$x = 6$$

He sold 6 pair of dress shoes, and he sold $9 - 6 = 3$ pairs of athletic shoes.

37.   MOVER'S PAY SCALE

	Hours worked	Hourly pay ($)	Total pay ($)
Regular rate	x	60	$60x$
Stair climbing rate	$20-x$	90	$90(20-x)$
Total	20		1,380

$$60x + 90(20-x) = 1,380$$
$$60x + 1,800 - 90x = 1,380$$
$$60x + 1,800 + (-90x) = 1,380$$
$$-30x + 1,800 = 1,380$$
$$-30x + 1,800 - 1,800 = 1,380 - 1,800$$
$$-30x = 1,380 + (-1,800)$$
$$-30x = -420$$
$$\frac{-30x}{-30} = \frac{-420}{-30}$$
$$x = 14$$

He worked 14 hours at regular pay and worked $20 - 14 = 6$ hours at stair climbing pay.

Writing

39. Answers will vary.

41. A son is 20 years younger than his father. If you combined their ages, it would total 50. Find the age of the father and son.

Review

43. Associative property of addition.

45. $-10^2 = -(10 \bullet 10)$
    $= -100$

47. Subtraction of a number is the same as **addition** of the opposite of that number.

49. $2 \bullet 2 \bullet 2 \bullet 5 \bullet 5 = 2^3 \bullet 5^2$

Key Concept Order of Operations

1. Perform all calculations within **parentheses** and other grouping symbols, following the order listen in Steps 2-4 below and working from the **innermost** pair to the **outermost** pair.

3. Perform all **multiplications** and **divisions** as they occur from left to right.

   If a fraction bar is present, evaluate the numerator and the denominator **separately**. Then do the division indicated by the fraction bar, if possible.

5.
$$-10+4-3^2 = -10+4-9$$
$$= -6-9$$
$$= -6+(-9)$$
$$= -15$$

7.
$$-2(-3)-12 \div 6 \bullet 3 = 6-12 \div 6 \bullet 3$$
$$= 6-2 \bullet 3$$
$$= 6-6$$
$$= 0$$

9.
$$2(4+3 \bullet 2)^2 -(-6) = 2(4+6)^2 -(-6)$$
$$= 2(10)^2 -(-6)$$
$$= 2(100)-(-6)$$
$$= 200-(-6)$$
$$= 200+6$$
$$= 206$$

11.
$$1-2\left|8w-w^3\right| \text{ for } w=-2$$
$$1-2\left|8(-2)-(-2)^3\right| = 1-2\left|8(-2)-(-8)\right|$$
$$= 1-2\left|-16-(-8)\right|$$
$$= 1-2\left|-16+8\right|$$
$$= 1-2\left|-8\right|$$
$$= 1-2(8)$$
$$= 1-16$$
$$= 1+(-16)$$
$$= -15$$

13.
$$P = 2l+2w \text{ for } l=30 \text{ and } w=16$$
$$P = 2(30)+2(16)$$
$$P = 60+32$$
$$P = 92 \text{ feet}$$

15.
$$(3x)4-2(5x)+x = 12x-10x+x$$
$$= 2x+x$$
$$= 3x$$

17. $15 - 3x = 23$ for $x = -3$

$15 - 3(-3) = 23$

$15 - (-9) = 23$

$15 + 9 = 23$

$24 \neq 23$ no.

19.   Undo subtraction first. Then undo the multiplication.

Chapter Three Review

Section 3.1 Variables and Algebraic Expressions

1. Brandon is closer by 250 miles.

2. $h+7$

3. $n-5$

4. $7x$

5. $\dfrac{6}{p}$

6. $s+(-15)$

7. $2l$

8. $D-100$

9. $r+2$

10. $\dfrac{45}{x}$

11. $\dfrac{c}{6}$ children in each room

12. $\$(1000-x)$

13. $x=$ hours wife drove
    $2x=$ hours husband drove

14. $w=$ width
    $w+3=$ length

15. $12x$

16. $\dfrac{d}{7}$

Section 3.2 Evaluating Algebraic Expressions and Formulas

17. RETAINING WALLS
    $h=$ height of wall
    $h-5=$ upper base
    $2h-3=$ lower base

18. upper base $=10-5=5$ feet
    lower base $=2(10)-3$
    $\qquad\qquad =20-3$
    $\qquad\qquad =17$ feet

19. $-2x + 6$ for $x = -3$

$$-2(-3) + 6 = 6 + 6$$
$$= 12$$

20. $\dfrac{6-a}{1+a}$ for $a = -2$

$$\dfrac{6-(-2)}{1+(-2)} = \dfrac{6+2}{-1}$$
$$= \dfrac{8}{-1}$$
$$= -8$$

21. $b^2 - 4ac$ for $a = 4$, $b = 6$, and $c = -4$

$$(6)^2 - 4(4)(-4) = 36 - 4(4)(-4)$$
$$= 36 - 16(-4)$$
$$= 36 - (-64)$$
$$= 36 + 64$$
$$= 100$$

22. $\dfrac{-2k^3}{1-2-3}$ for $k = -2$

$$\dfrac{-2(-2)^3}{1-2-3} = \dfrac{-2(-8)}{1+(-2)+(-3)}$$
$$= \dfrac{16}{-1+(-3)}$$
$$= \dfrac{16}{-4}$$
$$= -4$$

23. DISTANCE TRAVELED

	Rate (mph)	Time (hr)	Distance Traveled (mi)
Monorail	65	2	130
Subway	38	3	114
Train	x	6	$6x$
Bus	55	t	$55t$

24. SALE PRICE

$s = p - d$

$s = 315 - 37$

$s = 278$

The sale price is $278.

25. RETAIL PRICE

$r = p + m$

$r = 14,505 + 725$

$r = 15,230$

The retail price is $15,230.

26. 1998

27. Remember; $p = r - c$, 2000

28. The costs decreased.

29. TEMPERATURE CONVERSION

$$C = \frac{5(F-32)}{9} \text{ for } F = 77$$

$$C = \frac{5(77-32)}{9}$$

$$C = \frac{5(45)}{9}$$

$$C = \frac{225}{9}$$

$$C = 25$$

The pool is $2°C$ warmer.

30. DISTANCE FALLEN

$$d = 16t^2 \text{ for } t = 3$$

$$d = 16(3)^2$$

$$d = 16(9)$$

$$d = 144$$

The hammer falls 144 feet.

31. AVERAGE YEARS OF EXPERIENCE

Must realize there are 2 grandparents, 2 parents, and 1 grandson.

$$M = \frac{40+40+18+18+4}{5}$$

$$M = \frac{120}{5}$$

$$M = 24$$

The average number of years is 24 years.

Section 3.3 Simplifying Algebraic Expressions and the Distributive Property

32. $-2(5x) = -10x$

33. $-7(-6y) = 42xy$

34. $4d \cdot 3e \cdot 5 = 60de$

35. $(4s)8 = 32s$

36. $-1(-e)(2) = 2e$

37. $7x \cdot 7y = 49xy$

38. $4 \cdot 3k \cdot 7 = 84k$

39. $(-10t)(-10) = 100t$

40. $4(y+5) = 4y+20$

41. $-5(6t+9) = -30t-45$

42. $(-3-3x)7 = -21-21x$

43. $-3(4e-8x-1) = -12e+24x+3$

44. $-(6t-4) = -6t+4$

45. $-(5+x) = -5-x$

46. $-(6t-3s+1) = -6t+3s-1$

47. $-(-5a-3) = 5a+3$

Section 3.4 Combining Like Terms

48. $-4x, 8$

49. $-3y, 1$

50. factor

51. term

52. factor

53. term

54. yes

55. no

56. yes

57. no

58. $3x+4x = 7x$

59. $-3t-6t = -3t+(-6t)$
$$= -9t$$

60. $2z+(-5z) = -3z$

61. $6x-x = 5x$

62. $-6y-7y-(-y) = -6y+(-7y)+y$
$$= -13y+y$$
$$= -12y$$

63. $5w-8-4w+3 = w-5$

64. $-45d - 2a + 4a - d = 2a - 45d + (-d)$
$$= 2a - 46d$$

65. $5y + 8h - 3 + 7h + 5y + 2 = 15h + 10y - 1$

66. $7(y + 6) + 3(2y + 2) = 7y + 42 + 6y + 6$
$$= 13y + 48$$

67. $-4(t - 7) - (t + 6) = -4t + 28 - t - 6$
$$= -4t + (-t) + 22$$
$$= -5t + 22$$

68. $5x - 2(x - 6) = 5x - 2x + 12$
$$= 3x + 12$$

69. $6f + 7(12 - 8f) = 6f + 84 - 56f$
$$= 6f + (-56f) + 84$$
$$= -50f + 84$$

70. $0 - 14m = -14m$

71. $0 + 14m = 14m$

72. **HOLIDAY LIGHTS**

P_h = perimeter of house P_w = perimeter of window
$P_h = 35 + 42 + 35 + 42$ $P_w = 5 + 5 + 5 + 5$
$P_h = 154$ feet $P_w = 20$ feet

P_t = total feet of lights needed
$P_t = P_h + 2P_w$
$P_t = 154 + 2(20)$
$P_t = 154 + 40$
$P_t = 194$ feet

194 feet of lights are needed.

Section 3.5 Simplifying Expressions to Solve Equations

73. $-4x + 6 = 2(x + 12)$ for $x = -3$

$$-4(-3) + 6 = 2(-3 + 12)$$
$$12 + 6 = 2(9)$$
$$18 = 18$$

Yes, -3 is a solution because it makes a true statement when substituted into the equation.

74. $5a - 3a = -36$

$$2a = -36$$
$$\frac{2a}{2} = \frac{-36}{2}$$
$$a = -18$$

Check:

$$5(-18) - 3(-18) = -36$$
$$-90 - (-54) = -36$$
$$-90 + 54 = -36$$
$$-36 = -36$$

75. $3x - 4x = -8$

$$3x + (-4x) = -8$$
$$-x = -8$$
$$\frac{-x}{-1} = \frac{-8}{-1}$$
$$x = 8$$

Check:

$$3(8) - 4(8) = -8$$
$$24 - 32 = -8$$
$$24 + (-32) = -8$$
$$-8 = -8$$

76. $7x = 3x - 12$

$$7x - 3x = 3x - 12 - 3x$$
$$4x = -12$$
$$\frac{4x}{4} = \frac{-12}{4}$$
$$x = -3$$

Check:

$$7(-3) = 3(-3) - 12$$
$$-21 = -9 - 12$$
$$-21 = -9 + (-12)$$
$$-21 = -21$$

77. $5(y - 15) = 0$

$$5y - 75 = 0$$
$$5y - 75 + 75 = 0 + 75$$
$$5y = 75$$
$$\frac{5y}{5} = \frac{75}{5}$$
$$y = 15$$

Check:

$$5(15 - 15) = 0$$
$$5(0) = 0$$
$$0 = 0$$

78.
$$3a - (2a - 1) = -2$$
$$3a - 2a + 1 = -2$$
$$a + 1 = -2$$
$$a + 1 - 1 = -2 - 1$$
$$a = -2 + (-1)$$
$$a = -3$$

Check:
$$3(-3) - (2[-3] - 1) = -2$$
$$-9 - (-6 - 1) = -2$$
$$-9 - (-6 + [-1]) = -2$$
$$-9 - (-7) = -2$$
$$-9 + 7 = -2$$
$$-2 = -2$$

79.
$$15 = 5b + 1 + 2b$$
$$15 = 7b + 1$$
$$15 - 1 = 7b + 1 - 1$$
$$14 = 7b$$
$$\frac{14}{7} = \frac{7b}{7}$$
$$2 = b \ \ or \ \ b = 2$$

Check:
$$15 = 5(2) + 1 + 2(2)$$
$$15 = 10 + 1 + 4$$
$$15 = 15$$

80.
$$-6(2x + 3) = -(5x - 3)$$
$$-12x - 18 = -5x + 3$$
$$-12x - 18 + 12x = -5x + 3 + 12x$$
$$-18 = 7x + 3$$
$$-18 - 3 = 7x + 3 - 3$$
$$-18 + (-3) = 7x$$
$$-21 = 7x$$
$$\frac{-21}{7} = \frac{7x}{7}$$
$$-3 = x \ \ or \ \ x = -3$$

Check:
$$-6(2[-3] + 3) = -(5[-3] - 3)$$
$$-6(-6 + 3) = -(-15 - 3)$$
$$-6(-3) = -(-15 + [-3])$$
$$18 = -(-18)$$
$$18 = 18$$

81.
$$4 + 3(2x + 4) - 4 = -42$$
$$4 + 6x + 12 - 4 = -42$$
$$6x + 12 = -42$$
$$6x + 12 - 12 = -42 - 12$$
$$6x = -42 + (-12)$$
$$6x = -54$$
$$\frac{6x}{6} = \frac{-54}{6}$$
$$x = -9$$

Check:
$$4 + 3(2[-9] + 4) - 4 = -42$$
$$4 + 3(-18 + 4) - 4 = -42$$
$$4 + 3(-14) - 4 = -42$$
$$4 + (-42) - 4 = -42$$
$$-38 + (-4) = -42$$
$$-42 = -42$$

Section 3.6 Problem Solving

82.

Type of coin	Number	Value (¢)	Total value (¢)
Dime	6	10	60
Quarter	7	25	175
Penny	x	1	x
Nickel	n	5	$5n$

83. REFRESHMENTS

$56 - c$ cups left.

84. COLD STORAGE

h = hours needed to drop temp

$7h = 71 - 29$

$7h = 42$

$$\frac{7h}{7} = \frac{42}{7}$$

$h = 6$

It takes 6 hours.

85. FITNESS

d = distance she bikes

$d - 8$ = distance she jogs

$d + d - 8 = 18$

$2d - 8 = 18$

$2d - 8 + 8 = 18 + 8$

$2d = 26$

$$\frac{2d}{2} = \frac{26}{2}$$

$d = 13$

She bikes 13 miles.

86. CAR SHOWS

x = first day attendance

$2x$ = second day attendance

$x + 2x = 6,600$

$3x = 6,600$

$$\frac{3x}{3} = \frac{6,600}{3}$$

$x = 2,200$

2,200 people attended the first day.

87.   HEALTH FOOD

	Number of drinks sold	Cost of drink ($)	Total receipts ($)
$3 drinks	x	3	$3x$
$4 drinks	$50 - x$	4	$4(50 - x)$
Total	50		185

$$3x + 4(50 - x) = 185$$
$$3x + 200 - 4x = 185$$
$$3x + 200 + (-4x) = 185$$
$$-x + 200 = 185$$
$$-x + 200 - 200 = 185 - 200$$
$$-x = 185 + (-200)$$
$$-x = -15$$
$$\frac{-x}{-1} = \frac{-15}{-1}$$
$$x = 15$$

There were 15 $3 drinks sold and $50 - 15 = 35$ $4 drinks were sold.

Chapter Three Test

1a. $r - 2$

1b. $3xy$

1c. $x + 100$

2. $51,000 - e =$ husband's yearly earnings

3a. $\dfrac{x - 16}{x}$ for $x = 4$

$$\frac{4 - 16}{4} = \frac{4 + (-16)}{4}$$
$$= \frac{-12}{4}$$
$$= -3$$

3b. $2t^2 - 3(t - s)$ for $t = -2, s = 4$

$$2(-2)^2 - 3(-2 - 4)$$
$$= 2(-2)^2 - 3(-2 + [-4])$$
$$= 2(-2)^2 - 3(-6)$$
$$= 2(4) - 3(-6)$$
$$= 8 - (-18)$$
$$= 8 + 18$$
$$= 26$$

3c. $-a^2 + 10$ for $a = -3$

$$-(-3)^2 + 10 = -9 + 10$$
$$= 1$$

4. DISTANCE TRAVELED

Traveling from 9AM to noon means traveling for 3 hours.

$d = rt$
$d = (55)(3)$
$d = 165$

The motorist traveled 165 miles.

5.   PROFITS

$r = \text{total revenues} = 40,000 + 15,000 = 55,000$

$c = \text{total costs} = 13,000 + 5,000 = 18,000$

$p = r - c$

$p = 55,000 - 18,000$

$p = 37,000$

The total profits were \$37,000.

6.   FALLING OBJECTS

$d = 16t^2$

$d = 16(3)^2$

$d = 16(9)$

$d = 144$

In 3 seconds, the ball falls 144 feet, thus it is 56 feet short of hitting the ground.

7.   METER READINGS

$A = \text{average reading for period}$

$A = \dfrac{4 + 2 + (-4) + (-4) + 4 + 2 + (-4) + 2 + 4 + 4}{10}$

$A = \dfrac{10}{10}$

$A = 1$

The average meter value reading for the period is 1.

8.   LANDMARKS

$P_o = \text{perimeter of the outside of the "O"}$ $P_I = \text{perimeter of the inside of the "O"}$

$P_o = 40 + 40 + 40 + 40$ $P_I = 20 + 25 + 20 + 25$

$P_o = 160$ $P_I = 90$

$P = \text{total perimeter needed to be edged}$

$P = P_o + P_I$

$P = 160 + 90$

$P = 250$

They will need 250 feet of redwood edging.

9.   AIR CONDITIONING

$$C = \frac{5(F-32)}{9}$$

$$C = \frac{5(59-32)}{9}$$

$$C = \frac{5(27)}{9}$$

$$C = \frac{135}{9}$$

$$C = 15$$

It would be 15° C.

10a. $5(5x+1) = 25x+5$

10b. $-6(7-x) = -42+6x$

10c. $-(6y+4) = -6y-4$

10d. $3(2a+3b-7) = 6a+9b-21$

11a.   factor

11b.   term

12a. $-20y+6-8y+4 = -20y+(-8y)+10$
$$= -28y+10$$

12b. $-t-t-t = -t+(-t)+(-t)$
$$= -3t$$

13a. $8x^2, -4x, -6$

13b.   8

14a. $7x+4x = 11x$

14b. $3c \bullet 4e \bullet 2 = 24ce$

14c. $6x-x = 5x$

14d. $-5y(-6) = 30y$

14e. $0-7x = -7x$

14f. $0+9y = 9y$

15. $4(y+3)-5(2y+3)=4y+12-10y-15$

$$=4y+(-10y)+12+(-15)$$

$$=-6y-3$$

16. $6(-5)-8=12(-5-3)$

$-30-8=12(-5+[-3])$

$-30+(-8)=12(-8)$

$-38 \neq -96$

no

17. $5x-3x=-18$

$2x=-18$

$\dfrac{2x}{2}=\dfrac{-18}{2}$

$x=-9$

18. $6r=2r-12$

$6r-2r=2r-12-2r$

$4r=-12$

$\dfrac{4r}{4}=\dfrac{-12}{4}$

$r=-3$

19. $-45=3(1-4t)$

$-45=3-12t$

$-45-3=3-12t-3$

$-45+(-3)=-12t$

$-48=-12t$

$\dfrac{-48}{-12}=\dfrac{-12t}{-12}$

$4=t \ \ or \ \ t=4$

20. $6-(y-3)=19$

$6-y+3=19$

$9-y=19$

$9-y-9=19-9$

$-y=10$

$\dfrac{-y}{-1}=\dfrac{10}{-1}$

$y=-10$

21. $8+2(3x-4)=-60$

$8+6x-8=-60$

$6x=-60$

$\dfrac{6x}{6}=\dfrac{-60}{6}$

$x=-10$

22a. $10k$ 22b. $20(p+2)$ dollars

23.   DRIVING SCHOOLS

x = length of daily classroom session

$$4x + 2 = 14$$
$$4x + 2 - 2 = 14 - 2$$
$$4x = 12$$
$$\frac{4x}{4} = \frac{12}{4}$$
$$x = 3$$

Each class is 3 hours long.

24.   CABLE TELEVISION

t = time local shows aired

$t + 8$ = time national shows aired

$$t + t + 8 = 24$$
$$2t + 8 = 24$$
$$2t + 8 - 8 = 24 - 8$$
$$2t = 16$$
$$\frac{2t}{2} = \frac{16}{2}$$
$$t = 8$$

Local shows aired for 8 hours.

25.   Like terms are terms that have exactly the same variables and exponents.

26. t = length of trout
$t + 10$ = length of salmon
   or
s = length of salmon
$s - 10$ = length of trout

27.   Solve and simplify **do not** mean the same thing. We solve equations and we simplify expressions.

28.   Answers will vary. $4(7x - 5) = 28x - 20$

Chapter 1-3 Cumulative Review Exercises

1.   356,600,000 gallons

2.   50,000

3.
$$
\begin{array}{r}
38,908 \\
+15,696 \\
\hline
54,604
\end{array}
$$

4.
$$
\begin{array}{r}
9,700 \\
-5,491 \\
\hline
4,209
\end{array}
$$

5.
$$
\begin{array}{r}
345 \\
\times\ \ 67 \\
\hline
2,415 \\
20,700 \\
\hline
23,115
\end{array}
$$

6.
$$
\begin{array}{r}
87 \\
23{\overline{)2001}} \\
\underline{184} \\
161 \\
\underline{161} \\
0
\end{array}
$$

7. $683 + 459 = 1,142$

8.   The pattern is every 12 years.

2011 is the next year in the pattern.

9. $4 \bullet 5 = 4 + 4 + 4 + 4 + 4 = 20$

10.   ROOM DIVIDERS

There are 4 panels total that make up the divider, which makes 8 sides front and back.

To find the total area of fabric needed, the area of one panel will be found, then

multiplied by 8.

$A = (22)(62)$

$A = 1,364$

Total Area $= 8(1,364)$

Total Area $= 10,912$

$10,912$ in^2 of fabric will be needed.

11a.   1, 2, 3, 6, 9, 18

11b.   18

$$2 \bullet 3^2$$

12.   2, 3, 5, 7, 11, 13, 17, 19, 23, 29

13. $3 \bullet 9 = 27$, thus it is not prime

14. $$(9-2)^2 - 3^3 = (7)^2 - 3^3$$
$$= 49 - 27$$
$$= 22$$

15. $$250 = \frac{y}{2}$$
$$2(250) = 2\left(\frac{y}{2}\right)$$
$$500 = y \ \ or \ \ y = 500$$

16. $-(-6) = 6$

17.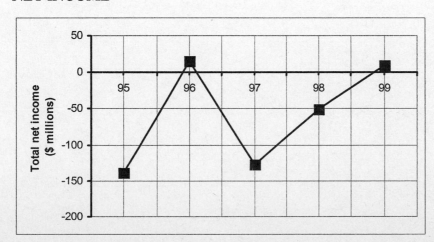

18. $|-5| = 5$

19.   False

20.   NET INCOME

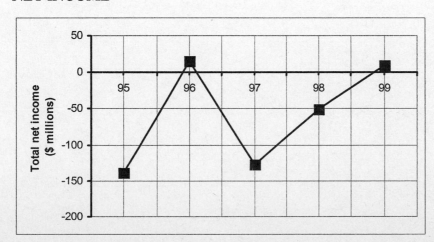

21. $-25 + 5 = -20$

22. $25 - (-5) = 25 + 5$
$$= 30$$

23. $-25(5)(-1) = -125(-1)$
$$= 125$$

24. $\dfrac{-25}{-5} = 5$

25. $\dfrac{(-6)^2 - 1^5}{-4 - 3} = \dfrac{36 - 1}{-4 + (-3)}$
$$= \dfrac{35}{-7}$$
$$= -5$$

26. $-3 + 3[-4 - 4 \bullet 2]^2 = -3 + 3[-4 - 8]^2$
$$= -3 + 3[-4 + (-8)]^2$$
$$= -3 + 3[-12]^2$$
$$= -3 + 3(144)$$
$$= -3 + 432$$
$$= 429$$

27. $-3^2 = -(3 \bullet 3) = -9$
$$(-3)^2 = (-3)(-3) = 9$$

28.   PLANETS

T = temperature range

$T = 810 - (-290)$

$T = 810 + 290$

$T = 1,100$

The temperature range is 1,100° F.

29. $-4x + 4 = -24$
$$-4x + 4 - 4 = -24 - 4$$
$$-4x = -24 + (-4)$$
$$-4x = -28$$
$$\dfrac{-4x}{-4} = \dfrac{-28}{-4}$$
$$x = 7$$

30. $-y = 10$
$$\dfrac{-y}{-1} = \dfrac{10}{-1}$$
$$y = -10$$

31. Division by zero, $\dfrac{12}{0}$;

Division of zero, $\dfrac{0}{12}$

Division by zero is always undefined.

32. commutative property of multiplication

33. $h+12$

34. TENNIS

$(26-x)$ inches

35. $x^2 = x \bullet x;$ while $2x = x+x$ or $2 \bullet x$

36. $x^2 - 2x + 1$ for $x = -5$

$$(-5)^2 - 2(-5) + 1 = 25 - 2(-5) + 1$$
$$= 25 - (-10) + 1$$
$$= 25 + 10 + 1$$
$$= 36$$

37.

	Rate (mph)	Time (hr)	Distance traveled (mi)
Truck	55	4	220

38. $5(2x-7) = 10x - 35$

39. $-6(-4t) = 24t$

40.

Term	Coefficient
$4a$	4
$-2y^2$	-2
x	1
$-m$	-1

41. Answers will vary

$x+9$, used as a term

$5x$, used as a factor

42. $25q$¢

43. $5b + 8 - 6b - 7 = 5b + (-6b) + 1$
$$= -b + 1$$

44.
$$8p+2p-1=-11$$
$$10p-1=-11$$
$$10p-1+1=-11+1$$
$$10p=-10$$
$$\frac{10p}{10}=\frac{-10}{10}$$
$$p=-1$$

Check:
$$8(-1)+2(-1)-1=-11$$
$$-8+(-2)-1=-11$$
$$-10-1=-11$$
$$-10+(-1)=-11$$
$$-11=-11$$

45.
$$7+2x=2-(4x+7)$$
$$7+2x=2-4x-7$$
$$7+2x=2-4x+(-7)$$
$$7+2x=-4x-5$$
$$7+2x+4x=-4x-5+4x$$
$$7+6x=-5$$
$$7+6x-7=-5-7$$
$$6x=-5+(-7)$$
$$6x=-12$$
$$\frac{6x}{6}=\frac{-12}{6}$$
$$x=-2$$

Check:
$$7+2(-2)=2-(4[-2]+7)$$
$$7+(-4)=2-(-8+7)$$
$$3=2-(-1)$$
$$3=2+1$$
$$3=3$$

46.   CLASS TIME

$t=$ time spent in lecture

$t-50=$ time spent in lab

$$t+t-50=300$$
$$2t-50=300$$
$$2t-50+50=300+50$$
$$2t=350$$
$$\frac{2t}{2}=\frac{350}{2}$$
$$t=175$$

The students spend 175 minutes in lecture.

Chapter 4 Fractions and Mixed Numbers

4.1 The Fundamental Property of Fractions

Vocabulary

1.   For the fraction $\frac{7}{8}$, 7 is the **numerator** and 8 is the **denominator**.

3.   A **proper** fraction is less than 1.

5.   Two fractions are **equivalent** if they have the same value.

7.   Multiplying the numerator and denominator of a fraction by a number to obtain an equivalent fraction that involves larger numbers is called expressing the fraction in **higher** terms or **building** up the fraction.

Concepts

9a.   2   9b.   3

9c.   5   9d.   7

11.   Equivalent fractions: $\frac{2}{6} = \frac{1}{3}$

13a.   In the first case, a common factor of 4 was found for the numbers 20 and 28. In the second, the numbers 20 and 28 were prime factored.

13b.   The results are the same. Either method can be used.

15.   The 2's in the numerator are not factors. Remember, factors are part of a multiplication, not an addition. **You cannot cancel common addends in a fraction ever!**

17a. $8 = \dfrac{8}{1}$ 17b. $-25 = -\dfrac{25}{1}$

17c. $x = \dfrac{x}{1}$ 17d. $7a = \dfrac{7a}{1}$

Notation

19. $$\frac{18}{24} = \frac{3 \bullet 3 \bullet 2}{3 \bullet 2 \bullet 2 \bullet 2}$$

$$= \frac{\overset{1}{\cancel{3}} \bullet 3 \bullet \overset{1}{\cancel{2}}}{\underset{1}{\cancel{3}} \bullet 2 \bullet 2 \bullet \underset{1}{\cancel{2}}}$$

$$= \frac{3}{4}$$

Practice

21. $$\frac{3}{9} = \frac{1 \bullet \cancel{3}}{3 \bullet \cancel{3}} = \frac{1}{3}$$

23. $$\frac{7}{21} = \frac{1 \bullet \cancel{7}}{3 \bullet \cancel{7}} = \frac{1}{3}$$

25. $$\frac{20}{30} = \frac{2 \bullet \cancel{10}}{3 \bullet \cancel{10}} = \frac{2}{3}$$

27. $$\frac{15}{6} = \frac{5 \bullet \cancel{3}}{2 \bullet \cancel{3}} = \frac{5}{2}$$

29. $$-\frac{28}{56} = -\frac{1 \bullet \cancel{28}}{2 \bullet \cancel{28}} = -\frac{1}{2}$$

31. $$-\frac{90}{105} = -\frac{6 \bullet \cancel{15}}{7 \bullet \cancel{15}} = -\frac{6}{7}$$

33. $$\frac{60}{108} = \frac{5 \bullet \cancel{12}}{9 \bullet \cancel{12}} = \frac{5}{9}$$

35. $$\frac{180}{210} = \frac{6 \bullet \cancel{30}}{7 \bullet \cancel{30}} = \frac{6}{7}$$

37. $$\frac{55}{67} = \text{already in lowest terms}$$

39. $$\frac{36}{96} = \frac{3 \bullet \cancel{12}}{8 \bullet \cancel{12}} = \frac{3}{8}$$

41. $$\frac{25x^2}{35x} = \frac{5 \bullet \cancel{5} \bullet \cancel{x} \bullet x}{7 \bullet \cancel{5} \bullet \cancel{x}} = \frac{5x}{7}$$

43. $$\frac{12t}{15t} = \frac{4 \bullet \cancel{3} \bullet \cancel{t}}{5 \bullet \cancel{3} \bullet \cancel{t}} = \frac{4}{5}$$

45. $$\frac{6a}{7a} = \frac{6 \bullet \cancel{a}}{7 \bullet \cancel{a}} = \frac{6}{7}$$

47. $$\frac{7xy}{8xy} = \frac{7 \bullet \cancel{x} \bullet \cancel{y}}{8 \bullet \cancel{x} \bullet \cancel{y}} = \frac{7}{8}$$

49. $-\dfrac{10rs}{30} = -\dfrac{1 \bullet \cancel{10} \bullet r \bullet s}{3 \bullet \cancel{10}} = -\dfrac{rs}{3}$

51. $\dfrac{15st^3}{25xt^3} = \dfrac{3 \bullet \cancel{5} \bullet s \bullet \cancel{t} \bullet \cancel{t} \bullet \cancel{t}}{5 \bullet \cancel{5} \bullet x \bullet \cancel{t} \bullet \cancel{t} \bullet \cancel{t}} = \dfrac{3s}{5x}$

53. $\dfrac{35r^2t}{28rt^2} = \dfrac{5 \bullet \cancel{7} \bullet r \bullet \cancel{r} \bullet \cancel{t}}{4 \bullet \cancel{7} \bullet \cancel{r} \bullet \cancel{t} \bullet t} = \dfrac{5r}{4t}$

55. $\dfrac{56p^4}{28p^6} = \dfrac{2 \bullet \cancel{28} \bullet \cancel{p} \bullet \cancel{p} \bullet \cancel{p} \bullet \cancel{p}}{1 \bullet \cancel{28} \bullet \cancel{p} \bullet \cancel{p} \bullet \cancel{p} \bullet \cancel{p} \bullet p \bullet p} = \dfrac{2}{p^2}$

57. $\dfrac{7}{8} = \dfrac{7 \bullet 5}{8 \bullet 5} = \dfrac{35}{40}$

59. $\dfrac{4}{5} = \dfrac{4 \bullet 7}{5 \bullet 7} = \dfrac{28}{35}$

61. $\dfrac{5}{6} = \dfrac{5 \bullet 9}{6 \bullet 9} = \dfrac{45}{54}$

63. $\dfrac{1}{2} = \dfrac{1 \bullet 15}{2 \bullet 15} = \dfrac{15}{30}$

65. $\dfrac{2}{7} = \dfrac{2 \bullet 2x}{7 \bullet 2x} = \dfrac{4x}{14x}$

67. $\dfrac{9}{10} = \dfrac{9 \bullet 6t}{10 \bullet 6t} = \dfrac{54t}{60t}$

69. $\dfrac{5}{4s} = \dfrac{5 \bullet 5}{4s \bullet 5} = \dfrac{25}{20s}$

71. $\dfrac{2}{15} = \dfrac{2 \bullet 3y}{15 \bullet 3y} = \dfrac{6y}{45y}$

73. $3 = \dfrac{3}{1} = \dfrac{3 \bullet 5}{1 \bullet 5} = \dfrac{15}{5}$

75. $6 = \dfrac{6}{1} = \dfrac{6 \bullet 8}{1 \bullet 8} = \dfrac{48}{8}$

77. $4a = \dfrac{4a}{1} = \dfrac{4a \bullet 9}{1 \bullet 9} = \dfrac{36a}{9}$

79. $-2t = -\dfrac{2t}{1} = -\dfrac{2t \bullet 2}{1 \bullet 2} = -\dfrac{4t}{2}$

Applications

81.   COMMUTING

He has made $\dfrac{3}{5}$ of the commute.

83.   SINKHOLES

$-\dfrac{15}{16}$ inches

85.   PERSONNEL RECORDS

Name	Total time to complete the job alone	Time worked alone	Amount of job completed
Bob	10 hours	7 hours	$\frac{7}{10}$
Ali	8 hours	1 hour	$\frac{1}{8}$

87.   MUSIC

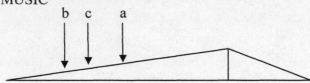

89.   MACHINERY

$\frac{1}{4}$ turn to the left, or $\frac{3}{4}$ turn to the right

91.   SUPERMARKET DISPLAYS

SNACKS

Potato

chips

Peanuts

Pretzels

Tortilla

chips

93.   CAMERAS

     $\frac{1}{250}$ will reduce chance of blur.

Writing

95. Equivalent fractions are fractions that have different numerators and denominators, but have the same value.

97. Three-fourths means 3 pieces out of 4 total, while three fifths means 3 pieces out of 5 total.

Review

99. $-5x + 1 = 16$

       $-5x + 1 - 1 = 16 - 1$

           $-5x = 15$

           $\dfrac{-5x}{-5} = \dfrac{15}{-5}$

             $x = -3$

101.   564,000

103. $10d$¢

4.2 Multiplying Fractions

Vocabulary

1. The word *of* in mathematics usually means **multiply**.

3. The result of a multiplication problem is called the **product**.

5. In the formula for the area of a triangle $A = \dfrac{1}{2}bh$, b stands for the length of the **base** and h

    stands for the **height**.

Concepts

7 $\dfrac{a}{b} \cdot \dfrac{c}{d} = \dfrac{ac}{bd}$

9a.

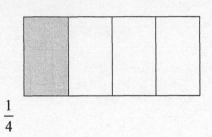

$\dfrac{1}{4}$

9b.

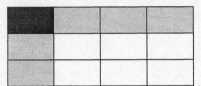

12 equal parts. 1 part is shaded twice.

$\dfrac{1}{3} \cdot \dfrac{1}{4} = \dfrac{1}{12}$

11a. negative

11b. positive

13a. true

13b. true

13c. false

13d. true

Notation

15.
$$\frac{5}{8} \cdot \frac{7}{15} = \frac{5 \cdot 7}{8 \cdot 15}$$
$$= \frac{5 \cdot 7}{8 \cdot 5 \cdot 3}$$
$$= \frac{\overset{1}{\cancel{5}} \cdot 7}{8 \cdot \underset{1}{\cancel{5}} \cdot 3}$$
$$= \frac{7}{24}$$

Practice

17.
$$\frac{1}{4} \cdot \frac{1}{2} = \frac{1}{8}$$

19.
$$\frac{3}{8} \cdot \frac{7}{16} = \frac{21}{128}$$

21.
$$\frac{2}{3} \cdot \frac{6}{7} = \frac{2}{3} \cdot \frac{2 \cdot 3}{7}$$
$$= \frac{2}{\cancel{3}} \cdot \frac{2 \cdot \cancel{3}}{7} = \frac{4}{7}$$

23.
$$\frac{14}{15} \cdot \frac{11}{8} = \frac{2 \cdot 7}{15} \cdot \frac{11}{2 \cdot 4}$$
$$= \frac{\cancel{2} \cdot 7}{15} \cdot \frac{11}{\cancel{2} \cdot 4} = \frac{77}{60}$$

25.
$$-\frac{15}{24} \cdot \frac{8}{25} = -\frac{3 \cdot 5}{8 \cdot 3} \cdot \frac{8}{5 \cdot 5}$$
$$= -\frac{\cancel{3} \cdot \cancel{5}}{\cancel{8} \cdot \cancel{3}} \cdot \frac{\cancel{8}}{5 \cdot \cancel{5}} = -\frac{1}{5}$$

27.
$$\left(-\frac{11}{21}\right)\left(-\frac{14}{33}\right) = \left(-\frac{11}{7 \cdot 3}\right)\left(-\frac{7 \cdot 2}{11 \cdot 3}\right)$$
$$= \left(-\frac{\cancel{11}}{\cancel{7} \cdot 3}\right)\left(-\frac{\cancel{7} \cdot 2}{\cancel{11} \cdot 3}\right) = \frac{2}{9}$$

29. $\dfrac{7}{10}\left(\dfrac{20}{21}\right)=\dfrac{7}{10}\left(\dfrac{10\bullet 2}{7\bullet 3}\right)$

$=\dfrac{\cancel{7}}{\cancel{10}}\left(\dfrac{\cancel{10}\bullet 2}{\cancel{7}\bullet 3}\right)=\dfrac{2}{3}$

31. $\dfrac{3}{4}\bullet\dfrac{4}{3}=\dfrac{3}{4}\bullet\dfrac{4}{3}$

$=\dfrac{\cancel{3}}{\cancel{4}}\bullet\dfrac{\cancel{4}}{\cancel{3}}=\dfrac{1}{1}=1$

33. $\dfrac{1}{3}\bullet\dfrac{15}{16}\bullet\dfrac{4}{25}=\dfrac{1}{3}\bullet\dfrac{5\bullet 3}{4\bullet 4}\bullet\dfrac{4}{5\bullet 5}$

$=\dfrac{1}{\cancel{3}}\bullet\dfrac{\cancel{5}\bullet\cancel{3}}{\cancel{4}\bullet 4}\bullet\dfrac{\cancel{4}}{\cancel{5}\bullet 5}=\dfrac{1}{20}$

35. $\left(\dfrac{2}{3}\right)\left(-\dfrac{1}{16}\right)\left(-\dfrac{4}{5}\right)=\left(\dfrac{2}{3}\right)\left(-\dfrac{1}{2\bullet 2\bullet 4}\right)\left(-\dfrac{4}{5}\right)$

$=\left(\dfrac{\cancel{2}}{3}\right)\left(-\dfrac{1}{2\bullet\cancel{2}\bullet\cancel{4}}\right)\left(-\dfrac{\cancel{4}}{5}\right)=\dfrac{1}{30}$

37. $\dfrac{5}{6}\bullet 18=\dfrac{5}{6}\bullet\dfrac{6\bullet 3}{1}$

$=\dfrac{5}{\cancel{6}}\bullet\dfrac{\cancel{6}\bullet 3}{1}=\dfrac{15}{1}=15$

39. $15\left(-\dfrac{4}{5}\right)=\dfrac{5\bullet 3}{1}\left(-\dfrac{4}{5}\right)$

$=\dfrac{\cancel{5}\bullet 3}{1}\left(-\dfrac{4}{\cancel{5}}\right)=-\dfrac{12}{1}=-12$

41. $\dfrac{5x}{12}\bullet\dfrac{1}{6x}=\dfrac{5\bullet x}{12}\bullet\dfrac{1}{6\bullet x}$

$=\dfrac{5\bullet\cancel{x}}{12}\bullet\dfrac{1}{6\bullet\cancel{x}}=\dfrac{5}{72}$

43. $\dfrac{b}{12}\bullet\dfrac{3}{10b}=\dfrac{b}{4\bullet 3}\bullet\dfrac{3}{10\bullet b}$

$=\dfrac{\cancel{b}}{4\bullet\cancel{3}}\bullet\dfrac{\cancel{3}}{10\bullet\cancel{b}}=\dfrac{1}{40}$

45. $\dfrac{1}{3}\bullet 3d=\dfrac{1}{3}\bullet\dfrac{3\bullet d}{1}$

$=\dfrac{1}{\cancel{3}}\bullet\dfrac{\cancel{3}\bullet d}{1}=\dfrac{d}{1}=d$

47. $\dfrac{2}{3}\bullet\dfrac{3s}{2}=\dfrac{2}{3}\bullet\dfrac{3\bullet s}{2}$

$=\dfrac{\cancel{2}}{\cancel{3}}\bullet\dfrac{\cancel{3}\bullet s}{\cancel{2}}=\dfrac{s}{1}=s$

49.

$$-\frac{5}{6} \bullet \frac{6}{5}c = -\frac{5}{6} \bullet \frac{6 \bullet c}{5}$$

$$= -\frac{\cancel{5}}{\cancel{6}} \bullet \frac{\cancel{6} \bullet c}{\cancel{5}} = -\frac{c}{1} = -c$$

51.

$$\frac{x}{2} \bullet \frac{4}{9x} = \frac{x}{2} \bullet \frac{2 \bullet 2}{9 \bullet x}$$

$$= \frac{\cancel{x}}{\cancel{2}} \bullet \frac{\cancel{2} \bullet 2}{9 \bullet \cancel{x}} = \frac{2}{9}$$

53.

$$4e \bullet \frac{e}{2} = \frac{2 \bullet 2 \bullet e}{1} \bullet \frac{e}{2}$$

$$= \frac{\cancel{2} \bullet 2 \bullet e}{1} \bullet \frac{e}{\cancel{2}} = \frac{2e^2}{1} = 2e^2$$

55.

$$\frac{5}{8x}\left(\frac{2x^3}{15}\right) = \frac{5}{2 \bullet 4 \bullet x}\left(\frac{2 \bullet x \bullet x \bullet x}{3 \bullet 5}\right)$$

$$= \frac{\cancel{5}}{\cancel{2} \bullet 4 \bullet \cancel{x}}\left(\frac{\cancel{2} \bullet \cancel{x} \bullet x \bullet x}{3 \bullet \cancel{5}}\right) = \frac{x^2}{12}$$

57.

$$-\frac{5c}{6cd^2} \bullet \frac{12d^4}{c} = -\frac{5 \bullet c}{6 \bullet c \bullet d \bullet d} \bullet \frac{6 \bullet 2 \bullet d \bullet d \bullet d \bullet d}{c}$$

$$= -\frac{5 \bullet \cancel{c}}{\cancel{6} \bullet c \bullet \cancel{d} \bullet \cancel{d}} \bullet \frac{\cancel{6} \bullet 2 \bullet \cancel{d} \bullet \cancel{d} \bullet d \bullet d}{\cancel{c}} = -\frac{10d^2}{c}$$

59.

$$-\frac{4h^2}{5}\left(-\frac{15}{16h^3}\right) = -\frac{4 \bullet h \bullet h}{5}\left(-\frac{5 \bullet 3}{4 \bullet 4 \bullet h \bullet h \bullet h}\right)$$

$$= -\frac{\cancel{4} \bullet \cancel{h} \bullet \cancel{h}}{\cancel{5}}\left(-\frac{\cancel{5} \bullet 3}{\cancel{4} \bullet 4 \bullet \cancel{h} \bullet \cancel{h} \bullet h}\right) = \frac{3}{4h}$$

61.

$$\frac{5}{6} \bullet x = \frac{5x}{6} \ or \ \frac{5}{6}x$$

63.

$$-\frac{8}{9} \bullet v = -\frac{8v}{9} \ or \ -\frac{8}{9}v$$

65.

$$\left(\frac{2}{3}\right)^2 = \frac{2}{3} \bullet \frac{2}{3} = \frac{4}{9}$$

67.

$$\left(-\frac{5}{9}\right)^2 = \left(-\frac{5}{9}\right)\left(-\frac{5}{9}\right) = \frac{25}{81}$$

69.

$$\left(\frac{4m}{3}\right)^2 = \left(\frac{4m}{3}\right)\left(\frac{4m}{3}\right) = \frac{16m^2}{9}$$

71.

$$\left(-\frac{3r}{4}\right)^2 = \left(-\frac{3r}{4}\right)\left(-\frac{3r}{4}\right)\left(-\frac{3r}{4}\right) = -\frac{27r^3}{64}$$

73.

$\bullet$	$\frac{1}{2}$	$\frac{1}{3}$	$\frac{1}{4}$	$\frac{1}{5}$	$\frac{1}{6}$
$\frac{1}{2}$	$\frac{1}{4}$	$\frac{1}{6}$	$\frac{1}{8}$	$\frac{1}{10}$	$\frac{1}{12}$
$\frac{1}{3}$	$\frac{1}{6}$	$\frac{1}{9}$	$\frac{1}{12}$	$\frac{1}{15}$	$\frac{1}{18}$
$\frac{1}{4}$	$\frac{1}{8}$	$\frac{1}{12}$	$\frac{1}{16}$	$\frac{1}{20}$	$\frac{1}{24}$
$\frac{1}{5}$	$\frac{1}{10}$	$\frac{1}{15}$	$\frac{1}{20}$	$\frac{1}{25}$	$\frac{1}{30}$
$\frac{1}{6}$	$\frac{1}{12}$	$\frac{1}{18}$	$\frac{1}{24}$	$\frac{1}{30}$	$\frac{1}{36}$

75.

$$A = \frac{1}{2}bh$$

$$A = \frac{1}{2}(10)(3)$$

$$A = \frac{1}{2}\left(\frac{10}{1}\right)\left(\frac{3}{1}\right)$$

$$A = \frac{1}{\cancel{2}}\left(\frac{5 \bullet \cancel{2}}{1}\right)\left(\frac{3}{1}\right) = \frac{15}{1} = 15$$

$$A = 15 \text{ ft}^2$$

77.

$$A = \frac{1}{2}bh$$

$$A = \frac{1}{2}(3)(5)$$

$$A = \frac{1}{2}\left(\frac{3}{1}\right)\left(\frac{5}{1}\right) = \frac{15}{2}$$

$$A = \frac{15}{2} \text{ yd}^2$$

Applications

79.   THE CONSTITUTION

v = votes required

$$v = \frac{2}{3}(435)$$

$$v = \frac{2}{3}\left(\frac{435}{1}\right)$$

$$v = \frac{2}{\cancel{3}}\left(\frac{\cancel{3} \bullet 145}{1}\right) = \frac{290}{1} = 290$$

290 votes are required.

81.   TENNIS BALLS

h_1 = height of first bounce

h_2 = height of second bounce

h_3 = height of third bounce

$$h_1 = \frac{1}{3}(54) \qquad\qquad h_2 = \frac{1}{3}(18) \qquad\qquad h_3 = \frac{1}{3}(6)$$

$$h_1 = \frac{1}{3}\left(\frac{54}{1}\right) \qquad h_2 = \frac{1}{3}\left(\frac{18}{1}\right) \qquad h_3 = \frac{1}{3}\left(\frac{6}{1}\right)$$

$$h_1 = \frac{1}{\cancel{3}}\left(\frac{18 \bullet \cancel{3}}{1}\right) = \frac{18}{1} = 18 \qquad h_2 = \frac{1}{\cancel{3}}\left(\frac{6 \bullet \cancel{3}}{1}\right) = \frac{6}{1} = 6 \qquad h_3 = \frac{1}{\cancel{3}}\left(\frac{2 \bullet \cancel{3}}{1}\right) = \frac{2}{1} = 2$$

18 inches on first bounce, 6 inches on second bounce, and 2 inches on third bounce.

83.   COOKING

Half of each amount is needed.

s = sugar required

$$s = \frac{1}{2} \bullet \frac{3}{4} = \frac{3}{8}$$

m = molasses required

$$m = \frac{1}{2} \bullet \frac{1}{3} = \frac{1}{6}$$

$\frac{3}{8}$ cup of sugar and $\frac{1}{6}$ cup of

molasses is required.

85.   BOTANY

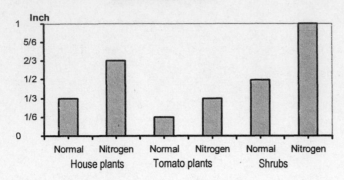

Growth Rate: June

87.   THE STARS AND STRIPES

A = area of the triangle

$A = \dfrac{1}{2}bh$, base = 22 in., height = 11 in.

$A = \dfrac{1}{2}(22)(11)$

$A = \dfrac{1}{2}\left(\dfrac{22}{1}\right)\left(\dfrac{11}{1}\right)$

$A = \dfrac{1}{\cancel{2}}\left(\dfrac{\cancel{2}\bullet 11}{1}\right)\left(\dfrac{11}{1}\right) = \dfrac{121}{1} = 121$

The area is 121 in^2.

89.   TILE DESIGNS

There are 4 triangular tiles total.

A = area of one triangle

$A = \dfrac{1}{2}bh$, base = 3 in., height = 3 in.

$A = \dfrac{1}{2}(3)(3)$

$A = \dfrac{1}{2}\left(\dfrac{3}{1}\right)\left(\dfrac{3}{1}\right) = \dfrac{9}{2}$

T = total area

$T = 4\bullet\dfrac{9}{2}$

$T = \dfrac{4}{1}\bullet\dfrac{9}{2}$

$T = \dfrac{\cancel{2}\bullet 2}{1}\bullet\dfrac{9}{\cancel{2}} = \dfrac{18}{1} = 18$

The area of tile that is blue is 18 in^2.

Writing

91. Answers will vary.

93. If the majority of the class voted to postpone a test that means that one more than half the number in the class voted to postpone a test. If there are 10 people in the class, it would take at least 6 votes (which is one more than half) to postpone the test.

Review

95. $2(x+7)=2x+14$

97. $2x+6=6$ *for* $x=-6$
$2(-6)+6=6$
$-12+6=6$
$-6\neq 6$ no

99.

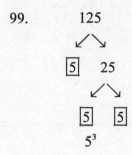

4.3 Dividing Fractions

Vocabulary

1.   Two numbers are called **<u>reciprocals</u>** if their product is 1.

Concepts

3. $\dfrac{1}{2} \div \dfrac{2}{3} = \dfrac{1}{2} \cdot \dfrac{3}{2}$

5.

$4 \div \dfrac{1}{3}$

$4 \div \dfrac{1}{3} = 12$

7. $\dfrac{4}{5} \cdot \dfrac{5}{4} = 1$

9a. $15 \div 3 = 5$

9b.
$$15 \div 3 = 15 \cdot \dfrac{1}{3}$$
$$= \dfrac{15}{1} \cdot \dfrac{1}{3}$$
$$= \dfrac{\cancel{3} \cdot 5}{1} \cdot \dfrac{1}{\cancel{3}} = \dfrac{5}{1} = 5$$

9c.   Division by 3 is the same as multiplication by $\dfrac{1}{3}$.

Notation

11.
$$\frac{25}{36} \div \frac{10}{9} = \frac{25}{36} \cdot \frac{9}{10}$$
$$= \frac{25 \cdot 9}{36 \cdot 10}$$
$$= \frac{5 \cdot 5 \cdot 9}{4 \cdot 9 \cdot 2 \cdot 5}$$
$$= \frac{\overset{1}{\cancel{5}} \cdot 5 \cdot \overset{1}{\cancel{9}}}{4 \cdot \underset{1}{\cancel{9}} \cdot 2 \cdot \underset{1}{\cancel{5}}}$$
$$= \frac{5}{8}$$

Practice

13.
$$\frac{1}{2} \div \frac{3}{5} = \frac{1}{2} \cdot \frac{5}{3}$$
$$= \frac{5}{6}$$

15.
$$\frac{3}{16} \div \frac{1}{9} = \frac{3}{16} \cdot \frac{9}{1}$$
$$= \frac{27}{16}$$

17.
$$\frac{4}{5} \div \frac{4}{5} = \frac{4}{5} \cdot \frac{5}{4}$$
$$= \frac{\cancel{4}}{\cancel{5}} \cdot \frac{\cancel{5}}{\cancel{4}} = 1$$

19.
$$\left(-\frac{7}{4}\right) \div \left(-\frac{21}{8}\right) = \left(-\frac{7}{4}\right) \cdot \left(-\frac{8}{21}\right)$$
$$= \left(-\frac{\cancel{7}}{\cancel{4}}\right) \cdot \left(-\frac{\cancel{4} \cdot 2}{\cancel{7} \cdot 3}\right)$$
$$= \frac{2}{3}$$

21.
$$3 \div \frac{1}{12} = \frac{3}{1} \cdot \frac{12}{1}$$
$$= \frac{36}{1} = 36$$

23.
$$120 \div \frac{12}{5} = \frac{120}{1} \cdot \frac{5}{12}$$
$$= \frac{\cancel{12} \cdot 10}{1} \cdot \frac{5}{\cancel{12}}$$
$$= \frac{50}{1} = 50$$

25.
$$-\frac{4}{5} \div (-6) = -\frac{4}{5} \bullet \left(-\frac{1}{6}\right)$$
$$= -\frac{\cancel{2} \bullet 2}{5} \bullet \left(-\frac{1}{\cancel{2} \bullet 3}\right)$$
$$= \frac{2}{15}$$

27.
$$\frac{15}{16} \div 180 = \frac{15}{16} \bullet \frac{1}{180}$$
$$= \frac{\cancel{15}}{16} \bullet \frac{1}{\cancel{15} \bullet 12}$$
$$= \frac{1}{192}$$

29.
$$-\frac{9}{10} \div \frac{4}{15} = -\frac{9}{10} \bullet \frac{15}{4}$$
$$= -\frac{9}{\cancel{5} \bullet 2} \bullet \frac{\cancel{5} \bullet 3}{4}$$
$$= -\frac{27}{8}$$

31.
$$\frac{9}{10} \div \left(-\frac{3}{25}\right) = \frac{9}{10} \bullet \left(-\frac{25}{3}\right)$$
$$= \frac{\cancel{3} \bullet 3}{\cancel{5} \bullet 2} \bullet \left(-\frac{\cancel{5} \bullet 5}{\cancel{3}}\right)$$
$$= -\frac{15}{2}$$

33.
$$-\frac{1}{8} \div 8 = -\frac{1}{8} \bullet \frac{1}{8}$$
$$= -\frac{1}{64}$$

35.
$$\frac{15}{32} \div \frac{15}{32} = \frac{15}{32} \bullet \frac{32}{15}$$
$$= \frac{\cancel{15}}{\cancel{32}} \bullet \frac{\cancel{32}}{\cancel{15}} = 1$$

37.
$$\frac{4a}{5} \div \frac{3}{2} = \frac{4a}{5} \bullet \frac{2}{3}$$
$$= \frac{8a}{15}$$

39.
$$\frac{t}{8} \div \frac{3}{4} = \frac{t}{8} \bullet \frac{4}{3}$$
$$= \frac{t}{\cancel{4} \bullet 2} \bullet \frac{\cancel{4}}{3}$$
$$= \frac{t}{6}$$

41.
$$\frac{13}{16b} \div \frac{1}{2} = \frac{13}{16b} \bullet \frac{2}{1}$$
$$= \frac{13}{\cancel{2} \bullet 8 \bullet b} \bullet \frac{\cancel{2}}{1}$$
$$= \frac{13}{8b}$$

43.
$$-\frac{15}{32y} \div \frac{3}{4} = -\frac{15}{32y} \bullet \frac{4}{3}$$
$$= -\frac{5 \bullet \cancel{3}}{\cancel{4} \bullet 8 \bullet y} \bullet \frac{\cancel{4}}{\cancel{3}}$$
$$= -\frac{5}{8y}$$

45.

$$a \div \frac{a}{b} = \frac{a}{1} \bullet \frac{b}{a}$$

$$= \frac{\cancel{a}}{1} \bullet \frac{b}{\cancel{a}}$$

$$= \frac{b}{1} = b$$

47.

$$\frac{x}{y} \div \frac{x}{y} = \frac{x}{y} \bullet \frac{y}{x}$$

$$= \frac{\cancel{x}}{\cancel{y}} \bullet \frac{\cancel{y}}{\cancel{x}} = 1$$

49.

$$\frac{2s}{3t} \div (-6) = \frac{2s}{3t} \bullet \left(-\frac{1}{6}\right)$$

$$= \frac{\cancel{2} \bullet s}{3t} \bullet \left(-\frac{1}{\cancel{2} \bullet 3}\right)$$

$$= -\frac{s}{9t}$$

51.

$$-\frac{9}{8}x \div \frac{3}{4x^2} = -\frac{9x}{8} \div \frac{4x^2}{3}$$

$$= -\frac{\cancel{3} \bullet 3 \bullet x}{\cancel{4} \bullet 2} \div \frac{\cancel{4} \bullet x \bullet x}{\cancel{3}}$$

$$= -\frac{3x^3}{2}$$

53.

$$-8x \div \left(-\frac{4x^3}{9}\right) = -\frac{8x}{1} \div \left(-\frac{9}{4x^3}\right)$$

$$= -\frac{\cancel{4} \bullet 2 \bullet \cancel{x}}{1} \div \left(-\frac{9}{\cancel{4} \bullet \cancel{x} \bullet x \bullet x}\right)$$

$$= \frac{18}{x^2}$$

55.

$$-\frac{x^2}{y^3} \div \frac{x}{y} = -\frac{x^2}{y^3} \bullet \frac{y}{x}$$

$$= -\frac{\cancel{x} \bullet x}{\cancel{y} \bullet y \bullet y} \bullet \frac{\cancel{y}}{\cancel{x}}$$

$$= -\frac{x}{y^2}$$

57.

$$-\frac{26x}{15} \div \frac{13}{45x} = -\frac{26x}{15} \bullet \frac{45x}{13}$$

$$= -\frac{\cancel{13} \bullet 2 \bullet x}{\cancel{15}} \bullet \frac{\cancel{15} \bullet 3 \bullet x}{\cancel{13}}$$

$$= -\frac{6x^2}{1} = -6x^2$$

Applications

59.   MARATHONS

n = number of laps needed

$$n = 26 \div \frac{1}{4}$$

$$n = \frac{26}{1} \bullet \frac{4}{1}$$

$$n = \frac{104}{1} = 104$$

It would take 104 laps.

61.   LASER TECHNOLOGIES

s = number of $\frac{1}{64}$ inch slices

$$s = \frac{7}{8} \div \frac{1}{64}$$

$$s = \frac{7}{8} \bullet \frac{64}{1}$$

$$s = \frac{7}{\cancel{8}} \bullet \frac{\cancel{8} \bullet 8}{1}$$

$$s = \frac{56}{1} = 56$$

You could make 56 slices.

63.   UNDERGROUND CABLES

Route 1

T_1 = time to install alonge Route 1
The total distance to be covered is 15 miles

$$T_1 = 15 \div \frac{3}{5}$$

$$T_1 = \frac{15}{1} \div \frac{5}{3}$$

$$T_1 = \frac{\cancel{3} \bullet 5}{1} \div \frac{5}{\cancel{3}}$$

$$T_1 = \frac{25}{1} = 25$$

Route 1 was takes less time to complete.

Route 2

T_2 = time to install alonge Route 2
The total distance to be covered is 12 miles

$$T_2 = 12 \div \frac{2}{5}$$

$$T_2 = \frac{12}{1} \div \frac{5}{2}$$

$$T_2 = \frac{\cancel{2} \bullet 6}{1} \div \frac{5}{\cancel{2}}$$

$$T_2 = \frac{30}{1} = 30$$

65. 3 × 5 CARDS

a) sixteen parts

b) $\dfrac{3}{4}$ inch thick

c) There are 90 cards in the stack.

t = thickness of one card

$$t = \frac{3}{4} \div 90$$

$$t = \frac{3}{4} \div \frac{90}{1}$$

$$t = \frac{\cancel{3}}{4} \bullet \frac{1}{\cancel{3} \bullet 30}$$

$$t = \frac{1}{120}$$

One card is $\frac{1}{120}$ inch thick.

67. FORESTRY

s = number of sections

$$s = 6{,}284 \div \frac{4}{5}$$

$$s = \frac{6{,}284}{1} \div \frac{4}{5}$$

$$s = \frac{6{,}284}{1} \bullet \frac{5}{4}$$

$$s = \frac{1{,}571 \bullet \cancel{4}}{1} \bullet \frac{5}{\cancel{4}}$$

$$s = \frac{7{,}855}{1} = 7{,}855$$

There are 7,855 sections.

Writing

69.   To divide two fractions, you must rewrite the division as multiplication using the following

conversion, $\dfrac{a}{b} \div \dfrac{c}{d} = \dfrac{a}{b} \bullet \dfrac{d}{c}$.

71.   Answers will vary. A 10 meter beam is to be cut into $\frac{1}{5}$ m size pieces. How many pieces

could be made?

Review

73.
$$4x + (-2) = -18$$
$$4x + (-2) + 2 = -18 + 2$$
$$4x = -16$$
$$\frac{4x}{4} = \frac{-16}{4}$$
$$x = -4$$

75. $p = r - c$

77.   False

79. $-3t + (-5T) + 4T + 8t = 5t - T$

4.4 Adding and Subtracting Fractions

Vocabulary

1. The <u>least</u> common denominator for a set of fractions is the smallest number each denominator will divide exactly.

3. To express a fraction in <u>higher</u> terms, we multiply the numerator and denominator by the same number.

Concepts

5. This rule tells us how to add fractions having like <u>denominators</u>. To find the sum, we add the <u>numerators</u> and then write that result over the <u>common </u>denominator.

7.   The denominators are unlike.

9a.   4   9b.   *c*

11a.   once   11b.   twice   11c.   three times

13. $2 \bullet 2 \bullet 3 \bullet 5 = 60$

15a.   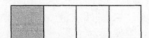 $\frac{1}{3}$ is larger

15b. $\frac{1}{4} = \frac{3}{12}$ and $\frac{1}{3} = \frac{4}{12}$
       therefore, $\frac{1}{3} > \frac{1}{4}$

Notation

17. $\dfrac{2}{5}+\dfrac{1}{3}=\dfrac{2\bullet 3}{5\bullet 3}+\dfrac{1\bullet 5}{3\bullet 5}$

$\phantom{\dfrac{2}{5}+\dfrac{1}{3}}=\dfrac{6}{15}+\dfrac{5}{15}$

$\phantom{\dfrac{2}{5}+\dfrac{1}{3}}=\dfrac{6+5}{15}$

$\phantom{\dfrac{2}{5}+\dfrac{1}{3}}=\dfrac{11}{15}$

Practice

19.   18

21.   24

23.   40

25.   60

27. $\dfrac{3}{7}+\dfrac{1}{7}=\dfrac{4}{7}$

29. $\dfrac{37}{103}-\dfrac{17}{103}=\dfrac{20}{103}$

31. $\dfrac{11}{25}-\dfrac{1}{25}=\dfrac{10}{25}$

$\phantom{\dfrac{11}{25}}=\dfrac{\cancel{5}\bullet 2}{\cancel{5}\bullet 5}$

$\phantom{\dfrac{11}{25}}=\dfrac{2}{5}$

33. $\dfrac{5}{d}+\dfrac{3}{d}=\dfrac{8}{d}$

35. $\dfrac{1}{4}+\dfrac{3}{8}=\dfrac{1\bullet 2}{4\bullet 2}+\dfrac{3}{8}$

$\phantom{\dfrac{1}{4}}=\dfrac{2}{8}+\dfrac{3}{8}$

$\phantom{\dfrac{1}{4}}=\dfrac{5}{8}$

37. $\dfrac{13}{20}-\dfrac{1}{5}=\dfrac{13}{20}-\dfrac{1\bullet 4}{5\bullet 4}$

$\phantom{\dfrac{13}{20}}=\dfrac{13}{20}-\dfrac{4}{20}$

$\phantom{\dfrac{13}{20}}=\dfrac{9}{20}$

39. $\dfrac{4}{5} + \dfrac{2}{3} = \dfrac{4 \bullet 3}{5 \bullet 3} + \dfrac{2 \bullet 5}{3 \bullet 5}$

             $= \dfrac{12}{15} + \dfrac{10}{15}$

             $= \dfrac{22}{15}$

41. $\dfrac{1}{8} + \dfrac{2}{7} = \dfrac{1 \bullet 7}{8 \bullet 7} + \dfrac{2 \bullet 8}{7 \bullet 8}$

             $= \dfrac{7}{56} + \dfrac{16}{56}$

             $= \dfrac{23}{56}$

43. $\dfrac{3}{4} - \dfrac{2}{3} = \dfrac{3 \bullet 3}{4 \bullet 3} - \dfrac{2 \bullet 4}{3 \bullet 4}$

             $= \dfrac{9}{12} - \dfrac{8}{12}$

             $= \dfrac{1}{12}$

45. $\dfrac{5}{6} - \dfrac{3}{4} = \dfrac{5 \bullet 2}{6 \bullet 2} - \dfrac{3 \bullet 3}{4 \bullet 3}$

             $= \dfrac{10}{12} - \dfrac{9}{12}$

             $= \dfrac{1}{12}$

47. $\dfrac{16}{25} - \left(-\dfrac{3}{10} \right) = \dfrac{16}{25} + \dfrac{3}{10}$

                   $= \dfrac{16 \bullet 2}{25 \bullet 2} + \dfrac{3 \bullet 5}{10 \bullet 5}$

                   $= \dfrac{32}{50} + \dfrac{15}{50}$

                   $= \dfrac{47}{50}$

49. $-\dfrac{7}{16} + \dfrac{1}{4} = -\dfrac{7}{16} + \dfrac{1 \bullet 4}{4 \bullet 4}$

               $= -\dfrac{7}{16} + \dfrac{4}{16}$

               $= -\dfrac{3}{16}$

51. $\dfrac{1}{12} - \dfrac{3}{4} = \dfrac{1}{12} - \dfrac{3 \bullet 3}{4 \bullet 3}$

            $= \dfrac{1}{12} - \dfrac{9}{12}$

            $= \dfrac{1}{12} + \left(-\dfrac{9}{12} \right)$

            $= -\dfrac{8}{12}$

            $= -\dfrac{\cancel{4} \bullet 2}{\cancel{4} \bullet 3} = -\dfrac{2}{3}$

53. $-\dfrac{5}{8} - \dfrac{1}{3} = -\dfrac{5 \bullet 3}{8 \bullet 3} - \dfrac{1 \bullet 8}{3 \bullet 8}$

            $= -\dfrac{15}{24} - \dfrac{8}{24}$

            $= -\dfrac{15}{24} + \left(-\dfrac{8}{24} \right)$

            $= -\dfrac{23}{24}$

55.

$$-3 + \frac{2}{5} = -\frac{3 \cdot 5}{1 \cdot 5} + \frac{2}{5}$$

$$= -\frac{15}{5} + \frac{2}{5}$$

$$= -\frac{13}{5}$$

57.

$$-\frac{3}{4} - 5 = -\frac{3}{4} - \frac{5 \cdot 4}{1 \cdot 4}$$

$$= -\frac{3}{4} - \frac{20}{4}$$

$$= -\frac{3}{4} + \left(-\frac{20}{4}\right)$$

$$= -\frac{23}{4}$$

59.

$$\frac{7}{8} - \frac{t}{7} = \frac{7 \cdot 7}{8 \cdot 7} - \frac{t \cdot 8}{7 \cdot 8}$$

$$= \frac{49}{56} - \frac{8t}{56}$$

$$= \frac{49 - 8t}{56}$$

61.

$$\frac{4}{5} - \frac{2b}{9} = \frac{4 \cdot 9}{5 \cdot 9} - \frac{2b \cdot 5}{9 \cdot 5}$$

$$= \frac{36}{45} - \frac{10b}{45}$$

$$= \frac{36 - 10b}{45}$$

63.

$$\frac{4}{7} - \frac{1}{r} = \frac{4 \cdot r}{7 \cdot r} - \frac{1 \cdot 7}{r \cdot 7}$$

$$= \frac{4r}{7r} - \frac{7}{7r}$$

$$= \frac{4r - 7}{7r}$$

65.

$$-\frac{5}{9} + \frac{1}{y} = -\frac{5 \cdot y}{9 \cdot y} + \frac{1 \cdot 9}{y \cdot 9}$$

$$= -\frac{5y}{9y} + \frac{9}{9y}$$

$$= \frac{-5y + 9}{9y}$$

67.

$$\frac{1}{3} + \frac{1}{4} + \frac{1}{5} = \frac{1 \cdot 20}{3 \cdot 20} + \frac{1 \cdot 15}{4 \cdot 15} + \frac{1 \cdot 12}{5 \cdot 12}$$

$$= \frac{20}{60} + \frac{15}{60} + \frac{12}{60}$$

$$= \frac{47}{60}$$

69.

$$-\frac{2}{3} + \frac{5}{4} + \frac{1}{6} = -\frac{2 \cdot 4}{3 \cdot 4} + \frac{5 \cdot 3}{4 \cdot 3} + \frac{1 \cdot 2}{6 \cdot 2}$$

$$= -\frac{8}{12} + \frac{15}{12} + \frac{2}{12}$$

$$= \frac{9}{12}$$

$$= \frac{\cancel{3} \cdot 3}{\cancel{3} \cdot 4} = \frac{3}{4}$$

71. $\dfrac{5}{24}+\dfrac{3}{16}=\dfrac{5\cdot 2}{24\cdot 2}+\dfrac{3\cdot 3}{16\cdot 3}$

$\qquad\qquad =\dfrac{10}{48}+\dfrac{9}{48}$

$\qquad\qquad =\dfrac{19}{48}$

73. $-\dfrac{11}{15}-\dfrac{2}{9}=-\dfrac{11\cdot 3}{15\cdot 3}-\dfrac{2\cdot 5}{9\cdot 5}$

$\qquad\qquad =-\dfrac{33}{45}-\dfrac{10}{45}$

$\qquad\qquad =-\dfrac{33}{45}+\left(-\dfrac{10}{45}\right)$

$\qquad\qquad =-\dfrac{43}{45}$

75. $\dfrac{7}{25}+\dfrac{1}{15}=\dfrac{7\cdot 3}{25\cdot 3}+\dfrac{1\cdot 5}{15\cdot 5}$

$\qquad\qquad =\dfrac{21}{75}+\dfrac{5}{75}$

$\qquad\qquad =\dfrac{26}{75}$

77. $\dfrac{4}{27}+\dfrac{1}{6}=\dfrac{4\cdot 2}{27\cdot 2}+\dfrac{1\cdot 9}{6\cdot 9}$

$\qquad\qquad =\dfrac{8}{54}+\dfrac{9}{54}$

$\qquad\qquad =\dfrac{17}{54}$

79. $\dfrac{11}{60}-\dfrac{2}{45}=\dfrac{11\cdot 3}{60\cdot 3}-\dfrac{2\cdot 4}{45\cdot 4}$

$\qquad\qquad =\dfrac{33}{180}-\dfrac{8}{180}$

$\qquad\qquad =\dfrac{25}{180}$

$\qquad\qquad =\dfrac{\cancel{5}\cdot 5}{\cancel{5}\cdot 36}=\dfrac{5}{36}$

81. $\dfrac{2}{15}-\dfrac{5}{12}=\dfrac{2\cdot 4}{15\cdot 4}-\dfrac{5\cdot 5}{12\cdot 5}$

$\qquad\qquad =\dfrac{8}{60}-\dfrac{25}{60}$

$\qquad\qquad =\dfrac{8}{60}+\left(-\dfrac{25}{60}\right)$

$\qquad\qquad =-\dfrac{17}{60}$

Applications

83a.   BOTANY

g = 2-year growth total

$$g = \frac{5}{32} + \frac{1}{16}$$

$$g = \frac{5}{32} + \frac{1 \bullet 2}{16 \bullet 2}$$

$$g = \frac{5}{32} + \frac{2}{32}$$

$$g = \frac{7}{32}$$

The growth was $\frac{7}{32}$ in.

83b.

d = difference in growth

$$d = \frac{5}{32} - \frac{1}{16}$$

$$d = \frac{5}{32} - \frac{1 \bullet 2}{16 \bullet 2}$$

$$d = \frac{5}{32} - \frac{2}{32}$$

$$d = \frac{3}{32}$$

The difference was $\frac{3}{32}$ in.

85.   FAMILY DINNER

L = total leftover pizza

$$L = \frac{3}{8} + \frac{2}{6}$$

$$L = \frac{3 \bullet 3}{8 \bullet 3} + \frac{2 \bullet 4}{6 \bullet 4}$$

$$L = \frac{9}{24} + \frac{8}{24}$$

$$L = \frac{17}{24}$$

There is $\frac{17}{24}$ of a pizza left. No the family could not have been fed by one.

87.   WEIGHTS AND MEASURES

e = error of the scale

$$e = \frac{3}{4} - \frac{11}{16}$$

$$e = \frac{3 \bullet 4}{4 \bullet 4} - \frac{11}{16}$$

$$e = \frac{12}{16} - \frac{11}{16}$$

$$e = \frac{1}{16}$$

The scale is off by $\frac{1}{16}$ lb. It is lower by $\frac{1}{16}$ lb, so they are undercharging.

89. **HIKING**

$$\frac{3}{4} = \frac{3 \bullet 10}{4 \bullet 10} = \frac{30}{40}, \quad \frac{4}{5} = \frac{4 \bullet 8}{5 \bullet 8} = \frac{32}{40},$$

$$\frac{5}{8} = \frac{5 \bullet 5}{8 \bullet 5} = \frac{25}{40}$$

Arranging from longest to shortest:

$$\frac{4}{5}, \frac{3}{4}, \frac{5}{8}$$

91. **STUDY HABITS**

s = students studying 2+ hours per day

$$s = \frac{2}{5} + \frac{3}{10}$$

$$s = \frac{2 \bullet 2}{5 \bullet 2} + \frac{3}{10}$$

$$s = \frac{4}{10} + \frac{3}{10}$$

$$s = \frac{7}{10}$$

$\frac{7}{10}$ of the students are studying 2 or

more hours daily.

93. **GARAGE DOOR OPENERS**

d = difference in strengths

$$d = \frac{1}{2} - \frac{1}{3}$$

$$d = \frac{1 \bullet 3}{2 \bullet 3} - \frac{1 \bullet 2}{3 \bullet 2}$$

$$d = \frac{3}{6} - \frac{2}{6}$$

$$d = \frac{1}{6}$$

The difference is $\frac{1}{6}$ horsepower.

Writing

95. They both use the fundamental property of fractions. For expressing fractions in lowest terms division is used, and for expressing fractions in higher terms, multiplication is used.

97. To compare 2 fractions with different denominators, they have to be converted to fractions with common denominators.

Review

99. $2(2+x)-3(x-1)=4+2x-3x+3$ 101. $x-5$

$\qquad\qquad\qquad\qquad = 7+2x+(-3x)$

$\qquad\qquad\qquad\qquad = 7-x \ \ or \ \ -x+7$

103. $P=2l+2w$ is the formula.

But remember, to find the perimeter you could also add all of the sides together.

The LCM and the GCF

1.　Since 3 and 5 are prime
$$LCM = 3 \bullet 5 = 15$$

3.

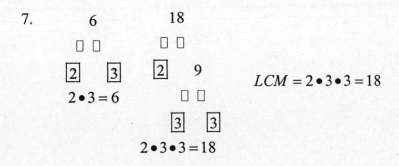

    8   14

$\boxed{2}$　4　　$\boxed{2}$　$\boxed{7}$

            $2 \bullet 7 = 14$

$\boxed{2}$　$\boxed{2}$

$2 \bullet 2 \bullet 2 = 8$

$$LCM = 2 \bullet 2 \bullet 2 \bullet 7 = 56$$

5.

  21   14

$\boxed{3}$　$\boxed{7}$　$\boxed{2}$　$\boxed{7}$
$3 \bullet 7 = 21$　$2 \bullet 7 = 14$

$$LCM = 2 \bullet 3 \bullet 7 = 42$$

7.

  6   18

$\boxed{2}$　$\boxed{3}$　$\boxed{2}$　9
$2 \bullet 3 = 6$

$\boxed{3}$　$\boxed{3}$

$2 \bullet 3 \bullet 3 = 18$

$$LCM = 2 \bullet 3 \bullet 3 = 18$$

9.

  44   60

$\boxed{11}$　4　　6　10

$\boxed{2}$　$\boxed{2}$　　$\boxed{2}$　$\boxed{3}$$\boxed{2}$　$\boxed{5}$
$2 \bullet 2 \bullet 11 = 44$　$2 \bullet 2 \bullet 3 \bullet 5 = 60$

$$LCM = 2 \bullet 2 \bullet 3 \bullet 5 \bullet 11 = 660$$

11.

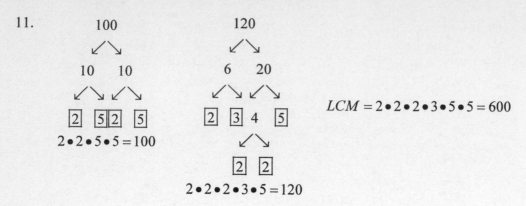

$LCM = 2 \bullet 2 \bullet 2 \bullet 3 \bullet 5 \bullet 5 = 600$

$2 \bullet 2 \bullet 5 \bullet 5 = 100$

$2 \bullet 2 \bullet 2 \bullet 3 \bullet 5 = 120$

13.

$2 \bullet 3 = 6$

$2 \bullet 2 \bullet 2 \bullet 3 = 24$

$2 \bullet 2 \bullet 3 \bullet 3 = 36$

$LCM = 2 \bullet 2 \bullet 2 \bullet 3 \bullet 3 = 72$

15.

$2 \bullet 3 \bullet 3 = 18$

$2 \bullet 3 \bullet 3 \bullet 3 = 54$

$3 \bullet 3 \bullet 7 = 63$

$LCM = 2 \bullet 3 \bullet 3 \bullet 3 \bullet 7 = 378$

17.

$GCF = 3$

$2 \bullet 3 = 6$ $3 \bullet 3 = 9$

19.

$GCF = 11$

$2 \bullet 11 = 22$ $3 \bullet 11 = 33$

21.

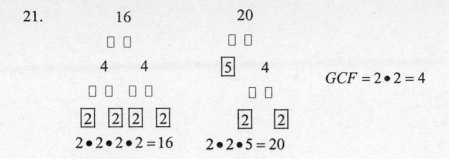

16   20

4   4   5   4 $GCF = 2 \cdot 2 = 4$

2 2 2 2   2   2

$2 \cdot 2 \cdot 2 \cdot 2 = 16$ $2 \cdot 2 \cdot 5 = 20$

23.

25   100

5   5   10   10 $GCF = 5 \cdot 5 = 25$

$5 \cdot 5 = 25$

2 5 2 5

$2 \cdot 2 \cdot 5 \cdot 5 = 100$

25.

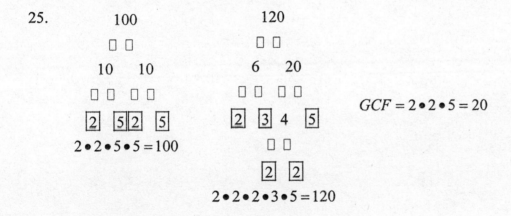

100   120

10   10   6   20

2 5 2 5   2 3 4 5 $GCF = 2 \cdot 2 \cdot 5 = 20$

$2 \cdot 2 \cdot 5 \cdot 5 = 100$ 2 2

$2 \cdot 2 \cdot 2 \cdot 3 \cdot 5 = 120$

27.

48   108

6   8   6   18

2 3 4 2   2 3 9 2 $GCF = 2 \cdot 2 \cdot 3 = 12$

2 2   3 3

$2 \cdot 2 \cdot 2 \cdot 2 \cdot 3 = 48$ $2 \cdot 2 \cdot 3 \cdot 3 \cdot 3 = 108$

29.

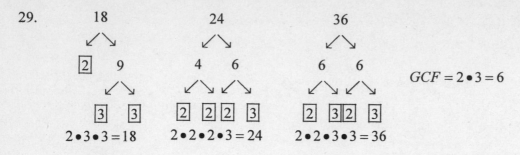

$$GCF = 2 \bullet 3 = 6$$

$2 \bullet 3 \bullet 3 = 18 \qquad 2 \bullet 2 \bullet 2 \bullet 3 = 24 \qquad 2 \bullet 2 \bullet 3 \bullet 3 = 36$

31.

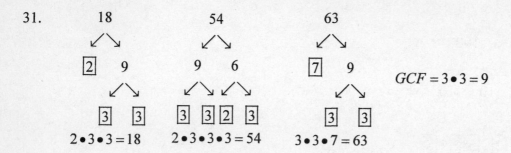

$$GCF = 3 \bullet 3 = 9$$

$2 \bullet 3 \bullet 3 = 18 \qquad 2 \bullet 3 \bullet 3 \bullet 3 = 54 \qquad 3 \bullet 3 \bullet 7 = 63$

33.   NURSING

2 hours = 120 minutes

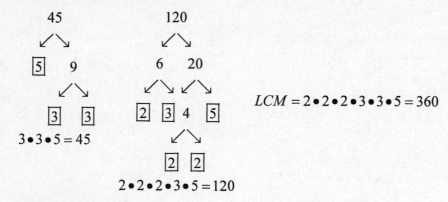

$$LCM = 2 \bullet 2 \bullet 2 \bullet 3 \bullet 3 \bullet 5 = 360$$

$3 \bullet 3 \bullet 5 = 45$

$2 \bullet 2 \bullet 2 \bullet 3 \bullet 5 = 120$

They will be in the same room again in 360 minutes or 6 hours.

4.5 Multiplying and Dividing Mixed Numbers

Vocabulary

1. A **mixed** number is the sum of a whole number and a proper fraction.

3. To **graph** a number means to locate its position on a number line and highlight it using a heavy dot.

Concepts

5a. $-5\dfrac{1}{2}°$

5b. $-6\dfrac{7}{8}$ in.

6a. $-2\dfrac{3}{10}$ in

6b. $-3\dfrac{1}{2}$ min

8a. $-2\dfrac{1}{2}$

8b. -2

9. $-\dfrac{4}{5}, -\dfrac{2}{5}, \dfrac{1}{5}$

11. DIVING

Forward $2\dfrac{1}{2}$ somersaults from the pike position.

13.

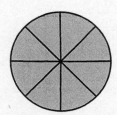 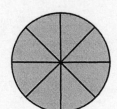 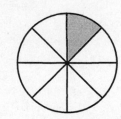

Notation

15.

$$-5\frac{1}{4} \cdot 1\frac{1}{7} = -\frac{21}{4} \cdot \frac{8}{7}$$

$$= -\frac{21 \cdot 8}{4 \cdot 7}$$

$$= -\frac{\overset{1}{\cancel{7}} \cdot 3 \cdot \overset{1}{\cancel{4}} \cdot 2}{\underset{1}{\cancel{4}} \cdot \underset{1}{\cancel{7}}}$$

$$= -\frac{6}{1}$$

$$= -6$$

Practice

17.

$$\frac{15}{4} = 3\frac{3}{4}$$

19.

$$\frac{29}{5} = 5\frac{4}{5}$$

21.

$$-\frac{20}{6} = -3\frac{2}{6} = -3\frac{\cancel{2} \cdot 1}{\cancel{2} \cdot 3} = -3\frac{1}{3}$$

23.

$$\frac{127}{12} = 10\frac{7}{12}$$

25.

$$6\frac{1}{2} = \frac{12+1}{2} = \frac{13}{2}$$

27.

$$20\frac{4}{5} = \frac{100+4}{5} = \frac{104}{5}$$

29.

$$-6\frac{2}{9} = -\frac{54+2}{9} = -\frac{56}{9}$$

31. $200\dfrac{2}{3} = \dfrac{600+2}{3} = \dfrac{602}{3}$

33.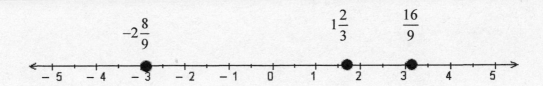

35.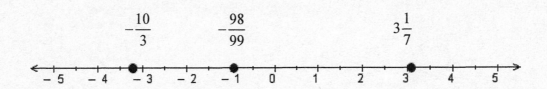

37. $1\dfrac{2}{3} \bullet 2\dfrac{1}{7} = \dfrac{5}{3} \bullet \dfrac{15}{7}$

   $\quad\quad = \dfrac{5}{\cancel{3}} \bullet \dfrac{5 \bullet \cancel{3}}{7}$

   $\quad\quad = \dfrac{25}{7} = 3\dfrac{4}{7}$

39. $-7\dfrac{1}{2}\left(-1\dfrac{2}{5}\right) = -\dfrac{15}{2}\left(-\dfrac{7}{5}\right)$

   $\quad\quad = -\dfrac{\cancel{5} \bullet 3}{2}\left(-\dfrac{7}{\cancel{5}}\right)$

   $\quad\quad = \dfrac{21}{2} = 10\dfrac{1}{2}$

41. $3\dfrac{1}{16} \bullet 4\dfrac{4}{7} = \dfrac{49}{16} \bullet \dfrac{32}{7}$

   $\quad\quad = \dfrac{7 \bullet \cancel{7}}{\cancel{16}} \bullet \dfrac{2 \bullet \cancel{16}}{\cancel{7}}$

   $\quad\quad = \dfrac{14}{1} = 14$

43. $-6 \bullet 2\dfrac{7}{24} = -\dfrac{6}{1} \bullet \dfrac{55}{24}$

   $\quad\quad = -\dfrac{\cancel{6}}{1} \bullet \dfrac{55}{4 \bullet \cancel{6}}$

   $\quad\quad = -\dfrac{55}{4} = -13\dfrac{3}{4}$

45. $2\dfrac{1}{2}\left(-3\dfrac{1}{3}\right) = \dfrac{5}{2}\left(-\dfrac{10}{3}\right)$

   $\quad\quad = \dfrac{5}{\cancel{2}}\left(-\dfrac{5 \bullet \cancel{2}}{3}\right)$

   $\quad\quad = -\dfrac{25}{3} = -8\dfrac{1}{3}$

47. $2\dfrac{5}{8} \bullet \dfrac{5}{27} = \dfrac{21}{8} \bullet \dfrac{5}{27}$

   $\quad\quad = \dfrac{7 \bullet \cancel{3}}{8} \bullet \dfrac{5}{9 \bullet \cancel{3}}$

   $\quad\quad = \dfrac{35}{72}$

49.

$$1\frac{2}{3} \cdot 6\left(-\frac{1}{8}\right) = \frac{5}{3} \cdot \frac{6}{1}\left(-\frac{1}{8}\right)$$

$$= \frac{5}{\cancel{3}} \cdot \frac{\cancel{2} \cdot \cancel{3}}{1}\left(-\frac{1}{4 \cdot \cancel{2}}\right)$$

$$= -\frac{5}{4} = -1\frac{1}{4}$$

51.

$$\left(1\frac{2}{3}\right)^2 = \left(1\frac{2}{3}\right)\left(1\frac{2}{3}\right)$$

$$= \left(\frac{5}{3}\right)\left(\frac{5}{3}\right)$$

$$= \frac{25}{9} = 2\frac{7}{9}$$

53.

$$\left(-1\frac{1}{3}\right)^3 = \left(-1\frac{1}{3}\right)\left(-1\frac{1}{3}\right)\left(-1\frac{1}{3}\right)$$

$$= \left(-\frac{4}{3}\right)\left(-\frac{4}{3}\right)\left(-\frac{4}{3}\right)$$

$$= -\frac{64}{27} = -2\frac{10}{27}$$

55.

$$3\frac{1}{3} \div 1\frac{5}{6} = \frac{10}{3} \div \frac{11}{6}$$

$$= \frac{10}{3} \cdot \frac{6}{11}$$

$$= \frac{10}{\cancel{3}} \cdot \frac{2 \cdot \cancel{3}}{11}$$

$$= \frac{20}{11} = 1\frac{9}{11}$$

57.

$$-6\frac{3}{5} \div 7\frac{1}{3} = -\frac{33}{5} \div \frac{22}{3}$$

$$= -\frac{33}{5} \cdot \frac{3}{22}$$

$$= -\frac{3 \cdot \cancel{11}}{5} \cdot \frac{3}{2 \cdot \cancel{11}}$$

$$= -\frac{9}{10}$$

59.

$$-20\frac{1}{4} \div \left(-1\frac{11}{16}\right) = -\frac{81}{4} \div \left(-\frac{27}{16}\right)$$

$$= -\frac{81}{4} \cdot \left(-\frac{16}{27}\right)$$

$$= -\frac{3 \cdot \cancel{27}}{\cancel{4}} \cdot \left(-\frac{4 \cdot \cancel{4}}{\cancel{27}}\right)$$

$$= \frac{12}{1} = 12$$

61.

$$6\frac{1}{4} \div 20 = \frac{25}{4} \div \frac{20}{1}$$

$$= \frac{25}{4} \cdot \frac{1}{20}$$

$$= \frac{5 \cdot \cancel{5}}{4} \cdot \frac{1}{4 \cdot \cancel{5}}$$

$$= \frac{5}{16}$$

63.

$$1\frac{2}{3} \div \left(-2\frac{1}{2}\right) = \frac{5}{3} \div \left(-\frac{5}{2}\right)$$

$$= \frac{5}{3} \cdot \left(-\frac{2}{5}\right)$$

$$= \frac{\cancel{5}}{3} \cdot \left(-\frac{2}{\cancel{5}}\right)$$

$$= -\frac{2}{3}$$

65.

$$8 \div 3\frac{1}{5} = \frac{8}{1} \div \frac{16}{5}$$

$$= \frac{8}{1} \cdot \frac{5}{16}$$

$$= \frac{\cancel{8}}{1} \cdot \frac{5}{2 \cdot \cancel{8}}$$

$$= \frac{5}{2} = 2\frac{1}{2}$$

67.

$$-4\frac{1}{2} \div 2\frac{1}{4} = -\frac{9}{2} \div \frac{9}{4}$$

$$= -\frac{9}{2} \cdot \frac{4}{9}$$

$$= -\frac{\cancel{9}}{\cancel{2}} \cdot \frac{2 \cdot \cancel{2}}{\cancel{9}}$$

$$= -\frac{2}{1} = -2$$

Applications

69. CALORIES

t = total calories

$$t = 3\frac{1}{5} \cdot 20$$

$$t = \frac{16}{5} \cdot \frac{20}{1}$$

$$t = \frac{16}{\cancel{5}} \cdot \frac{4 \cdot \cancel{5}}{1}$$

$$t = \frac{64}{1} = 64$$

There are 64 calories in a pack.

71. SHOPPING

c = cost of the fruit

$$c = 4\frac{1}{4} \cdot 84$$

$$c = \frac{17}{4} \cdot \frac{84}{1}$$

$$c = \frac{17}{\cancel{4}} \cdot \frac{21 \cdot \cancel{4}}{1}$$

$$c = \frac{357}{1} = 357 \text{ cents}$$

The cost is \$3.57.

73.   SUBDIVISIONS

There are 900 acres remaining.

L = number of lots

$$L = 900 \div 1\frac{1}{3}$$

$$L = \frac{900}{1} \div \frac{4}{3}$$

$$L = \frac{900}{1} \bullet \frac{3}{4}$$

$$L = \frac{225 \bullet \cancel{4}}{1} \bullet \frac{3}{\cancel{4}}$$

$$L = \frac{675}{1} = 675$$

There would be 675 lots created.

75.   GRAPH PAPER

L = length of paper W = width of paper

$$L = 11 \bullet \frac{1}{4} \qquad\qquad W = 5 \bullet \frac{1}{4}$$

$$L = \frac{11}{1} \bullet \frac{1}{4} \qquad\qquad W = \frac{5}{1} \bullet \frac{1}{4}$$

$$L = \frac{11}{4} = 2\frac{3}{4} \qquad\qquad W = \frac{5}{4} = 1\frac{1}{4}$$

The length is $2\frac{3}{4}$ in,

and the width is $1\frac{1}{4}$ in.

77. **EMERGENCY EXIT**

$$A = \frac{1}{2}bh, \text{ where } b = 8\frac{1}{4} \text{ and } h = 10\frac{1}{3}$$

$$A = \frac{1}{2}\left(8\frac{1}{4}\right)\left(10\frac{1}{3}\right)$$

$$A = \frac{1}{2}\left(\frac{33}{4}\right)\left(\frac{31}{3}\right)$$

$$A = \frac{1}{2}\left(\frac{11 \bullet \cancel{3}}{4}\right)\left(\frac{31}{\cancel{3}}\right)$$

$$A = \frac{341}{8} = 42\frac{5}{8}$$

The area of the sign is $42\frac{5}{8}$ in^2.

79. **FIRE ESCAPES**

It is important to realize that each story has 2 staircases.

N = number of steps per floor

$$N = 105 \div 7\frac{1}{2}$$

$$N = \frac{105}{1} \div \frac{15}{1}$$

$$N = \frac{105}{1} \bullet \frac{1}{15}$$

$$N = \frac{7 \bullet \cancel{15}}{1} \bullet \frac{1}{\cancel{15}}$$

$$N = \frac{7}{1} = 7$$

t = total steps in building

$$t = 2(7 \bullet 43)$$

$$t = 2(301)$$

$$t = 602$$

There are 602 steps total in the building.

81. **SHOPPING ON THE INTERNET**

She should get size 14 slim cut jeans.

Writing

83. $2\dfrac{3}{4}$ is a mixed number, while $2\left(\dfrac{3}{4}\right)$ is a multiplication problem.

85.   Divide the denominator into the numerator and the number of times it divides completely will be the whole number part of the mixed number, while the remainder will give the numerator of the fraction part.

Review

87. $3^2 \bullet 2^3 = 9 \bullet 8$

$= 72$

89. $8 + 8 + 8 + 8 = 4(8)$

91. $\dfrac{x}{2} = -12$

$2\left(\dfrac{x}{2}\right) = 2(-12)$

$x = -24$

93.   The variables are different.

4.6 Adding and Subtracting Mixed Numbers

Vocabulary

1. By the **commutative** property of addition, we can add numbers in any order.

3. To do the subtraction, we **borrow** 1 in the form of $\frac{3}{3}$.

Concepts

5a. Whole number part is 76 and the fractional part is $\frac{3}{4}$.

5b.
$$76\frac{3}{4} = 76 + \frac{3}{4}$$

7. The fundamental property of fractions.

9a.
$$9\frac{17}{16} = 10\frac{1}{16}$$

9b.
$$1,288\frac{7}{3} = 1,290\frac{1}{3}$$

9c.
$$16\frac{12}{8} = 17\frac{4}{8} = 17\frac{1 \bullet \cancel{4}}{2 \bullet \cancel{4}} = 17\frac{1}{2}$$

9d.
$$45\frac{24}{20} = 46\frac{4}{20} = 46\frac{\cancel{4} \bullet 1}{\cancel{4} \bullet 5} = 46\frac{1}{5}$$

Notation

11.
$$70\frac{3}{5} + 39\frac{2}{7} = 70 + \frac{3}{5} + 39 + \frac{2}{7}$$
$$= 70 + 39 + \frac{3}{5} + \frac{2}{7}$$
$$= 109 + \frac{3}{5} + \frac{2}{7}$$
$$= 109 + \frac{3 \bullet 7}{5 \bullet 7} + \frac{2 \bullet 5}{7 \bullet 5}$$
$$= 109 + \frac{21}{35} + \frac{10}{35}$$
$$= 109 + \frac{31}{35}$$
$$= 109\frac{31}{35}$$

Practice

13.
$$2\frac{1}{5}+2\frac{1}{5}=2+2+\frac{1}{5}+\frac{1}{5}$$
$$=4+\frac{2}{5}$$
$$=4\frac{2}{5}$$

15.
$$8\frac{2}{7}-3\frac{1}{7}=\frac{58}{7}-\frac{22}{7}$$
$$=\frac{36}{7}=5\frac{1}{7}$$

17.
$$3\frac{1}{4}+4\frac{1}{4}=\frac{13}{4}+\frac{17}{4}$$
$$=\frac{30}{4}=\frac{15\bullet\cancel{2}}{2\bullet\cancel{2}}$$
$$=\frac{15}{2}=7\frac{1}{2}$$

19.
$$4\frac{1}{6}+1\frac{1}{5}=\frac{25}{6}+\frac{6}{5}$$
$$=\frac{25\bullet5}{6\bullet5}+\frac{6\bullet6}{5\bullet6}$$
$$=\frac{125}{30}+\frac{36}{30}$$
$$=\frac{161}{30}=5\frac{11}{30}$$

21.
$$2\frac{1}{2}-1\frac{1}{4}=\frac{5}{2}-\frac{5}{4}$$
$$=\frac{5\bullet2}{2\bullet2}-\frac{5}{4}$$
$$=\frac{10}{4}-\frac{5}{4}$$
$$=\frac{5}{4}=1\frac{1}{4}$$

23.
$$2\frac{5}{6}-1\frac{3}{8}=\frac{17}{6}-\frac{11}{8}$$
$$=\frac{17\bullet4}{6\bullet4}-\frac{11\bullet3}{8\bullet3}$$
$$=\frac{68}{24}-\frac{33}{24}$$
$$=\frac{35}{24}=1\frac{11}{24}$$

25.
$$5\frac{1}{2}+3\frac{4}{5}=\frac{11}{2}+\frac{19}{5}$$
$$=\frac{11\bullet5}{2\bullet5}+\frac{19\bullet2}{5\bullet2}$$
$$=\frac{55}{10}+\frac{38}{10}$$
$$=\frac{93}{10}=9\frac{3}{10}$$

27.
$$7\frac{1}{2}-4\frac{1}{7}=\frac{15}{2}-\frac{29}{7}$$
$$=\frac{15\bullet7}{2\bullet7}-\frac{29\bullet2}{7\bullet2}$$
$$=\frac{105}{14}-\frac{58}{14}$$
$$=\frac{47}{14}=3\frac{5}{14}$$

29.
$$56\frac{2}{5} = 56\frac{2 \cdot 3}{5 \cdot 3} = 56\frac{6}{15}$$
$$+73\frac{1}{3} = +73\frac{1 \cdot 5}{3 \cdot 5} = +73\frac{5}{15}$$
$$129\frac{11}{15}$$

31.
$$380\frac{1}{6} = 380\frac{1 \cdot 2}{6 \cdot 2} = 380\frac{2}{12}$$
$$+17\frac{1}{4} = +17\frac{1 \cdot 3}{4 \cdot 3} = +17\frac{3}{12}$$
$$397\frac{5}{12}$$

33.
$$228\frac{5}{9} = 228\frac{5 \cdot 2}{9 \cdot 2} = 228\frac{10}{18}$$
$$+44\frac{2}{3} = +44\frac{2 \cdot 6}{3 \cdot 6} = +44\frac{12}{18}$$
$$= 272\frac{22}{18}$$
$$= 273\frac{4}{18} = 273\frac{2 \cdot 2}{9 \cdot 2} = 273\frac{2}{9}$$

35.
$$778\frac{5}{7} = 778\frac{5 \cdot 3}{7 \cdot 3} = 778\frac{15}{21}$$
$$-155\frac{1}{3} = -155\frac{1 \cdot 7}{3 \cdot 7} = -155\frac{7}{21}$$
$$623\frac{8}{21}$$

37.
$$140\frac{5}{6} = 140\frac{5 \cdot 5}{6 \cdot 5} = 140\frac{25}{30}$$
$$-129\frac{4}{5} = -129\frac{4 \cdot 6}{5 \cdot 6} = -129\frac{24}{30}$$
$$11\frac{1}{30}$$

39.
$$422\frac{13}{16} = 422\frac{13}{16} = 422\frac{13}{16}$$
$$-321\frac{3}{8} = -321\frac{3 \cdot 2}{8 \cdot 2} = -321\frac{6}{16}$$
$$101\frac{7}{16}$$

41.
$$16\frac{1}{4} = 15\frac{1}{4} + \frac{4}{4} = 15\frac{5}{4}$$
$$-13\frac{3}{4} = -13\frac{3}{4} = -13\frac{3}{4}$$
$$= 2\frac{2}{4}$$
$$= 2\frac{1 \cdot 2}{2 \cdot 2} = 2\frac{1}{2}$$

43.
$$76\frac{1}{6} = 76\frac{1 \cdot 4}{6 \cdot 4} = 76\frac{4}{24}$$
$$-49\frac{7}{8} = -49\frac{7 \cdot 3}{8 \cdot 3} = -49\frac{21}{24}$$
$$= 75\frac{4}{24} + \frac{24}{24} = 75\frac{28}{24}$$
$$= -49\frac{21}{24} = -49\frac{21}{24}$$
$$= 26\frac{7}{24}$$

45.

$$140\frac{3}{16} = 140\frac{3}{16} = 140\frac{3}{16}$$

$$-129\frac{3}{4} = -129\frac{3\bullet 4}{4\bullet 4} = -129\frac{12}{16}$$

$$= 139\frac{3}{16} + \frac{16}{16} = 139\frac{19}{16}$$

$$= -129\frac{12}{16} \qquad = -129\frac{12}{16}$$

$$= 10\frac{7}{16}$$

47.

$$334\frac{1}{9} = 334\frac{1\bullet 2}{9\bullet 2} = 334\frac{2}{18}$$

$$-13\frac{5}{6} = -13\frac{5\bullet 3}{6\bullet 3} = -13\frac{15}{18}$$

$$= 333\frac{2}{18} + \frac{18}{18} = 333\frac{20}{18}$$

$$= -13\frac{15}{18} \qquad = -13\frac{15}{18}$$

$$= 320\frac{5}{18}$$

49.

$$7 - \frac{2}{3} = \frac{7}{1} - \frac{2}{3}$$

$$= \frac{7\bullet 3}{1\bullet 3} - \frac{2}{3}$$

$$= \frac{21}{3} - \frac{2}{3}$$

$$= \frac{19}{3} = 6\frac{1}{3}$$

51.

$$9 - 8\frac{3}{4} = \frac{9}{1} - \frac{35}{4}$$

$$= \frac{9\bullet 4}{1\bullet 4} - \frac{35}{4}$$

$$= \frac{36}{4} - \frac{35}{4}$$

$$= \frac{1}{4}$$

53.

$$4\frac{1}{7} - \frac{4}{5} = \frac{29}{7} - \frac{4}{5}$$

$$= \frac{29\bullet 5}{7\bullet 5} - \frac{4\bullet 7}{5\bullet 7}$$

$$= \frac{145}{35} - \frac{28}{35}$$

$$= \frac{117}{35} = 3\frac{12}{35}$$

55.

$$6\frac{5}{8} - 3 = \frac{53}{8} - \frac{3}{1}$$

$$= \frac{53}{8} - \frac{3\bullet 8}{1\bullet 8}$$

$$= \frac{53}{8} - \frac{24}{8}$$

$$= \frac{29}{8} = 3\frac{5}{8}$$

57.

$$\frac{7}{3} + 2 = \frac{7}{3} + \frac{2}{1}$$

$$= \frac{7}{3} + \frac{2\bullet 3}{1\bullet 3}$$

$$= \frac{7}{3} + \frac{6}{3}$$

$$= \frac{13}{3} = 4\frac{1}{3}$$

59.

$$2 + 1\frac{7}{8} = \frac{2}{1} + \frac{15}{8}$$

$$= \frac{2\bullet 8}{1\bullet 8} + \frac{15}{8}$$

$$= \frac{16}{8} + \frac{15}{8}$$

$$= \frac{31}{8} = 3\frac{7}{8}$$

61.

$$12\frac{1}{2}+5\frac{3}{4}+35\frac{1}{6}=12+5+35+\frac{1}{2}+\frac{3}{4}+\frac{1}{6}$$

$$=52+\frac{1\bullet 6}{2\bullet 6}+\frac{3\bullet 3}{4\bullet 3}+\frac{1\bullet 2}{6\bullet 2}$$

$$=52+\frac{6}{12}+\frac{9}{12}+\frac{2}{12}$$

$$=52+\frac{17}{12}$$

$$=52+1\frac{5}{12}=53\frac{5}{12}$$

63.

$$58\frac{7}{8}+340+61\frac{1}{4}=58+340+61+\frac{7}{8}+\frac{1}{4}$$

$$=459+\frac{7}{8}+\frac{1\bullet 2}{4\bullet 2}$$

$$=459+\frac{7}{8}+\frac{2}{8}$$

$$=459+\frac{9}{8}$$

$$=459+1\frac{1}{8}=460\frac{1}{8}$$

65.

$$-3\frac{3}{4}+\left(-1\frac{1}{2}\right)=-\frac{15}{4}+\left(-\frac{3}{2}\right)$$

$$=-\frac{15}{4}+\left(-\frac{3\bullet 2}{2\bullet 2}\right)$$

$$=-\frac{15}{4}+\left(-\frac{6}{4}\right)$$

$$=-\frac{21}{4}=-5\frac{1}{4}$$

67.

$$-4\frac{5}{8}-1\frac{1}{4}=-\frac{37}{8}-\frac{5}{4}$$

$$=-\frac{37}{8}-\frac{5\bullet 2}{4\bullet 2}$$

$$=-\frac{37}{8}-\frac{10}{8}$$

$$=-\frac{37}{8}+\left(-\frac{10}{8}\right)$$

$$=-\frac{47}{8}=-5\frac{7}{8}$$

Applications

69.   FREEWAY TRAVEL

d = distance between exits

$$d = 3\frac{1}{2} - \frac{3}{4}$$

$$d = \frac{7}{2} - \frac{3}{4}$$

$$d = \frac{7 \bullet 2}{2 \bullet 2} - \frac{3}{4}$$

$$d = \frac{14}{4} - \frac{3}{4} = \frac{11}{4} = 2\frac{3}{4}$$

They are $2\frac{3}{4}$ miles apart.

71.   TRAIL MIX

t = total amount of trail mix

$$t = 2\frac{3}{4} + 2\left(\frac{1}{2}\right) + \frac{2}{3} + \frac{1}{3} + 2\frac{2}{3} + \frac{1}{4}$$

$$t = 2\frac{3}{4} + 1 + \frac{2}{3} + \frac{1}{3} + 2\frac{2}{3} + \frac{1}{4}$$

$$t = 2 + 1 + 2 + \frac{3}{4} + \frac{2}{3} + \frac{1}{3} + \frac{2}{3} + \frac{1}{4}$$

$$t = 5 + \frac{3 \bullet 3}{4 \bullet 3} + \frac{2 \bullet 4}{3 \bullet 4} + \frac{1 \bullet 4}{3 \bullet 4} + \frac{2 \bullet 4}{3 \bullet 4} + \frac{1 \bullet 3}{4 \bullet 3}$$

$$t = 5 + \frac{9}{12} + \frac{8}{12} + \frac{4}{12} + \frac{8}{12} + \frac{3}{12}$$

$$t = 5 + \frac{32}{12} = 5 + \frac{8 \bullet \cancel{4}}{3 \bullet \cancel{4}}$$

$$t = 5 + \frac{8}{3} = 5 + 2\frac{2}{3}$$

$$t = 7\frac{2}{3}$$

A total of $7\frac{2}{3}$ cups of trail mix will be made.

73.   HOSE REPAIRS

L = length of hose after repair

$$L = 50 - 1\frac{1}{2}$$

$$L = \frac{50}{1} - \frac{3}{2}$$

$$L = \frac{50 \bullet 2}{1 \bullet 2} - \frac{3}{2}$$

$$L = \frac{100}{2} - \frac{3}{2}$$

$$L = \frac{97}{2} = 48\frac{1}{2}$$

The hose is $48\frac{1}{2}$ feet long.

75a.   SHIPPING

	Rate (mph)	Time traveling	Distance traveled (mi)
Passenger ship	$16\frac{1}{2}$	1	$16\frac{1}{2}$
Cargo ship	$5\frac{1}{5}$	1	$5\frac{1}{5}$

75b d = distance between ships

$$d = 16\frac{1}{2} + 5\frac{1}{5}$$

$$d = 16 + 5 + \frac{1}{2} + \frac{1}{5}$$

$$d = 21 + \frac{1 \bullet 5}{2 \bullet 5} + \frac{1 \bullet 2}{5 \bullet 2}$$

$$d = 21 + \frac{5}{10} + \frac{2}{10}$$

$$d = 21 + \frac{7}{10} = 21\frac{7}{10}$$

The total distance between ships is $21\frac{7}{10}$ mi.

77a.   SERVICE STATIONS

p = difference in prices

$$p = 179\frac{9}{10} - 159\frac{9}{10}$$

$$179\frac{9}{10}$$

$$-159\frac{9}{10}$$

$$\overline{\phantom{-159\frac{9}{10}}}$$

$$20$$

$p = 20$

The price difference is 20 cents.

77b. f = difference between
full and self serve prices

$$f = 189\frac{9}{10} - 159\frac{9}{10}$$

$$189\frac{9}{10}$$

$$-159\frac{9}{10}$$

$$\overline{\phantom{-159\frac{9}{10}}}$$

$$30$$

$f = 30$

The price difference is 30 cents.

79.   WATER SLIDES

h = original height

$$h = 311\frac{5}{12} - 119\frac{2}{3}$$

$$311\frac{5}{12} = 311\frac{5}{12} = 311\frac{5}{12}$$

$$-119\frac{3}{4} = -119\frac{3 \bullet 3}{4 \bullet 3} = -119\frac{9}{12}$$

$$= 310\frac{5}{12} + \frac{12}{12} = 310\frac{17}{12}$$

$$= -119\frac{9}{12} = -119\frac{9}{12}$$

$$= 191\frac{8}{12}$$

$$h = 191\frac{2 \bullet \cancel{4}}{3 \bullet \cancel{4}} = 191\frac{2}{3}$$

The original height was $191\frac{2}{3}$ ft.

Writing

81. Answers will vary.

83. Convert the fraction with a 4 in the denominator to a denominator of 8 by multiplying the numerator and denominator by 2, then add the numerators, then convert to a mixed number if desired and reduce if necessary.

Review

85.
$$2x - 1 = 3x - 8$$
$$2x - 1 - 2x = 3x - 8 - 2x$$
$$-1 = x - 8$$
$$-1 + 8 = x - 8 + 8$$
$$7 = x \text{ or } x = 7$$

87.
$$8(x + 2) + 3(2 - x) = 8x + 16 + 6 - 3x$$
$$= 5x + 22$$

89.
$$-2 - (-8) = -2 + 8$$
$$= 6$$

91. Area measures the amount of surface a figure encloses.

4.7 Order of Operations and Complex Fractions

Vocabulary

1. $\dfrac{\dfrac{1}{2}}{\dfrac{3}{4}}$ is a **complex** fraction.

Concepts

3. $\dfrac{2}{3} \div \dfrac{1}{5}$ 5.   15

7.   negative   9. $LCD = 60$

Notation

11. $\dfrac{\dfrac{1}{8}}{\dfrac{3}{4}} = \dfrac{1}{8} \div \dfrac{3}{4}$

$= \dfrac{1}{8} \cdot \dfrac{4}{3}$

$= \dfrac{1 \cdot 4}{8 \cdot 3}$

$= \dfrac{1 \cdot \overset{1}{\cancel{4}}}{2 \cdot \underset{1}{\cancel{4}} \cdot 3}$

$= \dfrac{1}{6}$

Practice

13.
$$\frac{2}{3}\left(-\frac{1}{4}\right)+\frac{1}{2}=\frac{\cancel{2}}{3}\left(-\frac{1}{2\cdot\cancel{2}}\right)+\frac{1}{2}$$

$$=-\frac{1}{6}+\frac{1}{2}$$

$$=-\frac{1}{6}+\frac{1\cdot3}{2\cdot3}$$

$$=-\frac{1}{6}+\frac{3}{6}$$

$$=\frac{2}{6}=\frac{1\cdot\cancel{2}}{3\cdot\cancel{2}}$$

$$=\frac{1}{3}$$

15.
$$\frac{4}{5}-\left(-\frac{1}{3}\right)^{2}=\frac{4}{5}-\left(-\frac{1}{3}\right)\left(-\frac{1}{3}\right)$$

$$=\frac{4}{5}-\frac{1}{9}$$

$$=\frac{4\cdot9}{5\cdot9}-\frac{1\cdot5}{9\cdot5}$$

$$=\frac{36}{45}-\frac{5}{45}$$

$$=\frac{31}{45}$$

17.
$$-4\left(-\frac{1}{5}\right)-\left(\frac{1}{4}\right)\left(-\frac{1}{2}\right)=-\frac{4}{1}\left(-\frac{1}{5}\right)-\left(\frac{1}{4}\right)\left(-\frac{1}{2}\right)$$

$$=\frac{4}{5}-\left(-\frac{1}{8}\right)$$

$$=\frac{4}{5}+\frac{1}{8}$$

$$=\frac{4\cdot8}{5\cdot8}+\frac{1\cdot5}{8\cdot5}$$

$$=\frac{32}{40}+\frac{5}{40}$$

$$=\frac{37}{40}$$

19.
$$1\frac{3}{5}\left(\frac{1}{2}\right)^{2}\left(\frac{3}{4}\right)=\frac{8}{5}\left[\left(\frac{1}{2}\right)\left(\frac{1}{2}\right)\right]\left(\frac{3}{4}\right)$$

$$=\frac{8}{5}\left(\frac{1}{4}\right)\left(\frac{3}{4}\right)$$

$$=\frac{\cancel{2}\cdot\cancel{4}}{5}\left(\frac{1}{\cancel{4}}\right)\left(\frac{3}{\cancel{2}\cdot2}\right)$$

$$=\frac{3}{10}$$

21.

$$\frac{7}{8} - \left(\frac{4}{5} + 1\frac{3}{4}\right) = \frac{7}{8} - \left(\frac{4}{5} + \frac{7}{4}\right)$$

$$= \frac{7}{8} - \left(\frac{4 \bullet 4}{5 \bullet 4} + \frac{7 \bullet 5}{4 \bullet 5}\right)$$

$$= \frac{7}{8} - \left(\frac{16}{20} + \frac{35}{20}\right)$$

$$= \frac{7}{8} - \frac{51}{20}$$

$$= \frac{7 \bullet 5}{8 \bullet 5} - \frac{51 \bullet 2}{20 \bullet 2}$$

$$= \frac{35}{40} - \frac{102}{40}$$

$$= \frac{35}{40} + \left(-\frac{102}{40}\right)$$

$$= -\frac{67}{40} = -1\frac{27}{40}$$

23.

$$\left(\frac{9}{20} \div 2\frac{2}{5}\right) + \left(\frac{3}{4}\right)^2 = \left(\frac{9}{20} \div \frac{12}{5}\right) + \left(\frac{3}{4}\right)^2$$

$$= \left(\frac{9}{20} \bullet \frac{5}{12}\right) + \left(\frac{3}{4}\right)^2$$

$$= \left(\frac{3 \bullet \cancel{3}}{4 \bullet \cancel{5}} \bullet \frac{\cancel{5}}{4 \bullet \cancel{3}}\right) + \left(\frac{3}{4}\right)^2$$

$$= \frac{3}{16} + \left(\frac{3}{4}\right)^2$$

$$= \frac{3}{16} + \left(\frac{3}{4}\right)\left(\frac{3}{4}\right)$$

$$= \frac{3}{16} + \frac{9}{16}$$

$$= \frac{12}{16} = \frac{3 \bullet \cancel{4}}{4 \bullet \cancel{4}}$$

$$= \frac{3}{4}$$

25.
$$\left(-\frac{3}{4} \bullet \frac{9}{16}\right) + \left(\frac{1}{2} - \frac{1}{8}\right) = \left(-\frac{27}{64}\right) + \left(\frac{1 \bullet 4}{2 \bullet 4} - \frac{1}{8}\right)$$

$$= -\frac{27}{64} + \left(\frac{4}{8} - \frac{1}{8}\right)$$

$$= -\frac{27}{64} + \frac{3}{8}$$

$$= -\frac{27}{64} + \frac{3 \bullet 8}{8 \bullet 8}$$

$$= -\frac{27}{64} + \frac{24}{64}$$

$$= -\frac{3}{64}$$

27.
$$\left|\frac{2}{3} - \frac{9}{10}\right| \div \left(-\frac{1}{5}\right) = \left|\frac{2 \bullet 10}{3 \bullet 10} - \frac{9 \bullet 3}{10 \bullet 3}\right| \div \left(-\frac{1}{5}\right)$$

$$= \left|\frac{20}{30} - \frac{27}{30}\right| \div \left(-\frac{1}{5}\right)$$

$$= \left|\frac{20}{30} + \left(-\frac{27}{30}\right)\right| \div \left(-\frac{1}{5}\right)$$

$$= \left|-\frac{7}{30}\right| \div \left(-\frac{1}{5}\right)$$

$$= \frac{7}{30} \div \left(-\frac{1}{5}\right)$$

$$= \frac{7}{30} \bullet \left(-\frac{5}{1}\right)$$

$$= \frac{7}{6 \bullet \cancel{5}} \bullet \left(-\frac{\cancel{5}}{1}\right)$$

$$= -\frac{7}{6} = -1\frac{1}{6}$$

29.

$$\left(2-\frac{1}{2}\right)^2+\left(2+\frac{1}{2}\right)^2=\left(\frac{2}{1}-\frac{1}{2}\right)^2+\left(\frac{2}{1}+\frac{1}{2}\right)^2$$

$$=\left(\frac{2\bullet2}{1\bullet2}-\frac{1}{2}\right)^2+\left(\frac{2\bullet2}{1\bullet2}+\frac{1}{2}\right)^2$$

$$=\left(\frac{4}{2}-\frac{1}{2}\right)^2+\left(\frac{4}{2}+\frac{1}{2}\right)^2$$

$$=\left(\frac{3}{2}\right)^2+\left(\frac{5}{2}\right)^2$$

$$=\left(\frac{3}{2}\right)\left(\frac{3}{2}\right)+\left(\frac{5}{2}\right)\left(\frac{5}{2}\right)$$

$$=\frac{9}{4}+\frac{25}{4}$$

$$=\frac{34}{4}=\frac{17\bullet\cancel{2}}{2\bullet\cancel{2}}$$

$$=\frac{17}{2}=8\frac{1}{2}$$

31.

$$\text{half of}-7=\frac{1}{2}\left(-\frac{7}{1}\right)=-\frac{7}{2}$$

$$\text{squaring the result}=\left(-\frac{7}{2}\right)^2$$

$$=\left(-\frac{7}{2}\right)\left(-\frac{7}{2}\right)=\frac{49}{4}$$

33.

$$\text{half of}\frac{11}{2}=\frac{1}{2}\left(\frac{11}{2}\right)=\frac{11}{4}$$

$$\text{squaring the result}=\left(\frac{11}{4}\right)^2$$

$$=\left(\frac{11}{4}\right)\left(\frac{11}{4}\right)=\frac{121}{16}$$

35.

$$\frac{1}{3}\left(-\frac{1}{5}\right)^2+\left(-\frac{2}{3}\right)=\frac{1}{3}\left(\left[-\frac{1}{5}\right]\left[-\frac{1}{5}\right]\right)+\left(-\frac{2}{3}\right)$$

$$=\frac{1}{3}\left(\frac{1}{25}\right)+\left(-\frac{2}{3}\right)$$

$$=\frac{1}{75}+\left(-\frac{2}{3}\right)$$

$$=\frac{1}{75}+\left(-\frac{2\bullet25}{3\bullet25}\right)$$

$$=\frac{1}{75}+\left(-\frac{50}{75}\right)$$

$$=-\frac{49}{75}$$

37.

$$-1-\left(1\frac{3}{4}\bullet\left[-1\frac{2}{3}\right]\right)=-1-\left(\frac{7}{4}\bullet\left[-\frac{5}{3}\right]\right)$$

$$=-1-\left(-\frac{35}{12}\right)$$

$$=-\frac{12}{12}+\frac{35}{12}$$

$$=\frac{23}{12}=1\frac{11}{12}$$

39. P = perimeter of rectangle

$$P=2\left(2\frac{7}{8}\right)+2\left(1\frac{1}{4}\right)$$

$$P=\frac{2}{1}\left(\frac{23}{8}\right)+\frac{2}{1}\left(\frac{5}{4}\right)$$

$$P=\frac{\cancel{2}}{1}\left(\frac{23}{4\bullet\cancel{2}}\right)+\frac{\cancel{2}}{1}\left(\frac{5}{2\bullet\cancel{2}}\right)$$

$$P=\frac{23}{4}+\frac{5}{2}$$

$$P=\frac{23}{4}+\frac{5\bullet2}{2\bullet2}$$

$$P=\frac{23}{4}+\frac{10}{4}$$

$$P=\frac{33}{4}=8\frac{1}{4}\text{ in.}$$

41.

$$\frac{\dfrac{2}{3}}{\dfrac{4}{5}}=\frac{\dfrac{15}{1}\left(\dfrac{2}{3}\right)}{\dfrac{15}{1}\left(\dfrac{4}{5}\right)}=\frac{\dfrac{5\bullet\cancel{3}}{1}\left(\dfrac{2}{\cancel{3}}\right)}{\dfrac{\cancel{5}\bullet3}{1}\left(\dfrac{4}{\cancel{5}}\right)}=\frac{10}{12}=\frac{5\bullet\cancel{2}}{6\bullet\cancel{2}}$$

$$=\frac{5}{6}$$

43.

$$\frac{-\dfrac{14}{15}}{\dfrac{7}{10}}=\frac{\dfrac{30}{1}\left(-\dfrac{14}{15}\right)}{\dfrac{30}{1}\left(\dfrac{7}{10}\right)}=\frac{\dfrac{2\bullet\cancel{15}}{1}\left(-\dfrac{14}{\cancel{15}}\right)}{\dfrac{3\bullet\cancel{10}}{1}\left(\dfrac{7}{\cancel{10}}\right)}$$

$$=\frac{-28}{21}=-\frac{4\bullet\cancel{7}}{3\bullet\cancel{7}}=-\frac{4}{3}=-1\frac{1}{3}$$

45.

$$\frac{\dfrac{5}{1}}{\dfrac{10}{21}} = \frac{\dfrac{21}{1}\left(\dfrac{5}{1}\right)}{\dfrac{21}{1}\left(\dfrac{10}{21}\right)} = \frac{\dfrac{21}{1}\left(\dfrac{5}{1}\right)}{\dfrac{21}{1}\left(\dfrac{10}{\cancel{21}}\right)}$$

$$= \frac{105}{10} = \frac{21 \bullet \cancel{5}}{2 \bullet \cancel{5}} = \frac{21}{2} = 10\frac{1}{2}$$

47.

$$\frac{-\dfrac{5}{6}}{-1\dfrac{7}{8}} = \frac{-\dfrac{5}{6}}{-\dfrac{15}{8}} = \frac{\dfrac{24}{1}\left(-\dfrac{5}{6}\right)}{\dfrac{24}{1}\left(-\dfrac{15}{8}\right)}$$

$$= \frac{\dfrac{4 \bullet \cancel{6}}{1}\left(-\dfrac{5}{\cancel{6}}\right)}{\dfrac{3 \bullet \cancel{8}}{1}\left(-\dfrac{15}{\cancel{8}}\right)} = \frac{-20}{-45} = \frac{4 \bullet \cancel{5}}{9 \bullet \cancel{5}} = \frac{4}{9}$$

49.

$$\frac{\dfrac{1}{2}+\dfrac{1}{4}}{\dfrac{1}{2}-\dfrac{1}{4}} = \frac{\dfrac{4}{1}\left(\dfrac{1}{2}+\dfrac{1}{4}\right)}{\dfrac{4}{1}\left(\dfrac{1}{2}-\dfrac{1}{4}\right)} = \frac{\dfrac{4}{1}\left(\dfrac{1}{2}\right)+\dfrac{4}{1}\left(\dfrac{1}{4}\right)}{\dfrac{4}{1}\left(\dfrac{1}{2}\right)-\dfrac{4}{1}\left(\dfrac{1}{4}\right)}$$

$$= \frac{\dfrac{2 \bullet \cancel{2}}{1}\left(\dfrac{1}{\cancel{2}}\right)+\dfrac{\cancel{4}}{1}\left(\dfrac{1}{\cancel{4}}\right)}{\dfrac{2 \bullet \cancel{2}}{1}\left(\dfrac{1}{\cancel{2}}\right)-\dfrac{\cancel{4}}{1}\left(\dfrac{1}{\cancel{4}}\right)} = \frac{2+1}{2-1} = \frac{3}{1} = 3$$

51.

$$\frac{\dfrac{3}{8}+\dfrac{1}{4}}{\dfrac{3}{8}-\dfrac{1}{4}} = \frac{\dfrac{8}{1}\left(\dfrac{3}{8}+\dfrac{1}{4}\right)}{\dfrac{8}{1}\left(\dfrac{3}{8}-\dfrac{1}{4}\right)} = \frac{\dfrac{8}{1}\left(\dfrac{3}{8}\right)+\dfrac{8}{1}\left(\dfrac{1}{4}\right)}{\dfrac{8}{1}\left(\dfrac{3}{8}\right)-\dfrac{8}{1}\left(\dfrac{1}{4}\right)}$$

$$= \frac{\dfrac{\cancel{8}}{1}\left(\dfrac{3}{\cancel{8}}\right)+\dfrac{2 \bullet \cancel{4}}{1}\left(\dfrac{1}{\cancel{4}}\right)}{\dfrac{\cancel{8}}{1}\left(\dfrac{3}{\cancel{8}}\right)-\dfrac{2 \bullet \cancel{4}}{1}\left(\dfrac{1}{\cancel{4}}\right)} = \frac{3+2}{3-2} = \frac{5}{1} = 5$$

53.

$$\frac{\dfrac{1}{5}+3}{-\dfrac{4}{25}} = \frac{\dfrac{25}{1}\left(\dfrac{1}{5}+\dfrac{3}{1}\right)}{\dfrac{25}{1}\left(-\dfrac{4}{25}\right)} = \frac{\dfrac{25}{1}\left(\dfrac{1}{5}\right)+\dfrac{25}{1}\left(\dfrac{3}{1}\right)}{\dfrac{25}{1}\left(-\dfrac{4}{25}\right)}$$

$$= \frac{\dfrac{5 \bullet \cancel{5}}{1}\left(\dfrac{1}{\cancel{5}}\right)+\dfrac{25}{1}\left(\dfrac{3}{1}\right)}{\dfrac{\cancel{25}}{1}\left(-\dfrac{4}{\cancel{25}}\right)} = \frac{5+75}{-4}$$

$$= \frac{80}{-4} = -20$$

55.

$$\frac{5\dfrac{1}{2}}{-\dfrac{1}{4}+\dfrac{3}{4}} = \frac{\dfrac{11}{2}}{-\dfrac{1}{4}+\dfrac{3}{4}} = \frac{\dfrac{4}{1}\left(\dfrac{11}{2}\right)}{\dfrac{4}{1}\left(-\dfrac{1}{4}+\dfrac{3}{4}\right)}$$

$$= \frac{\dfrac{4}{1}\left(\dfrac{11}{2}\right)}{\dfrac{4}{1}\left(-\dfrac{1}{4}\right)+\dfrac{4}{1}\left(\dfrac{3}{4}\right)}$$

$$= \frac{\dfrac{2 \bullet \cancel{2}}{1}\left(\dfrac{11}{\cancel{2}}\right)}{\dfrac{\cancel{4}}{1}\left(-\dfrac{1}{\cancel{4}}\right)+\dfrac{\cancel{4}}{1}\left(\dfrac{3}{\cancel{4}}\right)} = \frac{22}{-1+3}$$

$$= \frac{22}{2} = 11$$

57.

$$\frac{\frac{1}{5}-\left(-\frac{1}{4}\right)}{\frac{1}{4}+\frac{4}{5}}=\frac{\frac{1}{5}+\frac{1}{4}}{\frac{1}{4}+\frac{4}{5}}=\frac{\frac{20}{1}\left(\frac{1}{5}+\frac{1}{4}\right)}{\frac{20}{1}\left(\frac{1}{4}+\frac{4}{5}\right)}$$

$$=\frac{\frac{20}{1}\left(\frac{1}{5}\right)+\frac{20}{1}\left(\frac{1}{4}\right)}{\frac{20}{1}\left(\frac{1}{4}\right)+\frac{20}{1}\left(\frac{4}{5}\right)}$$

$$=\frac{\frac{4\cdot 5}{1}\left(\frac{1}{5}\right)+\frac{5\cdot 4}{1}\left(\frac{1}{4}\right)}{\frac{5\cdot 4}{1}\left(\frac{1}{4}\right)+\frac{4\cdot 5}{1}\left(\frac{4}{5}\right)}=\frac{4+5}{5+16}$$

$$=\frac{9}{21}=\frac{3\cdot 3}{7\cdot 3}=\frac{3}{7}$$

59.

$$\frac{\frac{1}{3}+\left(-\frac{5}{6}\right)}{1\frac{1}{3}}=\frac{\frac{1}{3}+\left(-\frac{5}{6}\right)}{\frac{4}{3}}=\frac{\frac{6}{1}\left(\frac{1}{3}+\left[-\frac{5}{6}\right]\right)}{\frac{6}{1}\left(\frac{4}{3}\right)}$$

$$=\frac{\frac{6}{1}\left(\frac{1}{3}\right)+\frac{6}{1}\left(-\frac{5}{6}\right)}{\frac{6}{1}\left(\frac{4}{3}\right)}$$

$$=\frac{\frac{2\cdot 3}{1}\left(\frac{1}{3}\right)+\frac{6}{1}\left(-\frac{5}{6}\right)}{\frac{2\cdot 3}{1}\left(\frac{4}{3}\right)}=\frac{2+(-5)}{8}$$

$$=\frac{-3}{8}=-\frac{3}{8}$$

61.

$$\frac{-\frac{3}{4}+\frac{7}{8}}{2}=\frac{\frac{8}{1}\left(-\frac{3}{4}+\frac{7}{8}\right)}{\frac{8}{1}\left(\frac{2}{1}\right)}$$

$$=\frac{\frac{8}{1}\left(-\frac{3}{4}\right)+\frac{8}{1}\left(\frac{7}{8}\right)}{\frac{8}{1}\left(\frac{2}{1}\right)}$$

$$=\frac{\frac{2\cdot 4}{1}\left(-\frac{3}{4}\right)+\frac{8}{1}\left(\frac{7}{8}\right)}{\frac{8}{1}\left(\frac{2}{1}\right)}$$

$$=\frac{-6+7}{16}=\frac{1}{16}$$

63.

$$\left|\frac{2\left(-\dfrac{3}{4}\right)}{\dfrac{7}{8}-\left(-\dfrac{3}{4}\right)}\right| = \left|\frac{\dfrac{2}{1}\left(-\dfrac{3}{4}\right)}{\dfrac{7}{8}+\dfrac{3}{4}}\right| = \left|\frac{\dfrac{\cancel{2}}{1}\left(-\dfrac{3}{2\bullet\cancel{2}}\right)}{\dfrac{7}{8}+\dfrac{3}{4}}\right| = \left|\frac{-\dfrac{3}{2}}{\dfrac{7}{8}+\dfrac{3}{4}}\right| = \left|\frac{\dfrac{8}{1}\left(-\dfrac{3}{2}\right)}{\dfrac{8}{1}\left(\dfrac{7}{8}+\dfrac{3}{4}\right)}\right| = \left|\frac{\dfrac{8}{1}\left(-\dfrac{3}{2}\right)}{\dfrac{8}{1}\left(\dfrac{7}{8}\right)+\dfrac{8}{1}\left(\dfrac{3}{4}\right)}\right|$$

$$= \left|\frac{\dfrac{4\bullet\cancel{2}}{1}\left(-\dfrac{3}{\cancel{2}}\right)}{\dfrac{\cancel{8}}{1}\left(\dfrac{7}{\cancel{8}}\right)+\dfrac{2\bullet\cancel{4}}{1}\left(\dfrac{3}{\cancel{4}}\right)}\right| = \left|\frac{-12}{7+6}\right| = \left|-\frac{12}{13}\right| = \frac{12}{13}$$

Applications

65.   SANDWICH SHOPS

n = number of sandwiches

$$n = \left(1\frac{3}{4}+2\frac{1}{2}\right)\div\frac{1}{2}$$

$$n = \left(\frac{7}{4}+\frac{5}{2}\right)\div\frac{1}{2}$$

$$n = \left(\frac{7}{4}+\frac{5\bullet2}{2\bullet2}\right)\div\frac{1}{2}$$

$$n = \left(\frac{7}{4}+\frac{10}{4}\right)\div\frac{1}{2}$$

$$n = \frac{17}{4}\div\frac{1}{2}$$

$$n = \frac{17}{4}\bullet\frac{2}{1}$$

$$n = \frac{17}{2\bullet\cancel{2}}\bullet\frac{\cancel{2}}{1} = \frac{17}{2} = 8\frac{1}{2}$$

He can make $8\dfrac{1}{2}$ sandwiches.

67. PHYSICAL FITNESS

j = jogger distance

$$j = \left(2\frac{1}{2}\right)\left(1\frac{1}{2}\right)$$

$$j = \left(\frac{5}{2}\right)\left(\frac{3}{2}\right) = \frac{15}{4} = 3\frac{3}{4}$$

c = cyclist distance

$$c = \left(7\frac{1}{5}\right)\left(1\frac{1}{2}\right)$$

$$c = \left(\frac{36}{5}\right)\left(\frac{3}{2}\right)$$

$$c = \left(\frac{18 \bullet \cancel{2}}{5}\right)\left(\frac{3}{\cancel{2}}\right) = \frac{54}{5} = 10\frac{4}{5}$$

	Rate (mph)	Time (hr)	Distance (mi)
Jogger	$2\frac{1}{2}$	$1\frac{1}{2}$	$3\frac{3}{4}$
Cyclist	$7\frac{1}{5}$	$1\frac{1}{2}$	$10\frac{4}{5}$

d = distance apart

$$d = 3\frac{3}{4} + 10\frac{4}{5}$$

$$d = \frac{15}{4} + \frac{54}{5}$$

$$d = \frac{15 \bullet 5}{4 \bullet 5} + \frac{54 \bullet 4}{5 \bullet 4}$$

$$d = \frac{75}{20} + \frac{216}{20} = \frac{291}{20}$$

$$d = 14\frac{11}{20}$$

They are $14\frac{11}{20}$ mi apart.

69.   POSTAGE RATES

w = weight of the packages

$$w = \frac{1}{16} + \frac{5}{8} + \frac{1}{16} + \frac{1}{16} + \frac{1}{16}$$

$$w = \frac{1}{16} + \frac{5 \bullet 2}{8 \bullet 2} + \frac{1}{16} + \frac{1}{16} + \frac{1}{16}$$

$$w = \frac{1}{16} + \frac{10}{16} + \frac{1}{16} + \frac{1}{16} + \frac{1}{16} = \frac{14}{16}$$

$$w = \frac{7 \bullet \cancel{2}}{8 \bullet \cancel{2}} = \frac{7}{8}$$

Yes, the package is under the one ounce rate.

71.   PHYSICAL THERAPY

d = total distance walked

$$d = 7\left(\frac{1}{4}\right) + 7\left(\frac{1}{2}\right) + 7\left(\frac{3}{4}\right)$$

$$d = \frac{7}{1}\left(\frac{1}{4}\right) + \frac{7}{1}\left(\frac{1}{2}\right) + \frac{7}{1}\left(\frac{3}{4}\right)$$

$$d = \frac{7}{4} + \frac{7}{2} + \frac{21}{4}$$

$$d = \frac{7}{4} + \frac{7 \bullet 2}{2 \bullet 2} + \frac{21}{4}$$

$$d = \frac{7}{4} + \frac{14}{4} + \frac{21}{4} = \frac{42}{4}$$

$$d = \frac{21 \bullet \cancel{2}}{2 \bullet \cancel{2}} = \frac{21}{2} = 10\frac{1}{2}$$

The total distance was $10\frac{1}{2}$ miles.

73.   AMUSMEMENT PARKS

$$\frac{1}{\dfrac{1}{10} + \dfrac{1}{15}} = \frac{30(1)}{\dfrac{30}{1}\left(\dfrac{1}{10} + \dfrac{1}{15}\right)}$$

$$= \frac{30}{\dfrac{30}{1}\left(\dfrac{1}{10}\right) + \dfrac{30}{1}\left(\dfrac{1}{15}\right)}$$

$$= \frac{30}{\dfrac{3 \bullet \cancel{10}}{1}\left(\dfrac{1}{\cancel{10}}\right) + \dfrac{2 \bullet \cancel{15}}{1}\left(\dfrac{1}{\cancel{15}}\right)} = \frac{30}{3 + 2}$$

$$= \frac{30}{5} = 6$$

The time is 6 sec.

Writing

75. A complex fraction is a fraction in which the numerator, denominator, or both are fractions.

77. Answers will vary.

Review

79. $-4d - (-7d) = -4d + 7d$

$\qquad\qquad = 3d$

81. $2x(-x)$

83. $2 + 3\left[-3 - (-4 - 1)\right] = 2 + 3\left[-3 - (-4 + [-1])\right]$

$\qquad\qquad\qquad\qquad = 2 + 3\left[-3 - (-5)\right]$

$\qquad\qquad\qquad\qquad = 2 + 3\left[-3 + 5\right]$

$\qquad\qquad\qquad\qquad = 2 + 3\left[2\right]$

$\qquad\qquad\qquad\qquad = 2 + 6$

$\qquad\qquad\qquad\qquad = 8$

85. $3 \bullet 3 \bullet 3 \bullet x \bullet x \bullet x \bullet x \bullet x = 27x^5$

4.8 Solving Equations Containing Fractions

Vocabulary

1. To find the **reciprocal** of a fraction, invert the numerator and the denominator.

3. the **least** **common** **denominator** of a set of fractions is the smallest number each denominator will divide exactly.

Concepts

5.
$$\frac{5}{8}\left(\frac{40}{1}\right) = 25$$

$$\frac{5}{\cancel{8}}\left(\frac{\cancel{8} \bullet 5}{1}\right) = 25$$

$$\frac{25}{1} = 25$$

Yes, when substituted into the

equation the result is a true statement.

7.   1

9a. $\frac{4}{5}p$ 9b. $\frac{1}{4}t$

11.   Multiply both sides by $\frac{3}{2}$, or multiply both sides by 3 and then divide by 2.

Notation

13. $$\frac{7}{8}x = 21$$

$$\frac{8}{7}\left(\frac{7}{8}x\right) = \frac{8}{7}(21)$$

$$x = 24$$

15a. true

15b. false

15c. true

15d. true

Practice

17. $$\frac{4}{7}x = 16$$

$$\frac{7}{4}\left(\frac{4}{7}x\right) = \frac{7}{4}\left(\frac{16}{1}\right)$$

$$x = \frac{7}{\cancel{4}}\left(\frac{4 \cdot \cancel{4}}{1}\right)$$

$$x = \frac{28}{1} = 28$$

19. $$\frac{7}{8}t = -28$$

$$\frac{8}{7}\left(\frac{7}{8}t\right) = \frac{8}{7}\left(-\frac{28}{1}\right)$$

$$t = \frac{8}{\cancel{7}}\left(-\frac{4 \cdot \cancel{7}}{1}\right)$$

$$t = -\frac{32}{1} = -32$$

21. $$-\frac{3}{5}h = 4$$

$$-\frac{5}{3}\left(-\frac{3}{5}h\right) = -\frac{5}{3}\left(\frac{4}{1}\right)$$

$$t = -\frac{20}{3}$$

23. $$\frac{2}{3}x = \frac{4}{5}$$

$$\frac{3}{2}\left(\frac{2}{3}x\right) = \frac{3}{2}\left(\frac{4}{5}\right)$$

$$x = \frac{3}{\cancel{2}}\left(\frac{2 \cdot \cancel{2}}{5}\right)$$

$$x = \frac{6}{5}$$

25.
$$\frac{2}{5}y = 0$$
$$\frac{5}{2}\left(\frac{2}{5}y\right) = \frac{5}{2}(0)$$
$$y = 0$$

27.
$$-\frac{5c}{6} = -25$$
$$-\frac{6}{5}\left(-\frac{5c}{6}\right) = -\frac{6}{5}\left(-\frac{25}{1}\right)$$
$$c = -\frac{6}{\cancel{5}}\left(-\frac{5 \cdot \cancel{5}}{1}\right)$$
$$c = 30$$

29.
$$\frac{5f}{7} = -2$$
$$\frac{7}{5}\left(\frac{5f}{7}\right) = \frac{7}{5}\left(-\frac{2}{1}\right)$$
$$f = -\frac{14}{5}$$

31.
$$\frac{5}{8}y = \frac{1}{10}$$
$$\frac{8}{5}\left(\frac{5}{8}y\right) = \frac{8}{5}\left(\frac{1}{10}\right)$$
$$y = \frac{4 \cdot \cancel{2}}{5}\left(\frac{1}{5 \cdot \cancel{2}}\right)$$
$$y = \frac{4}{25}$$

33.
$$2x + 1 = 0$$
$$2x + 1 - 1 = 0 - 1$$
$$2x = -1$$
$$\frac{2x}{2} = \frac{-1}{2}$$
$$x = -\frac{1}{2}$$

35.
$$5x - 1 = 1$$
$$5x - 1 + 1 = 1 + 1$$
$$5x = 2$$
$$\frac{5x}{5} = \frac{2}{5}$$
$$x = \frac{2}{5}$$

37.
$$6x = 2x - 11$$
$$6x - 2x = 2x - 11 - 2x$$
$$4x = -11$$
$$\frac{4x}{4} = \frac{-11}{4}$$
$$x = -\frac{11}{4}$$

39.
$$2(y - 3) = 7$$
$$2y - 6 = 7$$
$$2y - 6 + 6 = 7 + 6$$
$$2y = 13$$
$$\frac{2y}{2} = \frac{13}{2}$$
$$y = \frac{13}{2}$$

41.
$$x - \frac{1}{9} = \frac{7}{9}$$
$$x - \frac{1}{9} + \frac{1}{9} = \frac{7}{9} + \frac{1}{9}$$
$$x = \frac{8}{9}$$

43.
$$x + \frac{1}{9} = \frac{4}{9}$$
$$x + \frac{1}{9} - \frac{1}{9} = \frac{4}{9} - \frac{1}{9}$$
$$x = \frac{3}{9} = \frac{1 \cdot \cancel{3}}{3 \cdot \cancel{3}}$$
$$x = \frac{1}{3}$$

45.
$$x - \frac{1}{6} = \frac{2}{9}$$
$$18\left(x - \frac{1}{6}\right) = 18\left(\frac{2}{9}\right)$$
$$18x - 3 = 4$$
$$18x - 3 + 3 = 4 + 3$$
$$18x = 7$$
$$\frac{18x}{18} = \frac{7}{18}$$
$$x = \frac{7}{18}$$

47.
$$y + \frac{7}{8} = \frac{1}{4}$$
$$8\left(y + \frac{7}{8}\right) = 8\left(\frac{1}{4}\right)$$
$$8y + 7 = 2$$
$$8y + 7 - 7 = 2 - 7$$
$$8y = -5$$
$$\frac{8y}{8} = \frac{-5}{8}$$
$$y = -\frac{5}{8}$$

49.
$$\frac{5}{4} + t = \frac{1}{4}$$
$$\frac{5}{4} + t - \frac{5}{4} = \frac{1}{4} - \frac{5}{4}$$
$$t = \frac{-4}{4}$$
$$t = -1$$

51.
$$x + \frac{3}{4} = -\frac{1}{2}$$
$$4\left(x + \frac{3}{4}\right) = 4\left(-\frac{1}{2}\right)$$
$$4x + 3 = -2$$
$$4x + 3 - 3 = -2 - 3$$
$$4x = -5$$
$$\frac{4x}{4} = \frac{-5}{4}$$
$$x = -\frac{5}{4}$$

53.

$$\frac{-x}{4}+1=10$$

$$4\left(\frac{-x}{4}+1\right)=4(10)$$

$$-x+4=40$$

$$-x+4-4=40-4$$

$$-x=36$$

$$\frac{-x}{-1}=\frac{36}{-1}$$

$$x=-36$$

55.

$$2x-\frac{1}{2}=\frac{1}{3}$$

$$6\left(2x-\frac{1}{2}\right)=6\left(\frac{1}{3}\right)$$

$$12x-3=2$$

$$12x-3+3=2+3$$

$$12x=5$$

$$\frac{12x}{12}=\frac{5}{12}$$

$$x=\frac{5}{12}$$

57.

$$\frac{1}{2}x-\frac{1}{9}=\frac{1}{3}$$

$$18\left(\frac{1}{2}x-\frac{1}{9}\right)=18\left(\frac{1}{3}\right)$$

$$9x-2=6$$

$$9x-2+2=6+2$$

$$9x=8$$

$$\frac{9x}{9}=\frac{8}{9}$$

$$x=\frac{8}{9}$$

59.

$$5+\frac{x}{3}=\frac{1}{2}$$

$$6\left(5+\frac{x}{3}\right)=6\left(\frac{1}{2}\right)$$

$$30+2x=3$$

$$30+2x-30=3-30$$

$$2x=-27$$

$$\frac{2x}{2}=\frac{-27}{2}$$

$$x=-\frac{27}{2}$$

61.
$$\frac{2}{5}x + 1 = \frac{1}{3} + x$$
$$15\left(\frac{2}{5}x + 1\right) = 15\left(\frac{1}{3} + x\right)$$
$$6x + 15 = 5 + 15x$$
$$6x + 15 - 6x = 5 + 15x - 6x$$
$$15 = 5 + 9x$$
$$15 - 5 = 5 + 9x - 5x$$
$$10 = 9x$$
$$\frac{10}{9} = \frac{9x}{9}$$
$$\frac{10}{9} = x \ or \ x = \frac{10}{9}$$

63.
$$\frac{x}{3} + \frac{x}{4} = -2$$
$$12\left(\frac{x}{3} + \frac{x}{4}\right) = 12(-2)$$
$$4x + 3x = -24$$
$$7x = -24$$
$$\frac{7x}{7} = \frac{-24}{7}$$
$$x = -\frac{24}{7}$$

65.
$$4 + \frac{s}{3} = 8$$
$$3\left(4 + \frac{s}{3}\right) = 3(8)$$
$$12 + s = 24$$
$$12 + s - 12 = 24 - 12$$
$$s = 12$$

67.
$$\frac{5h}{6} - 8 = 12$$
$$6\left(\frac{5h}{6} - 8\right) = 6(12)$$
$$5h - 48 = 72$$
$$5h - 48 + 48 = 72 + 48$$
$$\frac{5h}{5} = \frac{120}{5}$$
$$h = 24$$

69.
$$-4 + 9 + \frac{5t}{12} = 0$$
$$5 + \frac{5t}{12} = 0$$
$$5 + \frac{5t}{12} - 5 = 0 - 5$$
$$\frac{5t}{12} = -5$$
$$\frac{12}{5}\left(\frac{5t}{12}\right) = \frac{12}{\cancel{5}}\left(-\frac{\cancel{5}}{1}\right)$$
$$t = -\frac{12}{1} = -12$$

71.
$$-3 - 2 + \frac{4x}{15} = 0$$
$$-5 + \frac{4x}{15} = 0$$
$$-5 + \frac{4x}{15} + 5 = 0 + 5$$
$$\frac{4x}{15} = 5$$
$$\frac{15}{4}\left(\frac{4x}{15}\right) = \frac{15}{4}\left(\frac{5}{1}\right)$$
$$x = \frac{75}{4}$$

Applications

73.   TRANSMISSION REPAIRS

   Analyze the problem

   - Only $\frac{1}{3}$ of the customers needed new transmissions.

   - The shop installed **32** new transmissions last year.

   - Find the number of **customers** the shop had last year.

   Form an equation

   Let $x =$ **number of customers last year**.

   Key phrase: *one-third of*

   Translation: multiply

$\frac{1}{3}$ of the number of customers last year	was	32
$\frac{1}{3}x$	=	32

   Solve the equation

   $$\frac{1}{3}x = 32$$

   $$3\left(\frac{1}{3}x\right) = 3(32)$$

   $$x = 96$$

   State the conclusion **The shop had 96 customers last year**.

   Check the result

   If we find $\frac{1}{3}$ of 96, we get 32. The answer checks.

75.   TOOTH DEVELOPMENT

t = number of baby teeth child will have

$$\frac{4}{5}t = 16$$

$$\frac{5}{4}\left(\frac{4}{5}t\right) = \left(\frac{16}{1}\right)\frac{5}{4}$$

$$t = \left(\frac{4 \bullet \cancel{4}}{1}\right)\frac{5}{\cancel{4}} = \frac{20}{1} = 20$$

The child will have 20 baby teeth.

77.   TELEPHONE BOOKS

p = pages in the phone book

$$\frac{2}{3}p = 300$$

$$3\left(\frac{2}{3}p\right) = 3(300)$$

$$2p = 300$$

$$\frac{2p}{2} = \frac{300}{2}$$

$$p = 450$$

There are 450 pages total.

79.   HOME SALES

If $\frac{3}{4}$ are purchased, $\frac{1}{4}$ are left

n = number of homes

$$\frac{1}{4}n = 9$$

$$\frac{4}{1}\left(\frac{1}{4}n\right) = 4(9)$$

$$n = 36$$

There are 36 homes total.

81.   SAFETY REQUIREMENTS

w = width of the lights

$$A = lw$$

$$30 = 3\frac{3}{4}w$$

$$30 = \frac{15}{4}w$$

$$\frac{4}{15}\left(\frac{30}{1}\right) = \frac{4}{15}\left(\frac{15}{4}w\right)$$

$$\frac{4}{\cancel{15}}\left(\frac{2 \bullet \cancel{15}}{1}\right) = w$$

$$8 = w$$

The width should be 8 in.

83.   CPR CLASS

$$\frac{1}{4}+\frac{2}{3}=\frac{1\bullet 3}{4\bullet 3}+\frac{2\bullet 4}{3\bullet 4}$$

$$=\frac{3}{12}+\frac{8}{12}=\frac{11}{12}$$

There is only $\frac{1}{12}$ of the class left to occupy the 30 minutes for legal responsibilities.

t = time for the whole course

$$\frac{1}{12}t=30$$

$$\frac{12}{1}\left(\frac{1}{12}t\right)=12(30)$$

$$t=360$$

The course takes 360 min.

Writing

85.   Isolate means to perform mathematical operations on an equation until the variable is by itself on one side of the equals sign.

87.   The reciprocal method, because it would only take one step, but either method would work.

Review

89. $a(b+c)=ab+ac$

91. $C=\dfrac{5(F-32)}{9}$ *for* $F=41$

$$C=\frac{5(41-32)}{9}$$

$$C=\frac{5(\cancel{9})}{\cancel{9}}=5$$

41° F is the same as 5° C.

93.
$$5x - 3 = 2x + 12$$
$$5x - 3 - 2x = 2x + 12 - 2x$$
$$3x - 3 = 12$$
$$3x - 3 + 3 = 12 + 3$$
$$3x = 15$$
$$\frac{3x}{3} = \frac{15}{3}$$
$$x = 5$$

95   13,000,000

Key Concept The Fundamental Property of Fractions

1.   **Step 1:** The numerator and the denominator share a common

factor of 5.

Step 2: Apply the fundamental property of fractions. Divide $\dfrac{15}{25} = \dfrac{15 \div 5}{25 \div 5}$

the numerator and the denominator by a common factor 5.

Step 3: Perform the divisions to simplify the fraction. $= \dfrac{3}{5}$

2.   **Step 1:** Factor 15 as $5 \bullet 3$ and 25 as $5 \bullet 5$. $\dfrac{15}{25} = \dfrac{5 \bullet 3}{5 \bullet 5}$

Step 2: The slashes and small 1's indicate that the numerator $= \dfrac{\overset{1}{\cancel{5}} \bullet 3}{\underset{1}{\cancel{5}} \bullet 5}$

and the denominator have been divided by 5.

Step 3: Multiply in the numerator and denominator. $= \dfrac{3}{5}$

3.   **Step 1:** We must multiply 5 by 7 to obtain 35.

Step 2: Use the fundamental property of fractions. Multiply $\dfrac{1}{5} = \dfrac{1 \bullet 7}{5 \bullet 7}$

the numerator and the denominator by 7.

Step 3: Multiply the numerator and denominator. $= \dfrac{7}{35}$

4.   **Step 1:** We must multiply $3x$ by 6 to obtain $18x$.

Step 2: Use the fundamental property of fractions. Multiply $\dfrac{2}{3x} = \dfrac{2 \bullet 6}{3x \bullet 6}$

the numerator and the denominator by 6.

Step 3: Perform the multiplication in the numerator and the $= \dfrac{12}{18x}$

denominator.

Chapter Four Review

Section 4.1 The Fundamental Property of Fractions

1. $\dfrac{7}{24}$

2. The figure is not divided into equal parts.

3. $\dfrac{2}{-3} = \dfrac{-2}{3} = -\dfrac{2}{3}$

4. Equivalent fractions

   $\dfrac{6}{8} = \dfrac{3}{4}$

5. The numerator and denominator of the fraction are being divided by 2 using the fundamental property of fractions.

6. The slashes indicate that the numerator and denominator of the fractions are being divided by 2, the answer to these divisions is one.

7. $\dfrac{15}{45} = \dfrac{\cancel{15} \bullet 1}{\cancel{15} \bullet 3} = \dfrac{1}{3}$

8. $\dfrac{20}{48} = \dfrac{\cancel{4} \bullet 5}{\cancel{4} \bullet 12} = \dfrac{5}{12}$

9. $-\dfrac{63x^2}{84x} = -\dfrac{\cancel{21} \bullet 3 \bullet \cancel{x} \bullet x}{\cancel{21} \bullet 4 \bullet \cancel{x}} = -\dfrac{3x}{4}$

10. $\dfrac{66m^3n}{108m^4n} = \dfrac{\cancel{6} \bullet 11 \bullet \cancel{m} \bullet \cancel{m} \bullet \cancel{m} \bullet \cancel{n}}{\cancel{6} \bullet 18 \bullet \cancel{m} \bullet \cancel{m} \bullet \cancel{m} \bullet m \bullet \cancel{n}} = \dfrac{11}{18m}$

11. The numerator and denominator of the fraction are being multiplied by 2 using the fundamental property of fractions

12. $\dfrac{2}{3} = \dfrac{2 \bullet 6}{3 \bullet 6} = \dfrac{12}{18}$

13. $-\dfrac{3}{8} = -\dfrac{3 \bullet 2}{8 \bullet 2} = -\dfrac{6}{16}$

14. $\dfrac{7}{15} = \dfrac{7 \bullet 3a}{15 \bullet 3a} = \dfrac{21a}{45a}$

15. $\dfrac{4}{1} = \dfrac{4 \bullet 9}{1 \bullet 9} = \dfrac{36}{9}$

Section 4.2 Multiplying Fractions

16. $\dfrac{1}{2} \cdot \dfrac{1}{3} = \dfrac{1}{6}$

17. $\dfrac{2}{5}\left(-\dfrac{7}{9}\right) = -\dfrac{14}{45}$

18. $\dfrac{9}{16} \cdot \dfrac{20}{27} = \dfrac{\cancel{9}}{4 \cdot \cancel{4}} \cdot \dfrac{5 \cdot \cancel{4}}{3 \cdot \cancel{9}} = \dfrac{5}{12}$

19. $\dfrac{5}{6} \cdot \dfrac{1}{3} \cdot \dfrac{18}{25} = \dfrac{\cancel{5}}{\cancel{6}} \cdot \dfrac{1}{\cancel{3}} \cdot \dfrac{\cancel{3} \cdot \cancel{6}}{5 \cdot \cancel{5}} = \dfrac{1}{5}$

20. $\dfrac{3}{5} \cdot 7 = \dfrac{3}{5} \cdot \dfrac{7}{1} = \dfrac{21}{5}$

21. $-4\left(-\dfrac{9}{16}\right) = -\dfrac{\cancel{4}}{1}\left(-\dfrac{9}{4 \cdot \cancel{4}}\right) = \dfrac{9}{4}$

22. $3\left(\dfrac{1}{3}\right) = \dfrac{\cancel{3}}{1}\left(\dfrac{1}{\cancel{3}}\right) = 1$

23. $-\dfrac{6}{7}\left(-\dfrac{7}{6}\right) = 1$

24. true

25. false

26. $\dfrac{3t}{5} \cdot \dfrac{10}{27t} = \dfrac{\cancel{3} \cdot \cancel{t}}{\cancel{5}} \cdot \dfrac{2 \cdot \cancel{5}}{9 \cdot \cancel{3} \cdot \cancel{t}} = \dfrac{2}{9}$

27. $-\dfrac{2}{3}\left(\dfrac{4}{7}s\right) = -\dfrac{8}{21}s$

28. $\dfrac{4d^2}{9} \cdot \dfrac{3}{28d} = \dfrac{\cancel{4} \cdot \cancel{d} \cdot d}{3 \cdot \cancel{3}} \cdot \dfrac{\cancel{3}}{7 \cdot \cancel{4} \cdot \cancel{d}}$
$$= \dfrac{d}{21}$$

29. $9mn\left(-\dfrac{5}{81n^2}\right) = \dfrac{\cancel{9}m\cancel{n}}{1}\left(-\dfrac{5}{9 \cdot \cancel{9} \cdot n \cdot \cancel{n}}\right)$
$$= -\dfrac{5m}{9n}$$

30. $\left(\dfrac{3}{4}\right)^2 = \left(\dfrac{3}{4}\right)\left(\dfrac{3}{4}\right) = \dfrac{9}{16}$

31. $\left(-\dfrac{5}{2}\right)^3 = \left(-\dfrac{5}{2}\right)\left(-\dfrac{5}{2}\right)\left(-\dfrac{5}{2}\right) = -\dfrac{125}{8} = -15\dfrac{5}{8}$

32. $\left(\dfrac{x}{3}\right)^2 = \left(\dfrac{x}{3}\right)\left(\dfrac{x}{3}\right) = \dfrac{x^2}{9}$

33. $\left(-\dfrac{2c}{5}\right)^3 = \left(-\dfrac{2c}{5}\right)\left(-\dfrac{2c}{5}\right)\left(-\dfrac{2c}{5}\right) = -\dfrac{8c^3}{125}$

34. GRAVITY ON THE MOON

w = weight on moon

$$w = \frac{1}{6}(180)$$

$$w = \frac{1}{\cancel{6}}\left(\frac{30 \bullet \cancel{6}}{1}\right) = 30$$

He would weigh 30 lb.

35.

$$A = \frac{1}{2}bh$$

$$A = \frac{1}{2}(15)(8)$$

$$A = \frac{1}{\cancel{2}}\left(\frac{15}{1}\right)\left(\frac{4 \bullet \cancel{2}}{1}\right) = 60$$

The area is 60 sq in.

Section 4.3 Dividing Fractions

36. 8

37. $-\dfrac{12}{11}$

38. $\dfrac{1}{x}$

39. $\dfrac{c}{ab}$

40. $\dfrac{1}{6} \div \dfrac{11}{25} = \dfrac{1}{6} \bullet \dfrac{25}{11} = \dfrac{25}{66}$

41. $-\dfrac{7}{8} \div \dfrac{1}{4} = -\dfrac{7}{8} \bullet \dfrac{4}{1}$

$$= -\frac{7}{2 \bullet \cancel{4}} \bullet \frac{\cancel{4}}{1} = -\frac{7}{2}$$

42. $-\dfrac{15}{16} \div (-10) = -\dfrac{15}{16}\left(-\dfrac{1}{10}\right)$

$$= -\frac{3 \bullet \cancel{5}}{16}\left(-\frac{1}{2 \bullet \cancel{5}}\right) = \frac{3}{32}$$

43. $8 \div \dfrac{16}{5} = \dfrac{8}{1} \bullet \dfrac{5}{16}$

$$= \frac{\cancel{8}}{1} \bullet \frac{5}{2 \bullet \cancel{8}} = \frac{5}{2}$$

44. $\dfrac{t}{8} \div \dfrac{1}{4} = \dfrac{t}{8} \bullet \dfrac{4}{1}$

$$= \frac{t}{2 \bullet \cancel{4}} \bullet \frac{\cancel{4}}{1} = \frac{t}{2}$$

45. $\dfrac{4a}{5} \div \dfrac{a}{2} = \dfrac{4a}{5} \bullet \dfrac{2}{a}$

$$= \frac{4 \bullet \cancel{a}}{5} \bullet \frac{2}{\cancel{a}} = \frac{8}{5}$$

46.

$$-\frac{a}{b} \div \left(-\frac{b}{a}\right) = -\frac{a}{b}\left(-\frac{a}{b}\right) = \frac{a^2}{b^2}$$

47.

$$\frac{2}{3}x \div \left(-\frac{x^2}{9}\right) = \frac{2x}{3}\left(-\frac{9}{x^2}\right)$$

$$= \frac{2 \bullet \cancel{x}}{\cancel{3}}\left(-\frac{3 \bullet \cancel{3}}{x \bullet \cancel{x}}\right) = -\frac{6}{x}$$

48.   GOLD COINS

$n =$ number of coins

$$n = \frac{3}{4} \div \frac{1}{16}$$

$$n = \frac{3}{4} \bullet \frac{16}{1}$$

$$n = \frac{3}{\cancel{4}} \bullet \frac{4 \bullet \cancel{4}}{1} = 12$$

12 coins can be made.

Section 4.4 Adding and Subtracting Fractions

49.

$$\frac{2}{7} + \frac{3}{7} = \frac{5}{7}$$

50.

$$-\frac{3}{5} - \frac{3}{5} = -\frac{3}{5} + \left(-\frac{3}{5}\right) = -\frac{6}{5}$$

51.

$$\frac{3}{x} - \frac{1}{x} = \frac{2}{x}$$

52.

$$\frac{7}{8} + \frac{t}{8} = \frac{7+t}{8}$$

53.   The denominators are not the same.

54.

$$2 \bullet 3 \bullet 5 = 30 \qquad 3 \bullet 3 \bullet 5 = 45$$

$$LCD = 2 \bullet 3 \bullet 3 \bullet 5 = 90$$

55.
$$\frac{1}{6}+\frac{2}{3}=\frac{1}{6}+\frac{2\bullet 2}{3\bullet 2}$$
$$=\frac{1}{6}+\frac{4}{6}=\frac{5}{6}$$

56.
$$\frac{2}{5}+\left(-\frac{3}{8}\right)=\frac{2\bullet 8}{5\bullet 8}+\left(-\frac{3\bullet 5}{8\bullet 5}\right)$$
$$=\frac{16}{40}+\left(-\frac{15}{40}\right)=\frac{1}{40}$$

57.
$$-\frac{3}{8}-\frac{5}{6}=-\frac{3\bullet 3}{8\bullet 3}-\frac{5\bullet 4}{6\bullet 4}$$
$$=-\frac{9}{24}-\frac{20}{24}=-\frac{9}{24}+\left(-\frac{20}{24}\right)$$
$$=-\frac{29}{24}$$

58.
$$3-\frac{1}{7}=\frac{3}{1}-\frac{1}{7}=\frac{3\bullet 7}{1\bullet 7}-\frac{1}{7}$$
$$=\frac{21}{7}-\frac{1}{7}=\frac{20}{7}$$

59.
$$\frac{x}{25}-\frac{3}{10}=\frac{x\bullet 2}{25\bullet 2}-\frac{3\bullet 5}{10\bullet 5}$$
$$=\frac{2x}{50}-\frac{15}{50}=\frac{2x-15}{50}$$

60.
$$\frac{1}{3}+\frac{7}{y}=\frac{1\bullet y}{3\bullet y}+\frac{7\bullet 3}{y\bullet 3}$$
$$=\frac{y}{3y}+\frac{21}{3y}=\frac{y+21}{3y}$$

61.
$$\frac{13}{6}-6=\frac{13}{6}-\frac{6}{1}=\frac{13}{6}-\frac{6\bullet 6}{1\bullet 6}$$
$$=\frac{13}{6}-\frac{36}{6}=\frac{13}{6}+\left(-\frac{36}{6}\right)=-\frac{23}{6}$$

62.
$$\frac{1}{3}+\frac{1}{4}+\frac{1}{5}=\frac{1\bullet 20}{3\bullet 20}+\frac{1\bullet 15}{4\bullet 15}+\frac{1\bullet 12}{5\bullet 12}$$
$$=\frac{20}{60}+\frac{15}{60}+\frac{12}{60}=\frac{47}{60}$$

63. MACHINE SHOPS

m = amount that must be milled off

$$m=\frac{3}{4}-\frac{17}{32}$$
$$m=\frac{3\bullet 8}{4\bullet 8}-\frac{17}{32}$$
$$m=\frac{24}{32}-\frac{17}{32}$$
$$m=\frac{7}{32}$$

$\dfrac{7}{32}$ in must be milled off.

64. TELEMARKETING

Compare the two fractions.

$$\frac{2}{9}=\frac{2\bullet 11}{9\bullet 11}=\frac{22}{99}$$
$$\frac{3}{11}=\frac{3\bullet 9}{11\bullet 9}=\frac{27}{99}$$

Thus, the second hour she did better.

Section 4.5 Multiplying and Dividing Mixed Numbers

65. $2\dfrac{1}{6}$

66. $\dfrac{13}{6}$

67. $3\dfrac{1}{5}$

68. $-3\dfrac{11}{12}$

69. 1

70. $\dfrac{14}{6} = \dfrac{7 \bullet \cancel{2}}{3 \bullet \cancel{2}} = \dfrac{7}{3} = 2\dfrac{1}{3}$

71. $\dfrac{75}{8}$

72. $-\dfrac{11}{5}$

73. $\dfrac{201}{2}$

74. $\dfrac{199}{100}$

75.

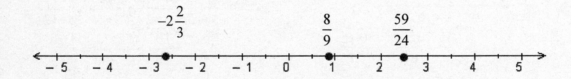

76. $-5\dfrac{1}{4} \bullet \dfrac{2}{35} = -\dfrac{21}{4} \bullet \dfrac{2}{35}$

$= -\dfrac{3 \bullet \cancel{7}}{2 \bullet \cancel{2}} \bullet \dfrac{\cancel{2}}{5 \bullet \cancel{7}} = -\dfrac{3}{10}$

77. $\left(-3\dfrac{1}{2}\right) \div \left(-3\dfrac{2}{3}\right) = \left(-\dfrac{7}{2}\right) \div \left(-\dfrac{11}{3}\right)$

$= \left(-\dfrac{7}{2}\right)\left(-\dfrac{3}{11}\right) = \dfrac{21}{22}$

78. $\left(-6\dfrac{2}{3}\right)(-6) = \left(-\dfrac{20}{3}\right)\left(-\dfrac{6}{1}\right)$

$= \left(-\dfrac{20}{\cancel{3}}\right)\left(-\dfrac{2 \bullet \cancel{3}}{1}\right) = 40$

79. $-8 \div 3\dfrac{1}{5} = -\dfrac{8}{1} \div \dfrac{16}{5} = -\dfrac{8}{1} \bullet \dfrac{5}{16}$

$= -\dfrac{\cancel{8}}{1} \bullet \dfrac{5}{2 \bullet \cancel{8}} = -\dfrac{5}{2} = -2\dfrac{1}{2}$

80.　CAMERA TRIPODS

L = length of leg when extended

$$L = 5\frac{1}{2}\left(8\frac{3}{4}\right)$$

$$L = \frac{11}{2}\left(\frac{35}{4}\right) = \frac{385}{8} = 48\frac{1}{8}$$

The legs will be $48\frac{1}{8}$ in. when extended.

Section 4.6　Adding and Subtracting Mixed Numbers

81.
$$1\frac{3}{8} + 2\frac{1}{5} = \frac{11}{8} + \frac{11}{5} = \frac{11\bullet5}{8\bullet5} + \frac{11\bullet8}{5\bullet8}$$
$$= \frac{55}{40} + \frac{88}{40} = \frac{143}{40} = 3\frac{23}{40}$$

82.
$$3\frac{1}{2} + 2\frac{2}{3} = \frac{7}{2} + \frac{8}{3} = \frac{7\bullet3}{2\bullet3} + \frac{8\bullet2}{3\bullet2}$$
$$= \frac{21}{6} + \frac{16}{6} = \frac{37}{6} = 6\frac{1}{6}$$

83.
$$2\frac{5}{6} - 1\frac{3}{4} = \frac{17}{6} - \frac{7}{4} = \frac{17\bullet2}{6\bullet2} - \frac{7\bullet3}{4\bullet3}$$
$$= \frac{34}{12} - \frac{21}{12} = \frac{13}{12} = 1\frac{1}{12}$$

84.
$$3\frac{7}{16} - 2\frac{1}{8} = \frac{55}{16} - \frac{17}{8} = \frac{55}{16} - \frac{17\bullet2}{8\bullet2}$$
$$= \frac{55}{16} - \frac{34}{16} = \frac{21}{16} = 1\frac{5}{16}$$

85.　PAINTING SUPPLIES

g = total gallons used by painter

$$g = 10\frac{3}{4} + 21\frac{1}{2} + 7\frac{2}{3}$$

$$g = 10 + 21 + 7 + \frac{3}{4} + \frac{1}{2} + \frac{2}{3}$$

$$g = 38 + \frac{3\bullet3}{4\bullet3} + \frac{1\bullet6}{2\bullet6} + \frac{2\bullet4}{3\bullet4}$$

The painter used $39\frac{11}{12}$ gallons.

$$g = 38 + \frac{9}{12} + \frac{6}{12} + \frac{8}{12}$$

$$g = 38 + \frac{23}{12}$$

$$g = 38 + 1\frac{11}{12} = 39\frac{11}{12}$$

86.

$$133\frac{1}{9} = 133\frac{1\bullet 2}{9\bullet 2} = 133\frac{2}{18}$$

$$+49\frac{1}{6} = +49\frac{1\bullet 3}{6\bullet 3} = +49\frac{3}{18}$$

$$= 182\frac{5}{18}$$

87.

$$98\frac{11}{20} = \quad 98\frac{11}{20} = \quad 98\frac{11}{20}$$

$$+14\frac{3}{5} = +14\frac{3\bullet 4}{5\bullet 4} = +14\frac{12}{20}$$

$$= 112\frac{23}{20}$$

$$= 112 + 1\frac{3}{20} = 113\frac{3}{20}$$

88.

$$50\frac{5}{8} = 50\frac{5\bullet 3}{8\bullet 3} = 50\frac{15}{24}$$

$$-19\frac{1}{6} = -19\frac{1\bullet 4}{6\bullet 4} = -19\frac{4}{24}$$

$$= 31\frac{11}{24}$$

89.

$$375\frac{3}{4}$$

$$-59$$

$$316\frac{3}{4}$$

90.

$$23\frac{1}{3} = \quad 23\frac{1\bullet 2}{3\bullet 2} \quad = 23\frac{2}{6}$$

$$-2\frac{5}{6} = -2\frac{5}{6} \quad = -2\frac{5}{6}$$

$$= \quad 22\frac{2}{6} + \frac{6}{6} = 22\frac{8}{6}$$

$$= -2\frac{5}{6} \quad = -2\frac{5}{6}$$

$$= 20\frac{3}{6}$$

$$= 20\frac{1\bullet \cancel{3}}{2\bullet \cancel{3}} = 20\frac{1}{2}$$

91.

$$39 = 38\frac{8}{8}$$

$$-4\frac{5}{8} = -4\frac{5}{8}$$

$$34\frac{3}{8}$$

Section 4.7 Order of Operations and Complex Fractions

92.

$$\frac{3}{4}+\left(-\frac{1}{3}\right)^2\left(\frac{5}{4}\right)=\frac{3}{4}+\left(\left[-\frac{1}{3}\right]\left[-\frac{1}{3}\right]\right)\left(\frac{5}{4}\right)$$

$$=\frac{3}{4}+\left(\frac{1}{9}\right)\left(\frac{5}{4}\right)$$

$$=\frac{3}{4}+\frac{5}{36}$$

$$=\frac{3\bullet9}{4\bullet9}+\frac{5}{36}$$

$$=\frac{27}{36}+\frac{5}{36}=\frac{32}{36}$$

$$=\frac{8\bullet\cancel{4}}{9\bullet\cancel{4}}=\frac{8}{9}$$

93.

$$\left(\frac{2}{3}\div\frac{16}{9}\right)-\left(1\frac{2}{3}\bullet\frac{1}{15}\right)=\left(\frac{2}{3}\bullet\frac{9}{16}\right)-\left(\frac{5}{3}\bullet\frac{1}{15}\right)$$

$$=\left(\frac{\cancel{2}}{\cancel{3}}\bullet\frac{3\bullet\cancel{3}}{8\bullet\cancel{2}}\right)-\left(\frac{\cancel{5}}{3}\bullet\frac{1}{3\bullet\cancel{5}}\right)$$

$$=\frac{3}{8}-\frac{1}{9}$$

$$=\frac{3\bullet9}{8\bullet9}-\frac{1\bullet8}{9\bullet8}$$

$$=\frac{27}{72}-\frac{8}{72}=\frac{19}{72}$$

94.

$$\frac{\frac{3}{5}}{-\frac{17}{20}} = \frac{\frac{20}{1}\left(\frac{3}{5}\right)}{\frac{20}{1}\left(-\frac{17}{20}\right)} = \frac{\frac{4\cdot\cancel{5}}{1}\left(\frac{3}{\cancel{5}}\right)}{\frac{\cancel{20}}{1}\left(-\frac{17}{\cancel{20}}\right)}$$

$$= \frac{12}{-17} = -\frac{12}{17}$$

95.

$$\frac{\frac{2}{3}-\frac{1}{6}}{-\frac{3}{4}-\frac{1}{2}} = \frac{\frac{12}{1}\left(\frac{2}{3}-\frac{1}{6}\right)}{\frac{12}{1}\left(-\frac{3}{4}-\frac{1}{2}\right)}$$

$$= \frac{\frac{12}{1}\left(\frac{2}{3}\right) - \frac{12}{1}\left(\frac{1}{6}\right)}{\frac{12}{1}\left(-\frac{3}{4}\right) - \frac{12}{1}\left(\frac{1}{2}\right)}$$

$$= \frac{\frac{4\cdot\cancel{3}}{1}\left(\frac{2}{\cancel{3}}\right) - \frac{2\cdot\cancel{6}}{1}\left(\frac{1}{\cancel{6}}\right)}{\frac{3\cdot\cancel{4}}{1}\left(-\frac{3}{\cancel{4}}\right) - \frac{6\cdot\cancel{2}}{1}\left(\frac{1}{\cancel{2}}\right)}$$

$$= \frac{8-2}{-9-6} = \frac{6}{-9+(-6)} = \frac{6}{-15}$$

$$= -\frac{2\cdot\cancel{3}}{5\cdot\cancel{3}} = -\frac{2}{5}$$

96.

$$\left(\frac{1}{8}\right)^2 - 2\left(-\frac{3}{4}\right) = \left(\frac{1}{8}\right)\left(\frac{1}{8}\right) - \frac{2}{1}\left(-\frac{3}{4}\right)$$

$$= \frac{1}{64} - \frac{\cancel{2}}{1}\left(-\frac{3}{2\cdot\cancel{2}}\right)$$

$$= \frac{1}{64} - \left(-\frac{3}{2}\right)$$

$$= \frac{1}{64} + \frac{3\cdot32}{2\cdot32}$$

$$= \frac{1}{64} + \frac{96}{64} = \frac{97}{64}$$

97.

$$-\left(-\frac{3}{4}\right)\left(\frac{1}{8}\right) + \left(-2\frac{1}{16}\right) = \frac{3}{32} + \left(-2\frac{1}{16}\right)$$

$$= \frac{3}{32} + \left(-\frac{33}{16}\right)$$

$$= \frac{3}{32} + \left(-\frac{33\cdot2}{16\cdot2}\right)$$

$$= \frac{3}{32} + \left(-\frac{66}{32}\right)$$

$$= -\frac{63}{32}$$

98.

$$-2\frac{1}{16} \div \left(-\frac{3}{4} \cdot \frac{1}{8}\right) = -\frac{33}{16} \div \left(-\frac{3}{32}\right)$$

$$= -\frac{33}{16}\left(-\frac{32}{3}\right)$$

$$= -\frac{11 \cdot \cancel{3}}{\cancel{16}}\left(-\frac{2 \cdot \cancel{16}}{\cancel{3}}\right)$$

$$= 22$$

99.

$$\frac{-\dfrac{3}{4}-\dfrac{1}{8}}{-2\dfrac{1}{16}} = \frac{-\dfrac{3}{4}-\dfrac{1}{8}}{-\dfrac{33}{16}} = \frac{\dfrac{16}{1}\left(-\dfrac{3}{4}-\dfrac{1}{8}\right)}{\dfrac{16}{1}\left(-\dfrac{33}{16}\right)}$$

$$= \frac{\dfrac{16}{1}\left(-\dfrac{3}{4}\right)-\dfrac{16}{1}\left(\dfrac{1}{8}\right)}{\dfrac{16}{1}\left(-\dfrac{33}{16}\right)}$$

$$= \frac{\dfrac{4 \cdot \cancel{16}}{1}\left(-\dfrac{3}{\cancel{4}}\right)-\dfrac{2 \cdot \cancel{8}}{1}\left(\dfrac{1}{\cancel{8}}\right)}{\dfrac{\cancel{16}}{1}\left(-\dfrac{33}{\cancel{16}}\right)}$$

$$= \frac{-12-2}{-33} = \frac{-12+(-2)}{-33} = \frac{-14}{-33}$$

$$= \frac{14}{33}$$

Section 4.8 Solving Equations Containing Fractions

100.

$$\frac{2}{3}x = 16$$

$$\frac{3}{2}\left(\frac{2}{3}x\right) = \frac{3}{2}\left(\frac{16}{1}\right)$$

$$x = \frac{3}{\cancel{2}}\left(\frac{8 \cdot \cancel{2}}{1}\right) = 24$$

Check:

$$\frac{2}{3}(24) = 16$$

$$\frac{2}{\cancel{3}}\left(\frac{8 \cdot \cancel{3}}{1}\right) = 16$$

$$16 = 16$$

101.

$$-\frac{7s}{4} = -49$$

$$-\frac{4}{7}\left(-\frac{7}{4}s\right) = -\frac{4}{7}\left(-\frac{49}{1}\right)$$

$$s = -\frac{4}{\cancel{7}}\left(-\frac{7 \cdot \cancel{7}}{1}\right) = 28$$

Check:

$$-\frac{7(28)}{4} = -49$$

$$-\frac{196}{4} = -49$$

$$-49 = -49$$

102.

$$\frac{y}{5} = -\frac{1}{15}$$

$$\frac{5}{1}\left(\frac{y}{5}\right) = \frac{5}{1}\left(-\frac{1}{15}\right)$$

$$y = \frac{\cancel{5}}{1}\left(-\frac{1}{3 \bullet \cancel{5}}\right) = -\frac{1}{3}$$

Check:

$$\frac{-\dfrac{1}{3}}{5} = -\frac{1}{15}$$

$$\frac{\dfrac{3}{1}\left(-\dfrac{1}{3}\right)}{3(5)} = -\frac{1}{15}$$

$$\frac{-1}{15} = -\frac{1}{15}$$

103.

$$2x - 3 = 8$$

$$2x - 3 + 3 = 8 + 3$$

$$2x = 11$$

$$\frac{2x}{2} = \frac{11}{2}$$

$$x = \frac{11}{2}$$

Check:

$$2\left(\frac{11}{2}\right) - 3 = 8$$

$$\frac{\cancel{2}}{1}\left(\frac{11}{\cancel{2}}\right) - 3 = 8$$

$$11 - 3 = 8$$

$$8 = 8$$

104.

$$\frac{c}{3} - \frac{c}{8} = 2$$

$$\frac{24}{1}\left(\frac{c}{3} - \frac{c}{8}\right) = 24(2)$$

$$8c - 3c = 48$$

$$5c = 48$$

$$\frac{5c}{5} = \frac{48}{5}$$

$$c = \frac{48}{5}$$

105.

$$\frac{5h}{9} - 1 = -3$$

$$\frac{9}{1}\left(\frac{5h}{9} - 1\right) = 9(-3)$$

$$5h - 9 = -27$$

$$5h - 9 + 9 = -27 + 9$$

$$5h = -18$$

$$\frac{5h}{5} = \frac{-18}{5}$$

$$h = -\frac{18}{5}$$

106.

$$4 - \frac{d}{4} = 0$$

$$\frac{4}{1}\left(4 - \frac{d}{4}\right) = 4(0)$$

$$16 - d = 0$$

$$16 - d - 16 = 0 - 16$$

$$-d = -16$$

$$\frac{-d}{-1} = \frac{-16}{-1}$$

$$d = 16$$

107.

$$\frac{t}{10} - \frac{2}{3} = \frac{1}{5}$$

$$\frac{30}{1}\left(\frac{t}{10} - \frac{2}{3}\right) = \frac{30}{1}\left(\frac{1}{5}\right)$$

$$3t - 20 = 6$$

$$3t - 20 + 20 = 6 + 20$$

$$3t = 26$$

$$\frac{3t}{3} = \frac{26}{3}$$

$$t = \frac{26}{3}$$

108. TEXTBOOKS

p = pages in the book

$$\frac{2}{3}p = 220$$

$$\frac{3}{1}\left(\frac{2}{3}p\right) = 3(220)$$

$$2p = 660$$

$$\frac{2p}{2} = \frac{660}{2}$$

$$p = 330$$

There are 330 pages in the book.

Chapter Four Test

1a. $\dfrac{4}{5}$

1b. $\dfrac{1}{5}$

2a. $\dfrac{27}{36} = \dfrac{3 \bullet \cancel{9}}{4 \bullet \cancel{9}} = \dfrac{3}{4}$

2b. $\dfrac{72n^2}{180n} = \dfrac{2 \bullet \cancel{36} \bullet \cancel{n} \bullet n}{5 \bullet \cancel{36} \bullet \cancel{n}} = \dfrac{2n}{5}$

3. $-\dfrac{3x}{4}\left(\dfrac{1}{5x^2}\right) = -\dfrac{3 \bullet \cancel{x}}{4}\left(\dfrac{1}{5 \bullet x \bullet \cancel{x}}\right)$

$= -\dfrac{3}{20x}$

4. COFFEE DRINKERS

c = morning coffee drinkers

$c = \dfrac{2}{5}(100)$

$c = \dfrac{2}{\cancel{5}}\left(\dfrac{20 \bullet \cancel{5}}{1}\right) = 40$

40 people of the 100 questioned started their morning with coffee.

5. $\dfrac{4a}{3} \div \dfrac{a^2}{9} = \dfrac{4a}{3} \bullet \dfrac{9}{a^2}$

$= \dfrac{4 \bullet \cancel{a}}{\cancel{3}} \bullet \dfrac{3 \bullet \cancel{3}}{a \bullet \cancel{a}} = \dfrac{12}{a}$

6. $\dfrac{x}{6} - \dfrac{4}{5} = \dfrac{x \bullet 5}{6 \bullet 5} - \dfrac{4 \bullet 6}{5 \bullet 6}$

$= \dfrac{5x}{30} - \dfrac{24}{30}$

$= \dfrac{5x - 24}{30}$

7. $\dfrac{7}{8} = \dfrac{7 \bullet 3a}{8 \bullet 3a} = \dfrac{21a}{24a}$

8.

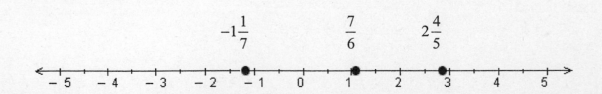

9. SPORTS CONTRACTS

$y =$ yearly salary

$$y = 13\frac{1}{2} \div 9$$

$$y = \frac{27}{2} \cdot \frac{1}{9}$$

$$y = \frac{3 \cdot \cancel{9}}{2} \cdot \frac{1}{\cancel{9}} = \frac{3}{2}$$

$$y = 1\frac{1}{2}$$

The salary is $\$1\frac{1}{2}$ million.

10.
$$-2\left(-2\frac{1}{12}\right)\left(\frac{2}{5}\right)^2 = -\frac{2}{1}\left(-\frac{25}{12}\right)\left(\left[\frac{2}{5}\right]\left[\frac{2}{5}\right]\right)$$

$$= -\frac{2}{1}\left(-\frac{25}{12}\right)\left(\frac{4}{25}\right)$$

$$= -\frac{2}{1}\left(-\frac{\cancel{25}}{3 \cdot \cancel{4}}\right)\left(\frac{\cancel{4}}{\cancel{25}}\right)$$

$$= \frac{2}{3}$$

11.
$$157\frac{5}{9} = 157\frac{5 \cdot 4}{9 \cdot 4} = 157\frac{20}{36}$$

$$+103\frac{3}{4} = +103\frac{3 \cdot 9}{4 \cdot 9} = +103\frac{27}{36}$$

$$= 260\frac{47}{36}$$

$$= 260 + 1\frac{11}{36} = 261\frac{11}{36}$$

12.
$$67\frac{1}{4} = 67\frac{1 \cdot 3}{4 \cdot 3} = 67\frac{3}{12}$$

$$-29\frac{5}{6} = -29\frac{5 \cdot 2}{6 \cdot 2} = -29\frac{10}{12}$$

$$= 66\frac{3}{12} + \frac{12}{12} = 66\frac{15}{12}$$

$$= -29\frac{10}{12} \qquad = -29\frac{10}{12}$$

$$= 37\frac{5}{12}$$

13a. 0

13b. $c =$ difference in chests

$$c = 42\frac{1}{4} - 39\frac{1}{2}$$

$$c = \frac{169}{4} - \frac{79}{2}$$

$$c = \frac{169}{4} - \frac{79 \cdot 2}{2 \cdot 2}$$

$$c = \frac{169}{4} - \frac{158}{4}$$

$$c = \frac{11}{4} = 2\frac{3}{4} \text{ in}$$

13c. $w =$ difference in waists

$$w = 31\frac{3}{4} - 28$$

$$w = 3\frac{3}{4} \text{ in}$$

14.
$$-\frac{3}{7}+2=-\frac{3}{7}+\frac{2}{1}=-\frac{3}{7}+\frac{2\bullet 7}{1\bullet 7}$$
$$=-\frac{3}{7}+\frac{14}{7}=\frac{11}{7}$$

15. SEWING

w = width of material

$$w=10\frac{1}{2}+\frac{5}{8}+\frac{5}{8}$$

$$w=10+\frac{1}{2}+\frac{5}{8}+\frac{5}{8}$$

$$w=10+\frac{1\bullet 4}{2\bullet 4}+\frac{5}{8}+\frac{5}{8}$$

$$w=10+\frac{4}{8}+\frac{5}{8}+\frac{5}{8}$$

$$w=10+\frac{14}{8}=10+\frac{7\bullet \cancel{2}}{4\bullet \cancel{2}}$$

$$w=10+\frac{7}{4}=10+1\frac{3}{4}=11\frac{3}{4}$$

The width should be $11\frac{3}{4}$ in.

16. P = perimeter of triangle

$$P=20+10\frac{2}{3}+22\frac{2}{3}$$

$$P=20+10+22+\frac{2}{3}+\frac{2}{3}$$

$$P=52+\frac{4}{3}=52+1\frac{1}{3}$$

$$P=53\frac{1}{3}\text{ in}$$

A = area of triangle

$$A=\frac{1}{2}bh$$

$$A=\frac{1}{2}\left(10\frac{2}{3}\right)(20)$$

$$A=\frac{1}{\cancel{2}}\left(\frac{32}{3}\right)\left(\frac{10\bullet \cancel{2}}{1}\right)$$

$$A=\frac{320}{3}=106\frac{2}{3}\text{ in}^2$$

17. NUTRITION

c = calories in box of tic tacs

$$c = 40\left(1\frac{1}{2}\right)$$

$$c = \frac{40}{1}\left(\frac{3}{2}\right)$$

$$c = \frac{\cancel{2} \bullet 20}{1}\left(\frac{3}{\cancel{2}}\right) = \frac{60}{1}$$

$$c = 60$$

There are 60 calories per box.

18. COOKING

n = number of servings total

$$n = 8 \div \frac{2}{3}$$

$$n = \frac{8}{1} \bullet \frac{3}{2}$$

$$n = \frac{\cancel{2} \bullet 4}{1} \bullet \frac{3}{\cancel{2}} = \frac{12}{1}$$

$$n = 12$$

There are 12 servings per roast.

19.

$$\left(\frac{2}{3} \bullet \frac{5}{16}\right) - \left(-1\frac{3}{5} \div 4\frac{4}{5}\right) = \left(\frac{\cancel{2}}{3} \bullet \frac{5}{8 \bullet \cancel{2}}\right) - \left(-\frac{8}{5} \div \frac{24}{5}\right)$$

$$= \left(\frac{5}{24}\right) - \left(-\frac{8}{5} \bullet \frac{5}{24}\right)$$

$$= \frac{5}{24} - \left(-\frac{\cancel{8}}{\cancel{5}} \bullet \frac{\cancel{5}}{3 \bullet \cancel{8}}\right)$$

$$= \frac{5}{24} - \left(-\frac{1}{3}\right)$$

$$= \frac{5}{24} + \frac{1 \bullet 8}{3 \bullet 8}$$

$$= \frac{5}{24} + \frac{8}{24} = \frac{13}{24}$$

20.

$$\frac{-\dfrac{5}{6}}{\dfrac{7}{8}} = \frac{\dfrac{24}{1}\left(-\dfrac{5}{6}\right)}{\dfrac{24}{1}\left(\dfrac{7}{8}\right)} = \frac{\dfrac{4\bullet\cancel{6}}{1}\left(-\dfrac{5}{\cancel{6}}\right)}{\dfrac{3\bullet\cancel{8}}{1}\left(\dfrac{7}{\cancel{8}}\right)}$$

$$= \frac{-20}{21} = -\frac{20}{21}$$

21.

$$\frac{\dfrac{1}{2}+\dfrac{1}{3}}{-\dfrac{1}{6}-\dfrac{1}{3}} = \frac{\dfrac{6}{1}\left(\dfrac{1}{2}+\dfrac{1}{3}\right)}{\dfrac{6}{1}\left(-\dfrac{1}{6}-\dfrac{1}{3}\right)}$$

$$= \frac{\dfrac{6}{1}\left(\dfrac{1}{2}\right)+\dfrac{6}{1}\left(\dfrac{1}{3}\right)}{\dfrac{6}{1}\left(-\dfrac{1}{6}\right)-\dfrac{6}{1}\left(\dfrac{1}{3}\right)}$$

$$= \frac{\dfrac{3\bullet\cancel{2}}{1}\left(\dfrac{1}{\cancel{2}}\right)+\dfrac{2\bullet\cancel{3}}{1}\left(\dfrac{1}{\cancel{3}}\right)}{\dfrac{\cancel{6}}{1}\left(-\dfrac{1}{\cancel{6}}\right)-\dfrac{2\bullet\cancel{3}}{1}\left(\dfrac{1}{\cancel{3}}\right)}$$

$$= \frac{3+2}{-1-2} = \frac{5}{-1+(-2)} = \frac{5}{-3}$$

$$= -\frac{5}{3}$$

22.

$$\frac{5}{2}(6) = 15$$

$$\frac{5}{\cancel{2}}\left(\frac{\cancel{2}\bullet 3}{1}\right) = 15$$

$$\frac{15}{1} = 15$$

Yes it is a solution because it makes a
true statement when substituted.

23.

$$\frac{x}{3} = 14$$

$$\frac{3}{1}\left(\frac{x}{3}\right) = 3(14)$$

$$x = 42$$

24.

$$-\frac{5}{2}t = 18$$

$$-\frac{2}{5}\left(-\frac{5}{2}t\right) = -\frac{2}{5}\left(\frac{18}{1}\right)$$

$$t = -\frac{36}{5}$$

25.

$$6x - 4 = -3$$

$$6x - 4 + 4 = -3 + 4$$

$$6x = 1$$

$$\frac{6x}{6} = \frac{1}{6}$$

$$x = \frac{1}{6}$$

26.
$$\frac{x}{6} - \frac{2}{3} = \frac{x}{12}$$
$$\frac{12}{1}\left(\frac{x}{6} - \frac{2}{3}\right) = \frac{12}{1}\left(\frac{x}{12}\right)$$
$$2x - 8 = x$$
$$2x - 8 - x = x - x$$
$$x - 8 = 0$$
$$x - 8 + 8 = 0 + 8$$
$$x = 8$$

27. JOB APPLICANTS

If $\frac{3}{4}$ had experience, then $\frac{1}{4}$ had no experience.

x = number that applied

$$\frac{1}{4}x = 36$$
$$\frac{4}{1}\left(\frac{1}{4}x\right) = 4(36)$$
$$x = 144$$

144 people applied.

28. The parts of a fraction are the numerator, fraction bar, and denominator. A fraction represents equal parts of a whole or a division.

29. When any number is multiplied by its reciprocal, the resulting product will always be one.

30a. Simplifying a fraction by dividing the numerator and denominator of a fraction by the same number.

30b. Equivalent fractions $\frac{1}{2} = \frac{2}{4}$

30c. Building a fraction by multiplying the numerator and denominator of a fraction by the same number.

Chapters 1-4 Cumulative Review Exercises

1.   5,434,700

2.   5,430,000

3.   THE STOCK MARKET

Approx 11,555 at 10:30 A.M.

4.   BANKS

The hundred billions place

5. $\begin{array}{r} 4{,}679 \\ +3{,}457 \\ \hline 8{,}136 \end{array}$

6. $\begin{array}{r} 7{,}897 \\ -4{,}378 \\ \hline 3{,}519 \end{array}$

7. $\begin{array}{r} 5{,}345 \\ \times\ \ \ 56 \\ \hline 32{,}070 \\ 267{,}250 \\ \hline 299{,}320 \end{array}$

8. $\begin{array}{r} 991 \\ 35\overline{)34685} = 991 \\ \underline{315} \\ 318 \\ \underline{315} \\ 35 \\ \underline{35} \\ 0 \end{array}$

9. P = perimeter of pool

$P = 2(150) + 2(75)$

$P = 300 + 150$

$P = 450$ ft

10. A = area of the pool

$A = (75)(150)$

$A = 11{,}250$ ft^2

11. $2^2 \bullet 3 \bullet 7$

12. $2 \bullet 3^2 \bullet 5^2$

13.

$$360$$

$$\boxed{2} \quad 180$$

$$9 \quad 20$$

$$\boxed{3} \quad \boxed{3} \, 4 \quad \boxed{5}$$

$$\boxed{2} \quad \boxed{2}$$

$$2^3 \bullet 3^2 \bullet 5$$

14.

$$3600$$

$$\boxed{2} \quad 1800$$

$$9 \quad 200$$

$$\boxed{3} \quad \boxed{3} \, 40 \quad \boxed{5}$$

$$4 \quad 10$$

$$\boxed{2} \quad \boxed{2}\boxed{2} \quad \boxed{5}$$

$$2^4 \bullet 3^2 \bullet 5^2$$

15.
$$6+(-2)(-5) = 6+10$$
$$= 16$$

16.
$$(-2)^3 - 3^3 = -8-27$$
$$= -8+(-27)$$
$$= -35$$

17.
$$\frac{2(-7)+3(2)}{2(-2)} = \frac{-14+6}{-4}$$
$$= \frac{-8}{-4} = 2$$

18.
$$\frac{2(3^2-4^2)}{-2(3)-1} = \frac{2(9-16)}{-6-1} = \frac{2(9+[-16])}{-6+(-1)}$$
$$= \frac{2(-7)}{-7} = \frac{-14}{-7} = 2$$

19. $x+15$

20. $x-8$

21. $4x$

22. $\dfrac{x}{10}$

23.
$$2(4)-1 = 8-1$$
$$= 7$$

24.
$$\frac{9(4)}{2}-4 = \frac{36}{2}-4$$
$$= 18-4 = 14$$

25. $3(4) - (4)^3 = 3(4) - 64$
    $ = 12 - 64$
    $ = 12 + (-64) = -52$

26. $4 + 2(4 - 7) = 4 + 2(4 + [-7])$
    $ = 4 + 2(-3)$
    $ = 4 + (-6)$
    $ = -2$

27. $-3(5x) = -15x$

28. $-4x(-7x) = 28x^2$

29. $-2(3x - 4) = -6x + 8$

30. $-5(3x - 2y + 4) = -15x + 10y - 20$

31. $-3x + 8x = 5x$

32. $4a^2 - (-3a^2) = 4a^2 + 3a^2 = 7a^2$

33. $4x - 3y - 5x + 2y = 4x + (-5x) - y$
    $ = -x - y$

34. $-2(3x - 4) + 2x = -6x + 8 + 2x$
    $ = -4x + 8$

35. $3x + 2 = -13$
    $3x + 2 - 2 = -13 - 2$
    $3x = -13 + (-2)$
    $3x = -15$
    $\dfrac{3x}{3} = \dfrac{-15}{3}$
    $x = -5$

    Check:

    $3(-5) + 2 = -13$
    $-15 + 2 = -13$
    $-13 = -13$

36. $-5z - 7 = 18$
    $-5z - 7 + 7 = 18 + 7$
    $-5z = 25$
    $\dfrac{-5z}{-5} = \dfrac{25}{-5}$
    $z = -5$

    Check:

    $-5(-5) - 7 = 18$
    $25 - 7 = 18$
    $18 = 18$

37.
$$\frac{y}{4} - 1 = -5$$
$$\frac{4}{1}\left(\frac{y}{4} - 1\right) = 4(-5)$$
$$y - 4 = -20$$
$$y - 4 + 4 = -20 + 4$$
$$y = -16$$

Check:

$$\frac{-16}{4} - 1 = -5$$
$$-4 - 1 = -5$$
$$-4 + (-1) = -5$$
$$-5 = -5$$

38.
$$\frac{n}{5} + 1 = 0$$
$$\frac{5}{1}\left(\frac{n}{5} + 1\right) = 5(0)$$
$$n + 5 = 0$$
$$n + 5 - 5 = 0 - 5$$
$$n = -5$$

Check:

$$\frac{-5}{5} + 1 = 0$$
$$-1 + 1 = 0$$
$$0 = 0$$

39.
$$6x - 12 = 2x + 4$$
$$6x - 12 - 2x = 2x + 4 - 2x$$
$$4x - 12 = 4$$
$$4x - 12 + 12 = 4 + 12$$
$$4x = 16$$
$$\frac{4x}{4} = \frac{16}{4}$$
$$x = 4$$

Check:

$$6(4) - 12 = 2(4) + 4$$
$$24 - 12 = 8 + 4$$
$$12 = 12$$

40.
$$3(2y - 8) = -2(y - 4)$$
$$6y - 24 = -2y + 8$$
$$6y - 24 + 2y = -2y + 8 + 2y$$
$$8y - 24 = 8$$
$$8y - 24 + 24 = 8 + 24$$
$$8y = 32$$
$$\frac{8y}{8} = \frac{32}{8}$$
$$y = 4$$

Check:

$$3(2[4] - 8) = -2(4 - 4)$$
$$3(8 - 8) = -2(0)$$
$$3(0) = 0$$
$$0 = 0$$

41. OBSERVATION HOURS

n = number of shifts needed

$$3n + 37 = 100$$
$$3n + 37 - 37 = 100 - 37$$
$$3n = 63$$
$$\frac{3n}{3} = \frac{63}{3}$$
$$n = 21$$

21 shifts are needed to complete.

42. GEOMETRY

w = width

Length is 4 times as long as width,

$4w$ = length

$$P = 2l + 2w$$
$$210 = 2(4w) + 2w$$
$$210 = 8w + 2w$$
$$210 = 10w$$
$$\frac{210}{10} = \frac{10w}{10}$$
$$21 = w \ or \ w = 21$$

width = 21 feet

length = $4w = 4(21) = 84$ feet

43. $\dfrac{21}{28} = \dfrac{3 \cdot \cancel{7}}{4 \cdot \cancel{7}} = \dfrac{3}{4}$

44 $\dfrac{40x^6y^4}{16x^3y^5} = \dfrac{5 \cdot \cancel{8} \cdot \cancel{x} \cdot \cancel{x} \cdot \cancel{x} \cdot x \cdot x \cdot x \cdot \cancel{y} \cdot \cancel{y} \cdot \cancel{y} \cdot \cancel{y}}{2 \cdot \cancel{8} \cdot \cancel{x} \cdot \cancel{x} \cdot \cancel{x} \cdot \cancel{y} \cdot \cancel{y} \cdot \cancel{y} \cdot \cancel{y} \cdot y} = \dfrac{5x^3}{2y}$

45. $\dfrac{6}{5}\left(-\dfrac{2}{3}\right) = \dfrac{2 \cdot \cancel{3}}{5}\left(-\dfrac{2}{\cancel{3}}\right) = -\dfrac{4}{5}$

46. $\dfrac{14p^2}{8} \div \dfrac{7p^3}{2} = \dfrac{14p^2}{8} \cdot \dfrac{2}{7p^3}$

$= \dfrac{\cancel{2} \cdot \cancel{7} \cdot \cancel{p} \cdot \cancel{p}}{\cancel{2} \cdot 2 \cdot \cancel{2}} \cdot \dfrac{\cancel{2}}{\cancel{7} \cdot \cancel{p} \cdot \cancel{p} \cdot p}$

$= \dfrac{1}{2p}$

47.
$$\frac{2}{3} + \frac{3}{4} = \frac{2 \bullet 4}{3 \bullet 4} + \frac{3 \bullet 3}{4 \bullet 3}$$
$$= \frac{8}{12} + \frac{9}{12}$$
$$= \frac{17}{12} = 1\frac{5}{12}$$

48.
$$\frac{4}{m} - \frac{3}{5} = \frac{4 \bullet 5}{m \bullet 5} - \frac{3 \bullet m}{5 \bullet m}$$
$$= \frac{20}{5m} - \frac{3m}{5m}$$
$$= \frac{20 - 3m}{5m}$$

49.
$$\frac{23}{6}$$

50.
$$-\frac{53}{8}$$

51.
$$4\frac{2}{3} + 5\frac{1}{4} = 4 + 5 + \frac{2}{3} + \frac{1}{4}$$
$$= 9 + \frac{2}{3} + \frac{1}{4}$$
$$= 9 + \frac{2 \bullet 4}{3 \bullet 4} + \frac{1 \bullet 3}{4 \bullet 3}$$
$$= 9 + \frac{8}{12} + \frac{3}{12}$$
$$= 9 + \frac{11}{12} = 9\frac{11}{12}$$

52.
$$14\frac{2}{5} - 8\frac{2}{3} = \frac{72}{5} - \frac{26}{3}$$
$$= \frac{72 \bullet 3}{5 \bullet 3} - \frac{26 \bullet 5}{3 \bullet 5}$$
$$= \frac{216}{15} - \frac{130}{15}$$
$$= \frac{86}{15} = 5\frac{11}{15}$$

53. **FIRE HAZARDS**

d = increase in distance

$$d = \frac{3}{4} - \frac{1}{16}$$
$$d = \frac{3 \bullet 4}{4 \bullet 4} - \frac{1}{16}$$
$$d = \frac{12}{16} - \frac{1}{16} = \frac{11}{16}$$

The distance was increased $\frac{11}{16}$ in.

54. **SHAVING**

t = time saved

$$t = \frac{1}{3} \bullet 90$$
$$t = \frac{1}{\cancel{3}} \bullet \frac{\cancel{3} \bullet 30}{1} = \frac{30}{1}$$
$$t = 30$$

He will save 30 seconds by using the lotion, and it will take him $90 - 30 = 60$ seconds to shave.

55. $\left(\dfrac{1}{4}-\dfrac{7}{8}\right)\div\left(-2\dfrac{3}{16}\right)=\left(\dfrac{1\bullet2}{4\bullet2}-\dfrac{7}{8}\right)\div\left(-\dfrac{35}{16}\right)$

$\qquad\qquad\qquad=\left(\dfrac{2}{8}-\dfrac{7}{8}\right)\div\left(-\dfrac{35}{16}\right)$

$\qquad\qquad\qquad=\left(\dfrac{2}{8}+\left[-\dfrac{7}{8}\right]\right)\div\left(-\dfrac{35}{16}\right)$

$\qquad\qquad\qquad=-\dfrac{5}{8}\left(-\dfrac{16}{35}\right)$

$\qquad\qquad\qquad=-\dfrac{\cancel{5}}{\cancel{8}}\left(-\dfrac{2\bullet\cancel{8}}{7\bullet\cancel{5}}\right)$

$\qquad\qquad\qquad=\dfrac{2}{7}$

56. $\dfrac{\dfrac{2}{3}-7}{4\dfrac{5}{6}}=\dfrac{\dfrac{2}{3}-7}{\dfrac{29}{6}}=\dfrac{\dfrac{6}{1}\left(\dfrac{2}{3}-7\right)}{\dfrac{6}{1}\left(\dfrac{29}{6}\right)}$

$\qquad=\dfrac{\dfrac{6}{1}\left(\dfrac{2}{3}\right)-6(7)}{\dfrac{6}{1}\left(\dfrac{29}{6}\right)}$

$\qquad=\dfrac{\dfrac{2\bullet\cancel{3}}{1}\left(\dfrac{2}{\cancel{3}}\right)-6(7)}{\dfrac{\cancel{6}}{1}\left(\dfrac{29}{\cancel{6}}\right)}$

$\qquad=\dfrac{4-42}{29}=\dfrac{4+(-42)}{29}=\dfrac{-38}{29}$

$\qquad=-\dfrac{38}{29}=-1\dfrac{9}{29}$

57.

$x+\dfrac{1}{5}=-\dfrac{14}{15}$

$\dfrac{15}{1}\left(x+\dfrac{1}{5}\right)=\dfrac{15}{1}\left(-\dfrac{14}{15}\right)$

$15x+3=-14$

$15x+3-3=-14-3$

$15x=-17$

$\dfrac{15x}{15}=\dfrac{-17}{15}$

$x=-\dfrac{17}{15}$

Check:

$-\dfrac{17}{15}+\dfrac{1}{5}=-\dfrac{14}{15}$

$-\dfrac{17}{15}+\dfrac{1\bullet3}{5\bullet3}=-\dfrac{14}{15}$

$-\dfrac{17}{15}+\dfrac{3}{15}=-\dfrac{14}{15}$

$-\dfrac{14}{15}=-\dfrac{14}{15}$

58.

$\dfrac{3}{4}x=\dfrac{5}{8}x+\dfrac{1}{2}$

$\dfrac{8}{1}\left(\dfrac{3}{4}x\right)=\dfrac{8}{1}\left(\dfrac{5}{8}x+\dfrac{1}{2}\right)$

$6x=5x+4$

$6x-5x=5x-4-5x$

$x=4$

Check:

$\dfrac{3}{4}\left(\dfrac{4}{1}\right)=\dfrac{5}{8}\left(\dfrac{4}{1}\right)+\dfrac{1}{2}$

$\dfrac{3}{\cancel{4}}\left(\dfrac{\cancel{4}}{1}\right)=\dfrac{5}{2\bullet\cancel{4}}\left(\dfrac{\cancel{4}}{1}\right)+\dfrac{1}{2}$

$3=\dfrac{5}{2}+\dfrac{1}{2}$

$3=\dfrac{6}{2}$

$3=3$

59.
$$\frac{2}{3}x = -10$$

$$\frac{3}{2}\left(\frac{2}{3}x\right) = \frac{3}{2}\left(-\frac{10}{1}\right)$$

$$x = \frac{3}{\cancel{2}}\left(-\frac{5 \bullet \cancel{2}}{1}\right) = -15$$

Check:

$$\frac{2}{3}\left(-\frac{15}{1}\right) = -10$$

$$\frac{2}{\cancel{3}}\left(-\frac{5 \bullet \cancel{3}}{1}\right) = -10$$

$$-10 = -10$$

60.
$$3y - 8 = 0$$

$$3y - 8 + 8 = 0 + 8$$

$$3y = 8$$

$$\frac{3y}{3} = \frac{8}{3}$$

$$y = \frac{8}{3}$$

Check:

$$3\left(\frac{8}{3}\right) - 8 = 0$$

$$\frac{\cancel{3}}{1}\left(\frac{8}{\cancel{3}}\right) - 8 = 0$$

$$8 - 8 = 0$$

$$0 = 0$$

61. An expression is a combination of numbers and/or variables with operation symbols. An equation contains an = symbol.

62. A variable is a letter that is used to stand for a number.

Chapter 5 Decimals

5.1 An Introduction to Decimals

Vocabulary

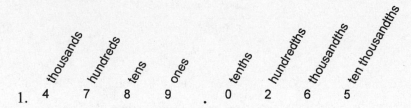

1.   4   7   8   9   .   0   2   6   5

3.   We can approximate a decimal number using the process called **rounding**.

Concepts

5a.   thirty-two and four hundred fifteen
      thousandths

5b.   32

5c. $\dfrac{415}{1,000}$

5d. $30 + 2 + \dfrac{4}{10} + \dfrac{1}{100} + \dfrac{5}{1,000}$

7.

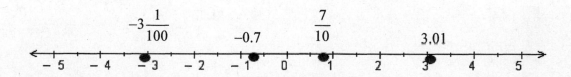

9a.   true

9b.   false

9c.   true

9d.   true

11. $\dfrac{47}{100}$, 0.47

13.

Notation

15.   9,816.0245

Practice

17.   fifty and one tenth;

$$50\frac{1}{10}$$

19.   negative one hundred thirty-seven ten-thousandths;

$$-\frac{137}{10,000}$$

21.   three hundred four and three ten-thousandths;

$$304\frac{3}{10,000}$$

23.   negative seventy-two and four hundred ninety-three thousandths;

$$-72\frac{493}{1,000}$$

25.   −0.39

27.   6.187

29.   506.1

31.   2.7

33.   −0.14

35.   33.00

37.   3.142

39.   1.414

41.   39

43.   2,988

45a. $3,090

45b. $3,090.30

47. $-23.45 < -23.1$

49. $-.065 > -.066$

51.   132.64, 132.6401, 132.6499

Applications

53.   WRITING CHECKS

$1,025.78

55.   INJECTIONS

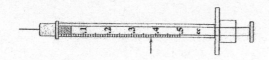

57a.   METRIC SYSTEM

0.30

57b.   1,609.34

57c.   453.59

57d.   3.79

59.   GEOLOGY

Sample	Location	Size (in.)	Classification
A	riverbank	0.009	sand
B	pond	0.0007	silt
C	NE corner	0.095	granule
D	dry lake	0.00003	clay

61.   AIR QUALITY

Texas City, Houston, Westport, Galveston, White Plains, Crestline

63.   THE OLYMPICS

Gold: Patterson; Silver: Khorkina; Bronze: Zhang

65.   E-COMMERCE

Largest gain was in 1Q, 2004; $0.25.

Largest loss was in 2Q, 2002; $0.25.

Writing

67.   A tenth is a fractional part of one, while ten is ten ones.

69.   Fractions and decimals are both a way to show parts of a whole.

71.   The notation is showing that the prices of the gas are really $1.799, $1.899, and $1.999

per gallon.

Review

73.
$$75\frac{3}{4}+88\frac{4}{5}=75+88+\frac{3}{4}+\frac{4}{5}$$
$$=163+\frac{3\bullet5}{4\bullet5}+\frac{4\bullet4}{5\bullet4}$$
$$=163+\frac{15}{20}+\frac{16}{20}$$
$$=163+\frac{31}{20}$$
$$=163+1\frac{11}{20}=164\frac{11}{20}$$

75.
$$5R-3(6-R)=5R-18+3R$$
$$=8R-18$$

77.
$$A=\frac{1}{2}bh$$
$$A=\frac{1}{2}(16)(9)$$
$$A=\frac{1}{\cancel{2}}\left(\frac{8\bullet\cancel{2}}{1}\right)\left(\frac{9}{1}\right)=72\text{ in}^2$$

79.
$$-2+(-3)+4=-5+4$$
$$=-1$$

5.2 Adding and Subtracting Decimals

Vocabulary

1.   The answer to an addition problem is called the **sum**.

Concepts

3.   To subtract signed decimals, add the **opposite** of the decimal that is being subtracted.

5.   Every whole number has an unwritten decimal **point** to its right.

Practice

7.
```
   32.5
 +  7.4
   39.9
```

9.
```
   21.6
 + 33.12
   54.72
```

11.
```
   12
 +  3.9
   15.9
```

13.
```
   0.03034
 + 0.2003
   0.23064
```

15.
```
   247.9
    40
 +   0.56
   288.46
```

17.
```
   45
    9.9
    0.12
 +  3.02
   58.04
```

19.
```
   12.98
 -  3.45
    9.53
```

21.
```
   78.1
 -  7.81
   70.29
```

23. $$\begin{array}{r} 5 \\ -\ 0.023 \\ \hline 4.977 \end{array}$$

25. $$\begin{array}{r} 24 \\ -\ 23.81 \\ \hline 0.19 \end{array}$$

27. $-45.6 + 34.7 = -10.9$

29. $46.09 + (-7.8) = 38.29$

31. $-7.8 + (-6.5) = -14.3$

33. $-0.0045 + (-0.031) = -0.0355$

35. $\begin{aligned} -9.5 - 7.1 &= -9.5 + (-7.1) \\ &= -16.6 \end{aligned}$

37. $\begin{aligned} 30.03 - (-17.88) &= 30.03 + 17.88 \\ &= 47.91 \end{aligned}$

39. $\begin{aligned} -2.002 - (-4.6) &= -2.002 + 4.6 \\ &= 2.598 \end{aligned}$

41. $\begin{aligned} -7 - (-18.01) &= -7 + 18.01 \\ &= 11.01 \end{aligned}$

43. $\begin{aligned} 3.4 - 6.6 + 7.3 &= 3.4 + (-6.6) + 7.3 \\ &= -3.2 + 7.3 \\ &= 4.1 \end{aligned}$

45. $\begin{aligned} (-9.1 - 6.05) - (-51) &= (-9.1 + [-6.05]) + 51 \\ &= -15.15 + 51 \\ &= 35.85 \end{aligned}$

47. $\begin{aligned} 16 - (67.2 + 6.27) &= 16 - (67.2 + 6.27) \\ &= 16 - 73.47 \\ &= 16 + (-73.47) \\ &= -57.47 \end{aligned}$

49. $(-7.2+6.3)-(-3.1-4) = -.9-(-3.1+[-4])$
$$= -.9-(-7.1)$$
$$= -.9+7.1$$
$$= 6.2$$

51. $|-14.1+6.9|+8 = |-7.2|+8$
$$= 7.2+8$$
$$= 15.2$$

53. $2.43+5.6 = 8.03$

Applications

55a. SPORTS PAGES

t = time for Italian team
$t = 53.03+0.014$
$t = 53.044$

The time was 53.044 sec.

55b.

p = 2nd place points
$p = 102.71-0.33$
$p = 102.38$

Second place had 102.38 points.

57. VEHICLE SPECIFICATIONS

w = wheelbase of car
$w = 187.8-43.5-40.9$
$w = 144.3-40.9$
$w = 103.4$

The wheelbase is 103.4 in.

59. BARAMETRIC PRESSURES

d = difference in pressures
$d = 30.7-28.9$
$d = 1.8$

The difference is 1.8. You would expect the weather to be fair around Texas and Oklahoma.

61. OFFSHORE DRILLING

	Pipe underwater (mi)	Pipe underground (mi)	Total pipe (mi)
Design1	1.74	2.32	4.06
Design2	2.90	0	2.90

63.   RECORD HOLDERS

   t = time difference

   $t = 53.52 - 10.49$

   $t = 43.03$

   She ran it 43.03 sec faster.

65.   DEPOSIT SLIPS

   subtotal $= 242.50 + 116.10 + 47.93 + 359.16$

   subtotal $= \$765.69$

   total deposit $= 765.69 - 25$

   total deposit $= \$740.69$

67a.   THE HOME SHOPPING NETWORK   67b.

   price difference $= 149.79 - 47.85$ total price $= 47.85 + 7.95$

   price difference $= \$101.94$ total price $= \$55.80$

69. $2,367.909 + 5,789.0253 = 8,156.9343$ 71. $9,000.09 - 7,067.445 = 1,932.645$

73. $3,434.768 - (908 - 2.3 + .0098)$

   $= 3,434.768 - 905.7098$

   $= 2,529.0582$

Writing

75.   To make sure that the correct place values are being added together.

77.   The number 37 is not lined up correctly.

Review

79.
$$44\frac{3}{8} + 66\frac{1}{5} = 44 + 66 + \frac{3}{8} + \frac{1}{5}$$
$$= 110 + \frac{3 \cdot 5}{8 \cdot 5} + \frac{1 \cdot 8}{5 \cdot 8}$$
$$= 110 + \frac{15}{40} + \frac{8}{40}$$
$$= 110 + \frac{23}{40} = 110\frac{23}{40}$$

81.
$$\frac{-15}{26} \cdot 1\frac{4}{9} = -\frac{15}{26} \cdot \frac{13}{9}$$
$$= -\frac{5 \cdot \cancel{3}}{2 \cdot \cancel{13}} \cdot \frac{\cancel{13}}{3 \cdot \cancel{3}}$$
$$= -\frac{5}{6}$$

5.3 Multiplying Decimals

Vocabulary

1. In the multiplication problem $2.89 \cdot 15.7$, the numbers 2.89 and 15.7 are called **factors**. The answer, 45.373, is called the **product**.

Concepts

3. To multiply decimals, multiply them as if they were **whole** numbers. The number of decimal places in the product is the same as the **sum** of the decimal places of the factors.

5a. $\dfrac{3}{10} \cdot \dfrac{7}{100} = \dfrac{21}{1,000}$

5b.
$$\begin{array}{r} .3 \\ \times .07 \\ \hline 0.021 \end{array}$$ They are the same number.

Notation

7.   2.3

Practice

9. $(0.4)(0.2) = 0.08$

11. $(-0.5)(0.3) = -0.15$

13. $(1.4)(0.7) = 0.98$

15. $(0.08)(0.9) = 0.072$

17. $(-5.6)(-2.2) = 12.32$

19. $(-4.9)(0.001) = -0.0049$

21. $(-0.35)(0.24) = -0.084$

23. $(-2.13)(4.05) = -8.6265$

25. $16 \cdot 0.6 = 9.6$

27. $-7(8.1) = -56.7$

29. $0.04(306) = 12.24$

31. $60.61(-0.3) = -18.183$

33.
$$-0.2(0.3)(-0.4) = -0.06(-0.4)$$
$$= 0.024$$

35.
$$5.5(10)(-0.3) = 55(-0.3)$$
$$= -16.5$$

37. $4.2 \bullet 10 = 42$

39. $67.164 \bullet 100 = 6,716.4$

41. $-0.056(10) = -0.56$

43. $1,000(8.05) = 8,050$

45. $0.098(10,000) = 980$

47. $-0.2 \bullet 1,000 = -200$

49.

Decimal	Its square
0.1	0.01
0.2	0.04
0.3	0.09
0.4	0.16
0.5	0.25
0.6	0.36
0.7	0.49
0.8	0.64
0.9	0.81

51.
$$(1.2)^2 = (1.2)(1.2) = 1.44$$

53.
$$(-1.3)^2 = (-1.3)(-1.3) = 1.69$$

55.
$$-4.6(23.4 - 19.6) = -4.6(3.8)$$
$$= -17.48$$

57.
$$(-0.2)^2 + 2(7.1) = (-0.2)(-0.2) + 2(7.1)$$
$$= 0.04 + 14.2$$
$$= 14.24$$

59.
$$(-0.7 - 0.5)(2.4 - 3.1) = (-0.7 + [-0.5])(2.4 + [-3.1])$$
$$= (-1.2)(-0.7)$$
$$= 0.84$$

61.
$$(0.5+0.6)^2(-3.2) = (1.1)^2(-3.2)$$
$$= (1.1 \bullet 1.1)(-3.2)$$
$$= (1.21)(-3.2)$$
$$= -3.872$$

63.
$$|-2.6| \bullet |-7.2| = 2.6 \bullet 7.2$$
$$= 18.72$$

65.
$$\left(|-2.6-6.7|\right)^2 = \left(|-2.6+[-6.7]|\right)^2$$
$$= \left(|-9.3|\right)^2$$
$$= (9.3)^2$$
$$= (9.3)(9.3) = 86.49$$

67.
$$3.14+2(1.2-[-6.7]) = 3.14+2(1.2+6.7)$$
$$= 3.14+2(7.9)$$
$$= 3.14+15.8$$
$$= 18.94$$

69.
$$-0.4+0.5(100)(-0.4)^2 = -0.4+0.5(100)(0.16)$$
$$= -0.4+50(0.16)$$
$$= -0.4+8$$
$$= 7.6$$

71.
$$10\left|(-1.1)^2-(2.2)^2\right| = 10|1.21-4.84|$$
$$= 10|1.21+(-4.84)|$$
$$= 10|-3.63|$$
$$= 10(3.63)$$
$$= 36.3$$

Applications

73a.   CONCERT SEATING

Ticket type	Price	Number sold	Receipts
Floor	$12.50	1,000	$12,500
Balcony	$15.75	100	$1,575

73b.

$$\text{total} = 12{,}500 + 1{,}575$$
$$\text{total} = \$14{,}075$$

75.   STORM DAMAGE

d = distance the house fell

$$d = 0.57 + 2(0.09)$$
$$d = 0.57 + 0.18$$
$$d = 0.75$$

The house fell 0.75 in total.

77.   WEIGHTLIFING

w = weight on barbell

$$w = 2(45.5 + 20.5 + 2.2)$$
$$w = 2(68.2)$$
$$w = 136.4$$

There are 136.4 lb on the barbell.

79.   BAKERY SUPPLIES

Type of nut	Price per pound	Pounds	Cost
Almonds	$5.95	16	$95.20
Walnuts	$4.95	25	$123.75
Peanuts	$3.85	x	$3.85x$

81.   POOL CONSTRUCTION

P = perimeter of pool

$$P = 2(30.3) + 2(50)$$
$$P = 60.6 + 100$$
$$P = 160.6$$

160.6 m of coping will be needed.

83.   BIOLOGY

$$34\text{Å} = 34(0.0000000004) = 0.000000136 \text{ in}$$

$$10\text{Å} = 10(0.0000000004) = 0.000000004 \text{ in}$$

$$3.4\text{Å} = 3.4(0.0000000004) = 0.0000000136 \text{ in}$$

85. $(-9.0089 + 10.0087)(15.3)$
$= (.9998)(15.3)$
$= 15.29694$

87. $(18.18 + 6.61)^2 + (5 - 9.09)^2$
$= (24.79)^2 + (-4.09)^2$
$= 614.5441 + 16.7281$
$= 631.2722$

89.   ELECTRIC BILLS

b = total bill
$b = 719(0.14277)$
$b = 102.65163$

The bill would be $102.65.

Writing

91.   Count the number of decimals involved in the multiplication, then count over that number to place the decimal.

93.   A decimal place is a place value for a digit in a number.

Review

95.
$$\frac{x}{2} - \frac{x}{3} = -2$$
$$\frac{6}{1}\left(\frac{x}{2} - \frac{x}{3}\right) = 6(-2)$$
$$3x - 2x = -12$$
$$x = -12$$

97.   the absolute value of negative three

99. $-\dfrac{8}{8} = -1$

5.4 Dividing Decimals

Vocabulary

1. In the division $2.5\overline{)4.075} = 1.63$, the decimal 4.075 is called the **dividend**, the decimal 2.5 is

    the **divisor**, and 1.63 is the **quotient**.

3. The **mean** of a list of values is the sum of those values divided by the number of values. The

    median of a list of values is the middle value.

5. The **range** of a list of values is the difference between the largest and smallest values.

Concepts

7. To divide by a decimal, move the decimal point of the divisor so that it becomes a **whole**

    number. The decimal point of the dividend is then moved the same number of places to the

    right. The decimal point in the quotient is written directly **above** the decimal point of the

    dividend.

9.   true   11.   10

13.   Check if $0.9 \bullet 2.13 = 1.917$ 15.   yes, it is correct.

17.
$$\text{mean} = \frac{2.3 + 2.3 + 3.3 + 3.8 + 4.5}{5}$$
median $= 3.6$ mode $= 2.3$ range $= 4.5 - 2.3$

range $= 2.2$

$$\text{mean} = \frac{16.2}{5} = 3.3$$

Notation

19.   moving the decimal point 2 places to the right in the divisor and the dividend

Practice

21.
$$
\begin{array}{r}
4.5 \\
8\overline{)36.0} \\
\underline{32} \\
40 \\
\underline{40} \\
0
\end{array}
$$

23. $-39 \div 4 = -9.75$

25. $49.6 \div 8 = 6.2$

27.
$$
\begin{array}{r}
32.1 \\
9\overline{)288.9} \\
\underline{27} \\
18 \\
\underline{18} \\
09 \\
\underline{9} \\
0
\end{array}
$$

29. $(-14.76) \div (-6) = 2.46$

31. $\dfrac{-55.02}{7} = -7.86$

33.
$$
\begin{array}{r}
2.66 \\
45\overline{)119.70} \\
\underline{90} \\
297 \\
\underline{270} \\
270 \\
\underline{270} \\
0
\end{array}
$$

35. $250.95 \div 35 = 7.17$

37.
$$41.6 \div 0.32 = 32\overline{)4160} \begin{array}{r} 130 \\ \end{array}$$

$$\begin{array}{r} \underline{32} \\ 96 \\ \underline{96} \\ 00 \\ \underline{0} \\ 0 \end{array}$$

39. $(-199.5) \div (-0.19) = 1,050$

41. $\dfrac{0.0102}{0.017} = \dfrac{102}{170} = \dfrac{6 \bullet \cancel{17}}{10 \bullet \cancel{17}} = \dfrac{6}{10} = 0.6$

43. $\dfrac{0.0186}{0.031} = \dfrac{186}{310} = \dfrac{6 \bullet 31}{10 \bullet 31} = \dfrac{6}{10} = 0.6$

45.
$$3\overline{)16.00} \approx 5.3 \begin{array}{r} 5.33 \\ \end{array}$$

$$\begin{array}{r} \underline{15} \\ 10 \\ \underline{9} \\ 10 \\ \underline{9} \\ 1 \end{array}$$

47. $-5.714 \div 2.4 \approx -2.4$

49. $12.243 \div 0.9 \approx 13.60$

51.
$$0.04\overline{)0.03164} = 4\overline{)3.164} \approx 0.79 \begin{array}{r} 0.791 \\ \end{array}$$

$$\begin{array}{r} \underline{28} \\ 36 \\ \underline{36} \\ 04 \\ \underline{4} \\ 0 \end{array}$$

53. $7.895 \div 100 = 0.07895$

55. $0.064 \div (-100) = -0.00064$

57. $1000\overline{)34.8} = 0.0348$

59. $\dfrac{45.04}{10} = 4.504$

61. $\dfrac{-1.2 - 3.4}{3(1.6)} = \dfrac{-1.2 + (-3.4)}{3(1.6)}$

$\qquad\qquad\quad = \dfrac{-4.6}{4.8} \approx -0.96$

63. $\dfrac{40.7(-5.3)}{0.4 - 0.61} = \dfrac{-215.71}{0.4 + (-0.61)}$

$\qquad\qquad\quad = \dfrac{-215.71}{-.21} \approx 1,027.19$

65. $\dfrac{5(48.38 - 32)}{9} = \dfrac{5(16.38)}{9}$

$\qquad\qquad\quad = \dfrac{81.9}{9} = 9.1$

67. $\dfrac{6.7 - (0.3)^2 + 1.6}{(0.3)^3} = \dfrac{6.7 - 0.09 + 1.6}{0.027}$

$\qquad\qquad\qquad = \dfrac{6.61 + 1.6}{0.027}$

$\qquad\qquad\qquad = \dfrac{8.21}{0.027} \approx 304.07$

Applications

69. **BUTCHER SHOPS**

n = number of slices

$n = 14 \div 0.05$

$n = 280$

280 slices would result.

71. **HIKING**

h = hours hiked

$h = 27.5 \div 2.5$

$h = 11$

The hiker walks for 11 more hours, which makes the arrival time 6P.M.

73. **SPRAY BOTTLES**

s = squeezes from bottle

$s = 8.5 \div 0.015$

$s = 566.\overline{6} \approx 567$

There are 567 squeezes per bottle.

75. **HOURLY PAY**

1998 average $= \dfrac{448.50}{34.5} = \$13.00$

2003 average $= \dfrac{517.36}{33.7} \approx \$15.35$

77. OIL WELLS

$$\text{avg} = \frac{0.68 + 0.36 + 0.44}{4}$$

$$\text{avg} = \frac{1.48}{4}$$

$$\text{avg} = 0.37$$

They must drill 0.37 mile per day.

79. OCTUPLETS

$$\text{mean} = \frac{24 + 27 + 28 + 26 + 11.2 + 17.5 + 28.5 + 18}{8}$$

$$\text{mean} = \frac{180.2}{8} = 22.525 \text{ oz}$$

$$\text{median} = 25 \text{ oz}$$

$$\text{range} = 28.5 - 11.2 = 17.3 \text{ oz}$$

81. COMPARISON SHOPPING

$$\text{mean} = \frac{4.29 + 3.98 + 3.89 + 4.19 + 4.29 + 4.19 + 4.09 + 4.39 + 4.24 + 3.97 + 3.99 + 4.29}{12}$$

$$\text{mean} = \frac{49.8}{12} = \$4.15$$

$$\text{median} = \$4.19$$

$$\text{mode} = \$4.29$$

$$\text{range} = 4.39 - 3.89 = \$0.50$$

83.
$$\frac{8.6 + 7.99 + (4.05)^2}{4.56} = \frac{8.6 + 7.99 + 16.4025}{4.56}$$

$$= \frac{32.9925}{4.56} \approx 7.24$$

85.
$$\left(\frac{45.9098}{-234.12}\right)^2 - 4 = (-.1961)^2 - 4$$

$$= .0385 - 4$$

$$= .0385 + (-4) \approx -3.96$$

Writing

87. Move the decimal point in the divisor all the way to the right and move the decimal the same number of places to the right in the dividend. Perform long division as you would normally making sure to transfer the decimal in the dividend to answer on top of the long division bar.

89. Equivalent means that the same outcome will occur from both divisions.

Review

91. $$\dfrac{\dfrac{7}{8}}{\dfrac{3}{4}} = \dfrac{\dfrac{8}{1}\left(\dfrac{7}{8}\right)}{\dfrac{8}{1}\left(\dfrac{3}{4}\right)} = \dfrac{\dfrac{\cancel{8}}{1}\left(\dfrac{7}{\cancel{8}}\right)}{\dfrac{2\bullet\cancel{4}}{1}\left(\dfrac{3}{\cancel{4}}\right)} = \dfrac{7}{6}$$

93. $\{\ldots,-3,-2,-1,0,1,2,3,\ldots\}$

95. $$-\dfrac{3}{4}A = -9$$

$$\left(-\dfrac{4}{3}\right)\left(-\dfrac{3}{4}A\right) = \left(-\dfrac{4}{3}\right)\left(-\dfrac{9}{1}\right)$$

$$A = \left(-\dfrac{4}{\cancel{3}}\right)\left(-\dfrac{3\bullet\cancel{3}}{1}\right) = 12$$

97. $5x - 6(x-1) - (-x) = 5x - 6x + 6 + x$

$$= 6$$

Estimation

1. The deluxe model is approximately $240 more expensive.

3. The economy model has approximately 2 cubic feet less storage.

5. She could purchase approximately 30 standard models.

7. Each would have to pay approximately $330.

9. There would be approximately $520 left to finance.

11. Answer is not reasonable.

13. Answer is reasonable.

15. Answer is reasonable.

17. Answer is not reasonable.

5.5 Fractions and Decimals

Vocabulary

1. The decimal form of the fraction $\frac{1}{3}$ is a **repeating** decimal, which is written $0.\overline{3}$ or $0.333. . . .$

3. The **decimal** equivalent of $\frac{1}{16}$ is 0.0625.

Concepts

5a. $7 \div 8$

5b.   To write a fraction as a decimal, divide the **numerator** of the fraction by the denominator.

7.   smaller

9.

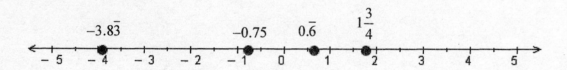

11a.   false   11b.   true

11c.   true   11d.   false

Notation

13a.   no   13b.   It is a repeating decimal.

Practice

15.
$$2\overline{)1.0}^{.5} \quad \text{so} \quad \frac{1}{2} = 0.5$$
$$\underline{10}$$
$$0$$

17.
$$8\overline{)5.000}^{.625} \quad \text{so} \quad -\frac{5}{8} = -0.625$$
$$\underline{48}$$
$$20$$
$$\underline{16}$$
$$40$$
$$\underline{40}$$
$$0$$

19.
$$\frac{9}{16} = 0.5625$$

21.
$$-\frac{17}{32} = -0.53125$$

23.
$$\frac{11}{20} = 0.55$$

25.
$$\frac{31}{40} = 0.775$$

27.
$$-\frac{3}{200} = -0.015$$

29.
$$\frac{1}{500} = 0.002$$

31.
$$\frac{2}{3} = 3\overline{)2.00}^{.66} = 0.\overline{6}$$
$$\underline{18}$$
$$20$$
$$\underline{18}$$
$$2$$

33.
$$\frac{5}{11} = 11\overline{)5.0000}^{.4545} = 0.\overline{45}$$
$$\underline{44}$$
$$60$$
$$\underline{55}$$
$$50$$
$$\underline{44}$$
$$60$$
$$\underline{55}$$
$$5$$

35.
$$-\frac{7}{12} = -0.58\overline{3}$$

37.
$$\frac{1}{30} = 0.0\overline{3}$$

39.
$$\frac{7}{30} = 30\overline{)7.000} \approx 0.23$$

$$\frac{.233}{7.000}$$
$$\underline{60}$$
$$100$$
$$\underline{90}$$
$$100$$
$$\underline{90}$$
$$10$$

41.
$$\frac{17}{45} = 45\overline{)17.000} \approx 0.38$$

$$\frac{.377}{17.000}$$
$$\underline{135}$$
$$350$$
$$\underline{315}$$
$$350$$
$$\underline{315}$$
$$35$$

43.
$$\frac{5}{33} = 33\overline{)5.0000} \approx 0.152$$

$$\frac{.1515}{5.0000}$$
$$\underline{33}$$
$$170$$
$$\underline{165}$$
$$50$$
$$\underline{33}$$
$$170$$
$$\underline{165}$$
$$5$$

45.
$$\frac{10}{27} = 27\overline{)10.0000} \approx 0.370$$

$$\frac{.3703}{10.0000}$$
$$\underline{81}$$
$$190$$
$$\underline{189}$$
$$10$$
$$\underline{0}$$
$$100$$
$$\underline{81}$$
$$19$$

47. $\dfrac{4}{3} \approx 1.33$

49. $-\dfrac{34}{11} \approx -3.09$

51. $3\dfrac{3}{4} = 3.75$

53. $-8\dfrac{2}{3} \approx -8.67$

55. $12\dfrac{11}{16} = 12.6875$

57. $203\dfrac{11}{15} \approx 203.73$

59. $\dfrac{7}{8} < 0.895$

61. $-\dfrac{11}{20} < -0.\overline{4}$

63.
$$\frac{1}{9} + 0.3 = \frac{1}{9} + \frac{3}{10}$$
$$= \frac{1 \bullet 10}{9 \bullet 10} + \frac{3 \bullet 9}{10 \bullet 9}$$
$$= \frac{10}{90} + \frac{27}{90}$$
$$= \frac{37}{90}$$

65.
$$0.9 - \frac{7}{12} = \frac{9}{10} - \frac{7}{12}$$
$$= \frac{9 \bullet 6}{10 \bullet 6} - \frac{7 \bullet 5}{12 \bullet 5}$$
$$= \frac{54}{60} - \frac{35}{60}$$
$$= \frac{19}{60}$$

67.
$$\frac{5}{11}(0.3) = \frac{5}{11}\left(\frac{3}{10}\right)$$
$$= \frac{\cancel{5}}{11}\left(\frac{3}{\cancel{5} \bullet 2}\right) = \frac{3}{22}$$

69.
$$\frac{1}{3}\left(-\frac{1}{15}\right)(0.5) = \frac{1}{3}\left(-\frac{1}{15}\right)\left(\frac{5}{10}\right)$$
$$= \frac{1}{3}\left(-\frac{1}{3 \bullet \cancel{5}}\right)\left(\frac{\cancel{5}}{10}\right) = -\frac{1}{90}$$

71.
$$\frac{1}{4}(0.25) + \frac{15}{16} = \frac{1}{4}\left(\frac{1}{4}\right) + \frac{15}{16}$$
$$= \frac{1}{16} + \frac{15}{16}$$
$$= \frac{16}{16} = 1$$

73.
$$0.24 + \frac{1}{3} = 0.24 + 0.33$$
$$= 0.57$$

75.
$$5.69 - \frac{5}{12} = 5.69 - 0.42$$
$$= 5.27$$

77.
$$\frac{3}{4}(0.43) - \frac{1}{12} = 0.75(0.43) - .08$$
$$= 0.32 - 0.08$$
$$= 0.24$$

79.
$$(3.5 + 6.7)\left(-\frac{1}{4}\right) = (10.2)(-0.25)$$
$$= -2.55$$

81.
$$\left(\frac{1}{5}\right)^2 (1.7) = (0.2)^2 (1.7)$$
$$= (0.04)(1.7)$$
$$= 0.068$$

83.

$$7.5 - (0.78)\left(\frac{1}{2}\right) = 7.5 - (0.78)(0.5)$$
$$= 7.5 - 0.39$$
$$= 7.11$$

85.

$$\frac{3}{8}(-3.2) + (4.5)\left(-\frac{1}{9}\right) = \frac{3}{\cancel{8}}\left(-\frac{\cancel{8} \bullet 0.4}{1}\right) + \left(\frac{\cancel{9} \bullet 0.5}{1}\right)\left(-\frac{1}{\cancel{9}}\right)$$
$$= 3(-0.4) + (0.5)(-1)$$
$$= -1.2 + (-0.5)$$
$$= -1.7$$

87.

$$\frac{3\frac{1}{5} + 2\frac{1}{2}}{5.69} + 3\frac{1}{4} = \frac{3.2 + 2.5}{5.69} + 3.25$$
$$= \frac{5.7}{5.69} + 3.25$$
$$= 1.00 + 3.25$$
$$= 4.25$$

89.

$$\frac{4}{3}(3.14)(3)^3 = \frac{4}{3}(3.14)(27)$$
$$= \frac{4}{\cancel{3}}(3.14)\left(\frac{9 \bullet \cancel{3}}{1}\right)$$
$$= 4(3.14)(9) = 113.04$$

91.

$$\frac{23}{101} = 0.\overline{2277}$$

93.

$$\frac{2,046}{55} = 37.2$$

Applications

95. DRAFTING

$$\frac{1}{16} \text{ in} = 0.0625 \text{ in}$$

$$\frac{6}{16} \text{ in} = 0.375 \text{ in}$$

$$\frac{9}{16} \text{ in} = 0.5625 \text{ in}$$

$$\frac{15}{16} \text{ in} = 0.9375 \text{ in}$$

97. GARDENING

nylon thickness $= 0.065$ in

trimmer thickness $= \dfrac{3}{40} = 0.075$ in

$\dfrac{3}{40}$ in thick brand is thicker.

99.   HORSE RACING

$$:23^2 = 23\frac{2}{5} = 23.4 \text{ sec}$$

$$:23^4 = 23\frac{4}{5} = 23.8 \text{ sec}$$

$$:24^1 = 24\frac{1}{5} = 24.2 \text{ sec}$$

$$:32^3 = 32\frac{3}{5} = 32.6 \text{ sec}$$

101.   WINDOW REPLACEMENTS

$$A = \frac{1}{2}bh$$

$$A = 0.5(6)(5.2)$$

$$A = 15.6 \text{ in}^2 \text{ for each window}$$

$$\text{total area} = 6(15.6) = 93.6 \text{ in}^2$$

Writing

103. Carry out long division on the fraction and round the decimal if necessary.

105. It is incorrect because the least number of repeating digits should always be used.

Review

107.
$$\begin{aligned} -2 + (-3) + 10 + (-6) &= -5 + 10 + (-6) \\ &= 5 + (-6) \\ &= -1 \end{aligned}$$

109.
$$\begin{aligned} 3T - 4T + 2(-4t) &= 3T - 4T + (-8t) \\ &= -1T + (-8T) \\ &= -T - 8t \end{aligned}$$

111. $4x^2 + 2x^2 = 6x^2$

5.6 Solving Equations Containing Decimals

Vocabulary

1. To <u>solve</u> an equation, we isolate the variable on one side of the equals symbol.

3. In the term $5.65t$, the number 5.65 is called the <u>coefficient</u>.

Concepts

5.
$$2.1(1.7) - 6.3 = -2.73$$
$$3.57 - 6.3 = -2.73$$
$$-2.73 = -2.73$$

7. $7.8x + 9.1 + 12.4$

9a. $0.25

9b. $0.01

9c. $2.50

9d. $0.99

11.   the distributive property

Notation

13.
$$0.6s - 2.3 = -1.82$$
$$0.6s - 2.3 + 2.3 = -1.82 + 2.3$$
$$0.6s = 0.48$$
$$\frac{0.6s}{0.6} = \frac{0.48}{0.6}$$
$$s = 0.8$$

Practice

15. $8.7x + 1.4x = 10.1x$

17. $0.05h - 0.03h = 0.02h$

19.
$$3.1r - 5.5r - 1.3r = -2.4 - 1.3r$$
$$= -3.7r$$

21. $3.2 - 8.78x + 9.1 = -8.78x + 12.3$

23. $5.6x - 8.3 - 6.1x + 12.2 = -0.5x + 3.9$

25. $0.05(100 - x) + 0.04x = 5 - 0.05x + 0.04x$

$$= -0.01x + 5$$

27. $x + 8.1 = 9.8$

$x + 8.1 - 8.1 = 9.8 - 8.1$

$x = 1.7$

29. $7.08 = t - 0.03$

$7.08 + 0.03 = t - 0.03 + 0.03$

$7.11 = t$ or $t = 7.11$

31. $-5.6 + h = -17.1$

$-5.6 + h + 5.6 = -17.1 + 5.6$

$h = -11.5$

33. $7.75 = t - (-7.85)$

$7.75 = t + 7.85$

$7.75 - 7.85 = t + 7.85 - 7.85$

$-0.1 = t$ or $t = -0.1$

35. $2x = -8.72$

$\dfrac{2x}{2} = \dfrac{-8.72}{2}$

$x = -4.36$

37. $-3.51 = -2.7x$

$\dfrac{-3.51}{-2.7} = \dfrac{-2.7x}{-2.7}$

$1.3 = x$ or $x = 1.3$

39. $\dfrac{x}{2.04} = -4$

$2.04\left(\dfrac{x}{2.04}\right) = 2.04(-4)$

$x = -8.16$

41. $\dfrac{-x}{5.1} = -4.4$

$5.1\left(\dfrac{-x}{5.1}\right) = 5.1(-4.4)$

$-x = -22.44$

$\dfrac{-x}{-1} = \dfrac{-22.44}{-1}$

$x = 22.44$

43. $\dfrac{1}{3}x = -7.06$

$3\left(\dfrac{1}{3}x\right) = 3(-7.06)$

$x = -21.18$

45. $\dfrac{x}{100} = 0.004$

$100\left(\dfrac{x}{100}\right) = 100(0.004)$

$x = 0.4$

47.
$$2x + 7.8 = 3.4$$
$$2x + 7.8 - 7.8 = 3.4 - 7.8$$
$$2x = -4.4$$
$$\frac{2x}{2} = \frac{-4.4}{2}$$
$$x = -2.2$$

49.
$$-0.8 = 5y + 9.2$$
$$-0.8 - 9.2 = 5y + 9.2 - 9.2$$
$$-10 = 5y$$
$$\frac{-10}{5} = \frac{5y}{5}$$
$$-2 = y \ \text{ or } \ y = -2$$

51.
$$0.3x - 2.1 = 7.2$$
$$0.3x - 2.1 + 2.1 = 7.2 + 2.1$$
$$0.3x = 9.3$$
$$\frac{0.3x}{0.3} = \frac{9.3}{0.3}$$
$$x = 31$$

53.
$$-1.5b + 2.7 = 1.2$$
$$-1.5b + 2.7 - 2.7 = 1.2 - 2.7$$
$$-1.5b = -1.5$$
$$\frac{-1.5b}{-1.5} = \frac{-1.5}{-1.5}$$
$$b = 1$$

55.
$$0.4a - 6 + 0.5a = -5.73$$
$$0.9a - 6 = -5.73$$
$$0.9a - 6 + 6 = -5.73 + 6$$
$$0.9a = 0.27$$
$$\frac{0.9a}{0.9} = \frac{0.27}{0.9}$$
$$a = 0.3$$

57.
$$2(t - 4.3) + 1.2 = -6.2$$
$$2t - 8.6 + 1.2 = -6.2$$
$$2t - 7.4 = -6.2$$
$$2t - 7.4 + 7.4 = -6.2 + 7.4$$
$$2t = 1.2$$
$$\frac{2t}{2} = \frac{1.2}{2}$$
$$t = 0.6$$

59.
$$1.2x - 1.3 = 2.4x + 0.02$$
$$1.2x - 1.3 - 1.2x = 2.4x + 0.02 - 1.2x$$
$$-1.3 = 1.2x + 0.02$$
$$-1.3 - 0.02 = 1.2x + 0.02 - 0.02$$
$$-1.32 = 1.2x$$
$$\frac{-1.32}{1.2} = \frac{1.2x}{1.2}$$
$$-1.1 = x \ \text{ or } \ x = -1.1$$

61.
$$53.7t - 10.1 = 46.3t + 4.7$$
$$53.7t - 10.1 - 46.3t = 46.3t + 4.7 - 46.3t$$
$$7.4t - 10.1 = 4.7$$
$$7.4t - 10.1 + 10.1 = 4.7 + 10.1$$
$$7.4t = 14.8$$
$$\frac{7.4t}{7.4} = \frac{14.8}{7.4}$$
$$t = 2$$

63.
$$2.1x - 4.6 = 7.3x - 11.36$$
$$2.1x - 4.6 - 2.1x = 7.3x - 11.36 - 2.1x$$
$$-4.6 = 5.2x - 11.36$$
$$-4.6 + 11.36 = 5.2x - 11.36 + 11.36$$
$$6.76 = 5.2x$$
$$\frac{6.76}{5.2} = \frac{5.2x}{5.2}$$
$$1.3 = x \quad \text{or} \quad x = 1.3$$

65.
$$0.06x + 0.09(100 - x) = 8.85$$
$$0.06x + 9 - 0.09x = 8.85$$
$$-0.03x + 9 = 8.85$$
$$-0.03x + 9 - 9 = 8.85 - 9$$
$$-0.03x = -0.15$$
$$\frac{-0.03x}{-0.03} = \frac{-0.15}{-0.03}$$
$$x = 5$$

Applications

67.   PETITION DRIVES

Analyze the problem

- Her base pay is <u>15</u> dollars a day.

- She makes <u>30</u> cents for each signature.

- She wants to make <u>60</u> dollars a day.

- Find the number of **signatures** she needs to get.

Form an equation

Let $x =$ <u>**the number of signatures she needs to collect**</u>.

We need to work in terms of the same units, so we write 30 cents as $0.30.

If we multiply the pay per signature by the number of signatures, we get the money she makes just from collecting signatures. Therefore, $0.30x =$ total amount in dollars made from collecting signatures.

Base pay	+	0.30	•	the number of signatures	is	60.
15	+	0.03x			=	60

Solve the equation

$$15 + 0.30x = 60$$
$$15 + 0.30x - 15 = 60 - 15$$
$$0.30x = 45$$
$$\frac{0.30x}{0.30} = \frac{45}{0.30}$$
$$x = 150$$

State the conclusion <u>**She needs to collect 150 signatures to make \$60**</u>.

Check the result

If she collects 150 signatures, she will make $0.30 \bullet 150 = 45$ dollars from the signatures. If we add this to \$15, we get \$60. The answer checks.

69. **DISASTER RELIEF**

m = money the county should ask for

$$6.8 + 12.5 + m = 27.9$$
$$19.3 + m = 27.9$$
$$19.3 + 3 - 19.3 = 27.9 - 19.3$$
$$m = 8.6$$

They should ask for $8.6 million.

71. **GRADE AVERAGES**

g = GPA at beginning of fall semester

$$g - 0.18 = 3.09$$
$$g - 0.18 + 0.18 = 3.09 + 0.18$$
$$g = 3.27$$

Her GPA was 3.27.

73. **POINTS PER GAME**

a = average as a junior

$$2a = 21.4$$
$$\frac{2a}{2} = \frac{21.4}{2}$$
$$a = 10.7$$

She averaged 10.7 points per game.

75. **FUEL EFFICIENCY**

a = average in 1960

$$a - 0.4 + 1.3 + 3.1 + 0.3 = 16.7$$
$$a + 4.3 = 16.7$$
$$a + 4.3 - 4.3 = 16.7 - 4.3$$
$$a = 12.4$$

The average was 12.4 mpg.

77. **CALLIGRAPHY**

n = number of words

$$0.15n + 20 = 50$$
$$0.15n + 20 - 20 = 50 - 20$$
$$0.15n = 30$$
$$\frac{0.15n}{0.15} = \frac{30}{0.15}$$
$$n = 200$$

They can get 200 words.

Writing

79. Answers will vary.

Review

81. $$-\frac{2}{3}+\frac{3}{4}=-\frac{2\bullet4}{3\bullet4}+\frac{3\bullet3}{4\bullet3}$$
$$=-\frac{8}{12}+\frac{9}{12}$$
$$=\frac{1}{12}$$

83. $$\left(-\frac{1}{2}\right)^{3}-\left(-1\right)^{3}=-\frac{1}{8}-\left(-1\right)$$
$$=-\frac{1}{8}+1$$
$$=-\frac{1}{8}+\frac{8}{8}=\frac{7}{8}$$

85. $$\frac{-3-3}{-3+4}=\frac{-3+\left(-3\right)}{1}=\frac{-6}{1}=-6$$

5.7 Square Roots

Vocabulary

1. When we find what number is squared to obtain a given number, we are finding the square **root** of the given number.

3. The symbol $\sqrt{}$ is called a **radical** symbol. It indicates that we are to find a **positive** square root.

5. In $\sqrt{26}$, the number 26 is called the **radicand**.

Concepts

7. The square of 5 is **25**, because $(5)^2 = 25$.

9. $\sqrt{49} = 7$, because $7^2 = 49$.

11. $\sqrt{\dfrac{9}{16}} = \dfrac{3}{4}$, because $\left(\dfrac{3}{4}\right)^2 = \dfrac{9}{16}$.

13. $\sqrt{6}, \sqrt{11}, \sqrt{23}, \sqrt{27}$

15a. $\sqrt{1} = 1$ 15b. $\sqrt{0} = 0$

17a. $\sqrt{6} \approx 2.4$ 17b. $(2.4)^2 = 5.76$ 17c. $6 - 5.76 = 0.24$

19.

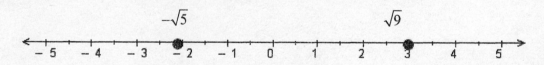

21a.   4 and 5.   21b.   9 and 10

Notation

23. $-\sqrt{49} + \sqrt{64} = -7 + 8$
$$= 1$$

Practice

25. $\sqrt{16} = 4$

27. $-\sqrt{121} = -11$

29. $-\sqrt{0.49} = -0.7$

31. $\sqrt{0.25} = 0.5$

33. $\sqrt{0.09} = 0.3$

35. $-\sqrt{\dfrac{1}{81}} = -\dfrac{1}{9}$

37. $-\sqrt{\dfrac{16}{9}} = -\dfrac{4}{3}$

39. $\sqrt{\dfrac{4}{25}} = \dfrac{2}{5}$

41. $5\sqrt{36} + 1 = 5(6) + 1$
$= 30 + 1 = 31$

43. $-4\sqrt{36} + 2\sqrt{4} = -4(6) + 2(2)$
$= -24 + 4 = -20$

45. $\sqrt{\dfrac{1}{16}} - \sqrt{\dfrac{9}{25}} = \dfrac{1}{4} - \dfrac{3}{5}$
$= \dfrac{1 \bullet 5}{4 \bullet 5} - \dfrac{3 \bullet 4}{5 \bullet 4}$
$= \dfrac{5}{20} - \dfrac{12}{20} = -\dfrac{7}{20}$

47. $5\left(\sqrt{49}\right)(-2) = 5(7)(-2)$
$= -70$

49. $\sqrt{0.04} + 2.36 = 0.2 + 2.36$
$= 2.56$

51. $-3\sqrt{1.44} = -3(1.2)$
$= -3.6$

53.

Number	Square root
1	1
2	1.414
3	1.732
4	2
5	2.236
6	2.449
7	2.646
8	2.828
9	3
10	3.162

55. $\sqrt{1,369} = 37$

57. $\sqrt{3,721} = 61$

59. $\sqrt{15} \approx 3.87$

61. $\sqrt{66} \approx 8.12$

63. $\sqrt{24.05} \approx 4.904$

65. $-\sqrt{11.1} \approx -3.332$

67. $\sqrt{24,000,201} = 4,899$

69. $-\sqrt{0.00111} = -0.0333$

Applications

71a. CARPENTRY

$\sqrt{25} = 5$ ft

71b.

$\sqrt{100} = 10$ ft

73. BASEBALL

$\sqrt{16,200} \approx 127.3$ ft

75. BIG-SCREEN TELEVISIONS

$\sqrt{1,764} = 42$ in

Writing

77.   The student mistook taking the square root of a number with taking half of the number.

79.   A nonterminating decimal is a decimal that never ends. Any square root that is not a perfect square of another number forms a nonterminating decimal, like $\sqrt{2}$, $\sqrt{10}$, etc .

81.   There is no number that when squared can give -4, or any negative number.

Review

83.   subtraction and multiplication

85.
$$5(-2)^2 - \frac{16}{4} = 5(4) - \frac{16}{4}$$
$$= 20 - 4$$
$$= 16$$

87. $\{0, 1, 2, 3, 4, 5, \ldots\}$

89.
$$8 + \frac{a}{5} = 14$$
$$5\left(8 + \frac{a}{5}\right) = 5(14)$$
$$40 + a = 70$$
$$40 + a - 40 = 70 - 40$$
$$a = 30$$

The Real Numbers

1.

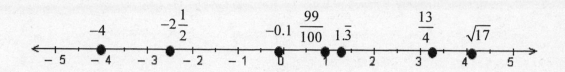

3.   Natural numbers

$\{1, 2, 3, 4, 5, \ldots\}$

5.   Integers

$\{\ldots -3, -2, -1, 0, 1, 2, 3, \ldots\}$

7.   Irrational numbers are the set of all numbers that **canno**t be expressed as fraction, terminating decimal, or repeating decimal.

9.   false

11.   false

13.   true

15.   false

17.   true

19.   There is no real number that, when squared, gives a negative number.

Key Concept Simplify and Solve

1a. $-2x+3+7x-11=5x-8$

1b. $-2x+3+7x-11=7$

$$5x-8=7$$

$$5x-8+8=7+8$$

$$5x=15$$

$$\frac{5x}{5}=\frac{15}{5}$$

$$x=3$$

3a. $3(0.2y-1.6)+0.6y=0.6y-4.8+0.6y$

$$=1.2y-4.8$$

3b. $3(0.2y-1.6)+0.6y=-6.6$

$$0.6y-4.8+0.6y=-6.6$$

$$1.2y-4.8=-6.6$$

$$1.2y-4.8+4.8=-6.6+4.8$$

$$1.2y=-1.8$$

$$\frac{1.2y}{1.2}=-\frac{1.8}{1.2}$$

$$y=-1.5$$

Chapter Five Review

Section 5.1 An Introduction to Decimals

1. $0.67, \dfrac{67}{100}$

2.
    0.8

3. $10 + 6 + \dfrac{4}{10} + \dfrac{5}{100} + \dfrac{2}{1,000} + \dfrac{3}{10,000}$

4. two and three tenths, $2\dfrac{3}{10}$

5. negative fifteen and fifty-nine hundredths, $-15\dfrac{59}{100}$

6. six hundred one ten thousandths, $\dfrac{601}{10,000}$

7. one one-hundred thousandth, $\dfrac{1}{100,000}$

8.
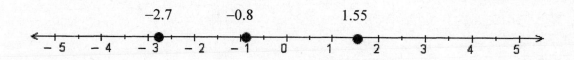

9. VALEDICTORIANS

    Washington, Diaz, Chou, Singh, Gerbac

10. true

11. $4.5 < 4.6$

12. $-2.35 > -2.53$

13. $10.90 = 10.9$

14. $0.027894 < 0.034$

15. 4.58

16. $3,706.090$

17. -0.1

18. 88.1

Section 5.2 Adding and Subtracting Decimals

19. $19.5 + 34.4 + 12.8 = 66.7$

20. $3.4 + 6.78 + 35 + 0.008 = 45.188$

21. $68.47 - 53.3 = 15.17$

22. $45.08 - 17.37 = 27.71$

23. $-16.1 + 8.4 = -7.7$

24. $-4.8 - (-7.9) = -4.8 + 7.9$
$$= 3.1$$

25. $-3.55 + (-1.25) = -4.8$

26. $-15.1 - 13.99 = -15.1 + (-13.99)$
$$= -29.09$$

27. $-8.8 + (-7.3 - 9.5) = -8.8 + (-7.3 + [-9.5])$
$$= -8.8 + (-16.8)$$
$$= -25.6$$

28. $(5 - 0.096) - (-0.035) = 4.904 + 0.035$
$$= 4.939$$

29.   SALE PRICES

s = sale price

$s = 52.20 - 3.99$

$s = 48.21$

The sale price is $48.21.

30.   MICROWAVE OVENS

w = height of window

$w = 13.4 - (2.5 + 2.75)$

$w = 13.4 - 5.25$

$w = 8.15$

The window is 8.15 in high.

Section 5.3 Multiplying Decimals

31. $(-0.6)(0.4) = -0.24$

32. $2.3 \bullet 0.9 = 2.07$

33. $5.5(-3.1) = -17.05$

34. $32.45(6.1) = 197.945$

35. $(-0.003)(-0.02) = 0.00006$

36. $7 \bullet 0.6 = 4.2$

37. $1,000(90.1452) = 90,145.2$

38. $(-10)(-2.897)(100) = 2,897$

39. $(0.2)^2 = 0.04$

40. $(-0.15)^2 = 0.0225$

41. $(3.3)^2 = 10.89$

42. $(0.1)^3 = 0.001$

43. $\begin{aligned}(0.6 + 0.7)^2 - 12.3 &= (1.3)^2 - 12.3 \\ &= 1.69 - 12.3 \\ &= 1.69 + (-12.3) \\ &= -10.61\end{aligned}$

44. $\begin{aligned}3(7.8) + 2(1.1)^2 &= 3(7.8) + 2(1.21) \\ &= 23.4 + 2.42 \\ &= 25.82\end{aligned}$

45. $\begin{aligned}2(3.14)(4)^2 - 8.1 &= 2(3.14)(16) - 8.1 \\ &= 6.28(16) - 8.1 \\ &= 100.48 - 8.1 \\ &= 92.38\end{aligned}$

46. WORD PROCESSORS

width of area $= 8.5 - (0.5 + 0.7)$

width of area $= 8.5 - 1.2$

width of area $= 7.3$ in

length of area $= 11 - (1.0 + 0.6)$

length of area $= 11 - 1.6$

length of area $= 9.4$ in

Area $= (9.4)(7.3)$

Area $= 68.62$ in^2

47. AUTO PAINTING

$t = $ thickness of finish

$t = 0.03 + 3(0.015) - 0.005$

$t = 0.03 + 0.045 - 0.005$

$t = 0.075 - 0.005$

$t = 0.07$ in

Section 5.4 Dividing Decimals

48.
$$
\begin{array}{r}
1.25 \\
12\overline{)15.00} \\
\underline{12} \\
30 \\
\underline{24} \\
60 \\
\underline{60} \\
0
\end{array}
$$

49. $-41.8 \div 4 = -10.45$

50. $\dfrac{-29.67}{-23} = 1.29$

51.
$$24.618 \div 6 = 6\overline{)24.618}$$
$$
\begin{array}{r}
4.103 \\
\underline{24} \\
06 \\
\underline{6} \\
01 \\
\underline{0} \\
18 \\
\underline{18} \\
0
\end{array}
$$

52. $12.47 \div (-4.3) = -2.9$

53. $\dfrac{0.0742}{1.4} = 0.053$

54. $\dfrac{15.75}{0.25} = 63$

55. $\dfrac{-0.03726}{-0.046} = 0.81$

56. $78.98 \div 6.1 \approx 12.9$

57. $\dfrac{-5.338}{0.008} \approx -667.3$

58. $\dfrac{5(68.4 - 32)}{9} = \dfrac{5(36.4)}{9}$

$= \dfrac{182}{9} \approx 20.22$

59. THANKSGIVING DINNER

$c =$ cost per person

$c = \dfrac{41.70}{5}$

$c = 8.34$

The cost per person was $8.34.

60. $89.76 \div 100 = 0.8976$

61. $\dfrac{0.0112}{-10} = -0.00112$

62. $\dfrac{(1.4)^2 + 2(4.6)}{0.5 + 0.3} = \dfrac{1.96 + 2(4.6)}{0.8}$

$= \dfrac{1.96 + 9.2}{0.8}$

$= \dfrac{11.16}{0.8} = 13.95$

63. SERVING SIZE

$s =$ servings per container

$s = \dfrac{15.5}{1.1}$

$s \approx 14.09$

There are just over 14 servings.

64. TELESCOPES

$r =$ number of revolutions needed

$r = \dfrac{0.2375}{0.025}$

$r = 9.5$

9.5 revolutions are required.

65. BLOOD SAMPLES

$\text{mean} = \dfrac{7.8 + 6.9 + 7.9 + 6.7 + 6.8 + 8.0 + 7.2 + 6.9 + 7.5}{9}$

mean $= 7.3$ microns

median $= 7.2$ microns

mode $= 6.9$ microns

range $= 8.0 - 6.7 = 1.3$ microns

66.   TOBACCO SETTLEMENTS

$$\text{median} = \frac{1.4 + 1.5}{2}$$

$$\text{median} = \frac{2.9}{2} = 1.45$$

The median payment was $1.45 billion.

Section 5.5 Fractions and Decimals

67. $\frac{7}{8} = 0.875$

68. $-\frac{2}{5} = -0.4$

69. $\frac{9}{16} = 0.5625$

70. $\frac{3}{50} = 0.06$

71. $\frac{6}{11} = 0.\overline{54}$

72. $-\frac{2}{3} = -0.\overline{6}$

73. $\frac{19}{33} \approx 0.58$

74. $\frac{31}{30} \approx 1.03$

75. $\frac{13}{25} > 0.499$

76. $-0.\overline{26} > -\frac{4}{15}$

77.

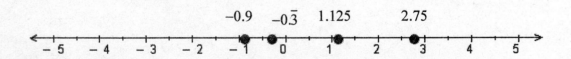

78.

$$\frac{1}{3} + 0.4 = \frac{1}{3} + \frac{4}{10} = \frac{1}{3} + \frac{2 \bullet \cancel{2}}{5 \bullet \cancel{2}}$$

$$= \frac{1}{3} + \frac{2}{5}$$

$$= \frac{1 \bullet 5}{3 \bullet 5} + \frac{2 \bullet 3}{5 \bullet 3}$$

$$= \frac{5}{15} + \frac{6}{15} = \frac{11}{15}$$

79.

$$\frac{4}{5}(-7.8) = 0.8(-7.8)$$

$$= -6.24$$

80.

$$\frac{1}{2}(9.7 + 8.9)(10) = \frac{1}{2}(18.6)\left(\frac{10}{1}\right)$$

$$= \frac{1}{\cancel{2}}(18.6)\left(\frac{5 \bullet \cancel{2}}{1}\right)$$

$$= 93$$

81.

$$\frac{1}{3}(3.14)(3)^2(4.2) = \frac{1}{3}(3.14)(9)(4.2)$$

$$= \frac{1}{\cancel{3}}(3.14)\left(\frac{3 \bullet \cancel{3}}{1}\right)(4.2)$$

$$= 39.564$$

82.

$$\frac{4}{3}(3.14)(2)^3 = \frac{4}{3}(3.14)(8)$$

$$= \frac{100.48}{3}$$

$$\approx 33.49$$

83. ROADSIDE EMERGENCIES

$$A = \frac{1}{2}(6.4)(10.9)$$

$$A = \frac{69.76}{2}$$

$$A = 34.88 \text{ in}^2$$

Section 5.6 Solving Equations Containing Decimals

84.

$$y + 12.4 = -6.01$$

$$y + 12.4 - 12.4 = -6.01 - 12.4$$

$$y = -6.01 + (-12.4)$$

$$y = -18.41$$

85.

$$0.23 + x = 5$$

$$0.23 + x - 0.23 = 5 - 0.23$$

$$x = 4.77$$

86.

$$\frac{x}{1.78} = -3$$

$$1.78\left(\frac{x}{1.78}\right) = 1.78(-3)$$

$$x = -5.34$$

87.

$$-16.1b = -27.37$$

$$\frac{-16.1b}{-16.1} = \frac{-27.37}{-16.1}$$

$$b = 1.7$$

88.
$$-1.3 + 1.2(-1.1) = 2.4(-1.1) + 0.02$$
$$-1.3 + (-1.32) = -2.64 + 0.02$$
$$-2.62 = -2.62$$

yes

89. $5.7a - 12.4 - 2.9a = 2.8a - 12.4$

90.
$$2(0.3t - 0.4) + 3(0.8t - 0.2)$$
$$= 0.6t - 0.8 + 2.4t - 0.6$$
$$= 3t - 1.4$$

91.
$$1.7y + 1.24 = -1.4y - 0.62$$
$$1.7y + 1.24 + 1.4y = -1.4y - 0.62 + 1.4y$$
$$3.1y + 1.24 = -0.62$$
$$3.1y + 1.24 - 1.24 = -0.62 - 1.24$$
$$3.1y = -1.86$$
$$\frac{3.1y}{3.1} = \frac{-1.86}{3.1}$$
$$y = -0.6$$

92.
$$0.05(1,000 - x) + 0.9x = 60.2$$
$$50 - 0.05x + 0.9x = 60.2$$
$$50 + 0.85x = 60.2$$
$$50 + 0.85x - 50 = 60.2 - 50$$
$$0.85x = 10.2$$
$$\frac{0.85x}{0.85} = \frac{10.2}{0.85}$$
$$x = 12$$

93.   BOWLING

g = number of games
$$0.95g + 1.45 = 10$$
$$0.95g + 1.45 - 1.45 = 10 - 1.45$$
$$0.95g = 8.55$$
$$\frac{0.95g}{0.95} = \frac{8.55}{0.95}$$
$$g = 9$$

9 games can be bowled.

Section 5.7 Square Roots

94a.   The symbol $\sqrt{}$ is called a __radical__ symbol.

94b. $\sqrt{64} = 8$, because $8^2 = 64$.

95. $\sqrt{49} = 7$

96. $-\sqrt{16} = -4$

97. $\sqrt{100} = 10$

98. $\sqrt{0.09} = 0.3$

99. $\sqrt{\dfrac{64}{25}} = \dfrac{8}{5}$

100. $\sqrt{0.81} = 0.9$

101. $-\sqrt{\dfrac{1}{36}} = -\dfrac{1}{6}$

102. $\sqrt{0} = 0$

103. 9 and 10

104. $\sqrt{11} \approx 3.3$

$(3.3)^2 = 10.89$

It differs by 0.11.

105.

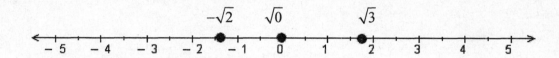

106. $-3\sqrt{100} = -3(10) = -30$

107. $5\sqrt{0.25} = 5(0.5) = 2.5$

108. $-3\sqrt{49} - \sqrt{36} = -3(7) - 6$
$= -21 - 6$
$= -27$

109. $\sqrt{\dfrac{9}{100}} + \sqrt{1.44} = \dfrac{3}{10} + 1.2$
$= 0.3 + 1.2$
$= 1.5$

110. $\sqrt{19} \approx 4.36$

Chapter Five Test

1. $\dfrac{79}{100}$, 0.79

2. WATER PURITY

   Selway, Monroe, Paston, Covington, Cadia

3a. SKATEBOARDING

   sixty-two and fifty-five hundredths;

   $62\dfrac{55}{100}$

3b. MONEY

   eight thousand thirteen one hundred-

   thousandths; $\dfrac{8,013}{100,000}$

4. 33.050

5. SKATING RECEIPTS

   t = total receipts

   $t = 30.25 + 62.25 + 40.50 + 75.75$

   $t = 208.75$

   The total receipts were $208.75.

6a. $567.909 \div 1,000 = 0.567909$

6b. $0.00458 \bullet 100 = 0.458$

7. EARTHQUAKE FAULT LINES

   d = distance the ground dropped

   $d = 0.83 + 0.19$

   $d = 1.02$

   The ground dropped 1.02 in total.

8. $2 + 4.56 + 0.89 + 3.3 = 10.75$

9. $45.2 - 39.079 = 6.121$

10. $\left(0.32\right)^2 = 0.1024$

11. $-6.7\left(-2.1\right) = 14.07$

12. NEW YORK CITY

$A = (2.5)(0.5)$

$A = 1.25 \text{ mi}^2$

13. TELEPHONE BOOKS

t = thickness of paper

$t = \dfrac{2.3}{565}$

$t \approx 0.004$

Each page is about 0.004 in thick.

14. $4.1 - (3.2)(0.4)^2 = 4.1 - (3.2)(0.16)$

$= 4.1 - .512$

$= 3.588$

15a. $\dfrac{17}{50} = 0.34$

15b $\dfrac{5}{12} = 0.41\overline{6}$

16. $\dfrac{12.146}{-5.3} \approx -2.29$

17.

$$
\begin{array}{r}
1.1818 \\
11\overline{)13.0000} \\
\underline{11} \\
20 \\
\underline{11} \\
90 \\
\underline{88} \\
20 \\
\underline{11} \\
90 \\
\underline{88} \\
2
\end{array}
$$

$= 1.\overline{18}$

18.   RATINGS

$$\text{mean} = \frac{4.5 + 4.4 + 3.9 + 3.6 + 3.1 + 3.1 + 2.9}{7}$$

$$\text{mean} = \frac{25.5}{7} \approx 3.6$$

$$\text{median} = 3.6$$

$$\text{mode} = 3.1$$

$$\text{range} = 4.5 - 2.9 = 1.6$$

19.   STATISTICS

Median means that half the families had more debt and half had less debt.

20.

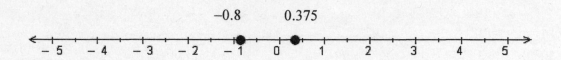

21. $\dfrac{2}{3} + 0.7 = \dfrac{2}{3} + \dfrac{7}{10} = \dfrac{2 \bullet 10}{3 \bullet 10} + \dfrac{7 \bullet 3}{10 \bullet 3}$ 22. $6.18s + 8.9 - 1.22s - 6.6 = 4.96s + 2.3$

$\qquad = \dfrac{20}{30} + \dfrac{21}{30} = \dfrac{41}{30}$

23. $2.1(x - 3) + 3.1(x - 4) = 2.1x - 6.3 + 3.1x - 12.4$

$\qquad\qquad\qquad\qquad\quad = 5.2x - 18.7$

24. $-2.4t = 16.8$ 25. $-0.008 + x = 6$

$\qquad \dfrac{-2.4t}{-2.4} = \dfrac{16.8}{-2.4}$ $-0.008 + x + 0.008 = 6 + 0.008$

$\qquad\qquad t = -7$ $x = 6.008$

26.
$$0.3x - 0.53 = 0.0225 + 1.6x$$
$$0.3x - 0.53 - 0.3x = 0.0225 + 1.6x - 0.3x$$
$$-0.53 = 0.0225 + 1.3x$$
$$-0.53 - 0.0225 = 0.0225 + 1.3x - 0.0225$$
$$-0.5525 = 1.3x$$
$$\frac{-0.5525}{1.3} = \frac{1.3x}{1.3}$$
$$-0.425 = x \text{ or } x = -0.425$$

27. CHEMISTRY

$w =$ missing weight
$$1.86 + 2.09 + w = 4.37$$
$$3.95 + w = 4.37$$
$$3.95 + w - 3.95 = 4.37 - 3.95$$
$$w = 0.42$$

The missing weight is 0.42 g.

28. WEDDING COSTS

$n =$ number of announcements
$$0.95n + 24 = 100$$
$$0.95n + 24 - 24 = 100 - 24$$
$$0.95n = 76$$
$$\frac{0.95n}{0.95} = \frac{76}{0.95}$$
$$n = 80$$

They could get 80 announcements.

29. $\sqrt{144} = 12$, because $12^2 = 144$.

30.

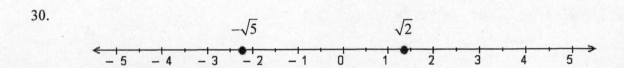

31.
$$-2\sqrt{25} + 3\sqrt{49} = -2(5) + 3(7)$$
$$= -10 + 21$$
$$= 11$$

32.
$$\sqrt{\frac{1}{36}} - \sqrt{\frac{1}{25}} = \frac{1}{6} - \frac{1}{5} = \frac{1 \bullet 5}{6 \bullet 5} - \frac{1 \bullet 6}{5 \bullet 6}$$
$$= \frac{5}{30} - \frac{6}{30} = -\frac{1}{30}$$

33a. $-6.78 > -6.79$

33b. $\dfrac{3}{8} > 0.3$

33c. $\sqrt{\dfrac{16}{81}} > \dfrac{16}{81}$

33d. $0.\overline{45} > 0.45$

34a. $-\sqrt{0.04} = -0.2$

34b. $\sqrt{1.69} = 1.3$

35.   It is smaller because the decimal part of the number, 0.2999 is less than 0.3, so the number must be smaller no matter how many decimal places it has.

36.   A repeating decimal is a repeating number (or numbers) that occurs as the result of division. For instance $\dfrac{1}{3} = 0.33333\ldots$ which can be written with an over bar instead of writing all of the repeating threes as such $0.\overline{3}$.

Chapters 1-5 Cumulative Review Exercises

1. THE EXECUTIVE BRANCH

    We must find the difference for 1 year,

    then multiply it by 4.

    $d = $ difference for 4 years

    $d = 4(400,000 - 203,000)$

    $d = 4(197,000)$

    $d = 788,000$

    The difference is \$788,000.

2. $x + (y + z) = (x + y) + z$

3.
    $$43\overline{)1203} = 27\,R\,42$$
    $$\begin{array}{r} 27 \\ \hline 86 \\ \hline 343 \\ 301 \\ \hline 42 \end{array}$$

4. 1,000 thousands are in 1 million.

5.
    $$220$$
    $$\boxed{2} \quad 110$$
    $$\boxed{2} \quad 55$$
    $$\boxed{11} \quad \boxed{5}$$
    $$2^2 \bullet 5 \bullet 11$$

6. 1, 2, 4, 5, 10, 20

7. {0, 1, 2, 3, 4, 5, . . .}

8. $-8 + (-5) = -13$

9.   Subtraction is the same as **adding** the opposite.

10.

$$(-6)^2 - 2(5 - 4 \bullet 2) = (-6)^2 - 2(5 - 8)$$
$$= (-6)^2 - 2(-3)$$
$$= 36 - 2(-3)$$
$$= 36 - (-6)$$
$$= 36 + 6$$
$$= 42$$

11. $-15 = -5 \bullet 3$

12. $(-1)^5 = -1$

13.

$$8 - 2d = -5 - 5$$
$$8 - 2d = -10$$
$$8 - 2d - 8 = -10 - 8$$
$$-2d = -18$$
$$\frac{-2d}{-2} = \frac{-18}{-2}$$
$$d = 9$$

14.

$$0 = 6 + \frac{c}{-5}$$
$$0 - 6 = 6 + \frac{c}{-5} - 6$$
$$-6 = \frac{c}{-5}$$
$$-5(-6) = -5\left(\frac{c}{-5}\right)$$
$$30 = c \text{ or } c = 30$$

15. $\left|-7(5)\right| = \left|-35\right| = 35$

16.   102

17. $3x$ ft

18.   CHECKING ACCOUNTS

b = balance before deposit
$$b + 995 = -105$$
$$b + 995 - 995 = -105 - 995$$
$$b = -105 + (-995)$$
$$b = -1,100$$

The balance was $\$-1,100$.

19a. $(k + 1)$ in

19b. $(m - 1)$ in

20. $2(4x+5)$

21. 3 terms

22. $3w - 8w = 3w + (-8w)$
$$= -5w$$

23. $-(5x-4) + 6(2x-7) = -3$
$$-5x + 4 + 12x - 42 = -3$$
$$7x - 38 = -3$$
$$7x - 38 + 38 = -3 + 38$$
$$7x = 35$$
$$\frac{7x}{7} = \frac{35}{7}$$
$$x = 5$$

24. $\dfrac{6}{13}$

25. equivalent fractions

26. $\dfrac{90x^2}{126x} = \dfrac{\cancel{18} \bullet 5 \bullet \cancel{x} \bullet x}{\cancel{18} \bullet 7 \bullet \cancel{x}} = \dfrac{5x}{7}$

27. $\dfrac{3}{8} \bullet \dfrac{7}{16} = \dfrac{21}{128}$

28. $-\dfrac{15}{8y} \div \dfrac{10}{y^3} = -\dfrac{15}{8y} \bullet \dfrac{y^3}{10}$
$$= -\dfrac{3 \bullet \cancel{5}}{8 \cancel{y}} \bullet \dfrac{y \bullet y \bullet \cancel{y}}{2 \bullet \cancel{5}} = -\dfrac{3y^2}{16}$$

29. $\dfrac{4}{m} + \dfrac{2}{7} = \dfrac{4 \bullet 7}{m \bullet 7} + \dfrac{2 \bullet m}{7 \bullet m}$
$$= \dfrac{28}{7m} + \dfrac{2m}{7m} = \dfrac{2m+8}{7m}$$

30. $-4\dfrac{1}{4}\left(-4\dfrac{1}{2}\right) = -\dfrac{17}{4}\left(-\dfrac{9}{2}\right)$
$$= \dfrac{153}{8} = 19\dfrac{1}{8}$$

31. $76\dfrac{1}{6} = \quad 76\dfrac{1 \bullet 4}{6 \bullet 4} = \quad 76\dfrac{4}{24}$

$\underline{-49\dfrac{7}{8}} = \underline{-49\dfrac{7 \bullet 3}{8 \bullet 3}} = \quad \underline{-49\dfrac{21}{24}}$

$75\dfrac{4}{24} + \dfrac{24}{24} \ = 75\dfrac{28}{24}$

$\underline{-49\dfrac{21}{24}} \qquad = \underline{-49\dfrac{21}{24}}$

$$= 26\dfrac{7}{24}$$

32.

$$\dfrac{\dfrac{5}{27}}{-\dfrac{5}{9}} = \dfrac{\dfrac{27}{1}\left(\dfrac{5}{27}\right)}{\dfrac{27}{1}\left(-\dfrac{5}{9}\right)} = \dfrac{\dfrac{\cancel{27}}{1}\left(\dfrac{5}{\cancel{27}}\right)}{\dfrac{3\bullet\cancel{9}}{1}\left(-\dfrac{5}{\cancel{9}}\right)}$$

$$= \dfrac{5}{-15} = -\dfrac{1\bullet\cancel{5}}{3\bullet\cancel{5}} = -\dfrac{1}{3}$$

33.

$$\dfrac{2}{3}y = -30$$

$$\dfrac{3}{2}\left(\dfrac{2}{3}y\right) = \dfrac{3}{2}\left(-\dfrac{30}{1}\right)$$

$$y = \dfrac{3}{\cancel{2}}\left(-\dfrac{15\bullet\cancel{2}}{1}\right) = -45$$

34.

$$\dfrac{d}{6} - \dfrac{2}{3} = \dfrac{d}{12}$$

$$12\left(\dfrac{d}{6} - \dfrac{2}{3}\right) = 12\left(\dfrac{d}{12}\right)$$

$$2d - 8 = d$$

$$2d - 8 - d = d - d$$

$$d - 8 = 0$$

$$d - 8 + 8 = 0 + 8$$

$$d = 8$$

35. KITE

Look at the kite as 2 triangles placed side by side. We need to find the area of one triangle, then double that for the total area.

A = total area

$$A = 2\left(\dfrac{1}{2}[21]\left[7\dfrac{1}{2}\right]\right)$$

$$A = 2\left(\dfrac{1}{2}\left[\dfrac{21}{1}\right]\left[\dfrac{15}{2}\right]\right)$$

$$A = 2\left(\dfrac{315}{4}\right)$$

$$A = \dfrac{\cancel{2}}{1}\left(\dfrac{315}{2\bullet\cancel{2}}\right) = \dfrac{315}{2} = 157\dfrac{1}{2}$$

The area is $157\dfrac{1}{2}$ in^2.

36.

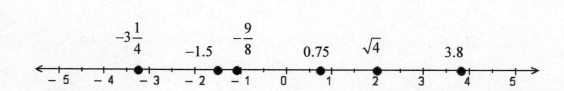

37. GLASS

    0.001 in

38. $356.1978 < 356.22$

39. $-1.8(4.52) = -8.136$

40. $\dfrac{-21.28}{-3.8} = 5.6$

41. $56.012(100) = 5{,}601.2$

42. $\dfrac{0.897}{10{,}000} = 0.0000897$

43. $-9.1 - (-6.05 - 51) = -9.1 - (-6.05 + [-51])$
    $$= -9.1 - (-57.05)$$
    $$= -9.1 + 57.05$$
    $$= 47.95$$

44. WEEKLY SCHEDULES

    t = time spent watching t.v.
    $$48.3 + 34.5 + t + 21 + 3.1 + 27.5 = 168$$
    $$t + 134.4 = 168$$
    $$t + 134.4 - 134.4 = 168 - 134.4$$
    $$t = 33.6$$

    33.6 hours per week were spent watching television.

45. LITERATURE

    $$C = \frac{5(451 - 32)}{9}$$
    $$C = \frac{5(419)}{9}$$
    $$C = \frac{2{,}095}{9} \approx 232.8°C$$

46.   TEAM GPA

$$\text{mean} = \frac{2.20 + 2.25 + 2.56 + 2.75 + 2.75 + 2.99 + 3.02 + 3.04 + 3.23 + 3.87 + 4.00}{12}$$

$$\text{mean} = \frac{36.24}{12} = 3.02$$

$$\text{median} = \frac{2.99 + 3.02}{2} = \frac{6.01}{2} = 3.005$$

$$\text{mode} = 2.75$$

47. $\dfrac{5}{12} = 0.41\overline{6}$

48.
$$0.2t - 3 = 0.7t + 1.5$$
$$0.2t - 3 - 0.2t = 0.7t + 1.5 - 0.2t$$
$$-3 = 0.5t + 1.5$$
$$-3 - 1.5 = 0.5t$$
$$-4.5 = 0.5t$$
$$\frac{-4.5}{0.5} = \frac{0.5t}{0.5}$$
$$-9 = t \ \text{ or } \ t = -9$$

49.   CONCESSIONAIRES

$b = $ bags of nuts she needs to sell
$$0.35b + 22 = 50$$
$$0.35b + 22 - 22 = 50 - 22$$
$$0.35b = 28$$
$$\frac{0.35b}{0.35} = \frac{28}{0.35}$$
$$b = 80$$

She needs to sell 80 bags of nuts.

50. $-4\sqrt{36} + 2\sqrt{81} = -4(6) + 2(9)$
$$= -24 + 18$$
$$= -6$$

Chapter 6: Graphing, Exponents, and Polynomials

Section 6.1 The Rectangular Coordinate System

Vocabulary

1. The pair $(2, 5)$ is called an **ordered** pair.

3. Since the ordered pair $(1, 3)$ satisfies $x + y = 4$, we say that $(1, 3)$ is a **solution** of the equation.

5. The rectangular coordinate system is sometimes called the **Cartesian** coordinate system.

7. The point where the *x*- and *y*- axes cross is called the **origin**.

Concepts

9. BURNING CALORIES

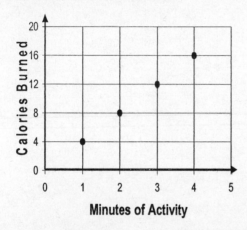

11. $2x + 3y = 14 \quad where \quad x = 2, y = 3$

$$2(2) + 3(3) = 14$$
$$4 + 9 = 14$$
$$13 = 14, \, no$$

13. To plot the point with coordinates $(3, -4)$, we start at the **origin** and move 3 units to the **right** and then move 4 units **down**.

Notation

15. $4x + 3y = 14$ 17.   yes

        $4(2) + 3y = 14$

        $8 + 3y = 14$

        $3y = 6$

        $y = 2$

Practice

19a. If $x = 0$, then $y = 12$.   19b. If $y = 0$, then $x = 4$.   19c. If $x = 2$, then $y = 6$.

21a. $(0, 8)$ 21b. $(4, 0)$ 21c. $(3, 2)$

23.

x	y	(x, y)
0	−5	$(0, -5)$
4	0	$(4, 0)$
8	5	$(8, 5)$

25.

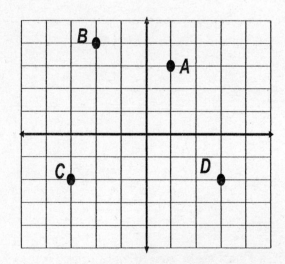

27.

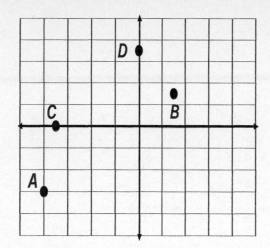

29. *A* (2,4), *B* (-3, 3),   31. *A* (-3, -4), *B* (2.5, 3.5),
     C (-2, -3), *D* (4, -3)   *C* (-2.5, 0), *D* (2.5, 0)

Applications

33.   MAPS

Rockford (5, B), Mount Carroll (1, C), Harvard (7, A), intersection (5, E)

35a.   EARTHQUAKES   35b. no   35c. yes

$(2, -1)$

37.   THE GLOBE

New Delhi, Kampala, Coats Land, Reykjavik, Buenos Aires, Havana

39.   COOKING

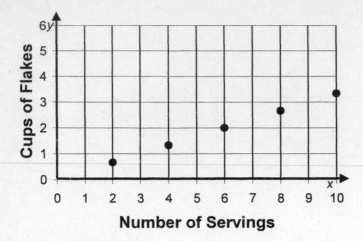

Writing

41.   The *x* and *y* values in each coordinate are different, thus the points must be different.

43.   To plot (4, -3), start at the origin (0, 0), count to the right 4 places on the *x*-axis (right because 4 is positive), then count down 3 places on the *y*-axis (down because 3 is negative) and mark the point.

Review

45. $(-8-5)-3 = (-8+[-5])-3$
$$= -13-3$$
$$= -13+(-3)$$
$$= -16$$

47. $(-4)^2 - 3^2 = 16-9$
$$= 7$$

49.
$$\frac{x}{3} + 3 = 10$$

$$\frac{x}{3} + 3 - 3 = 10 - 3$$

$$\frac{x}{3} = 7$$

$$(3)\frac{x}{3} = 7(3)$$

$$x = 21$$

Check:

$$\frac{21}{3} + 3 = 10$$

$$7 + 3 = 10$$

$$10 = 10$$

51.
$$5 - (7 - x) = -5$$

$$5 - 7 + x = -5$$

$$5 + (-7) + x = -5$$

$$-2 + x = -5$$

$$-2 + 2 + x = -5 + 2$$

$$x = -3$$

Check:

$$5 - (7 - [-3]) = -5$$

$$5 - (7 + 3) = -5$$

$$5 - 10 = -5$$

$$5 + (-10) = -5$$

$$-5 = -5$$

53.
$$\left(4^2\right)^4 = 65{,}536$$

6.2 Graphing Linear Equations

Vocabulary

1.   The graph of a linear equation is a **line**.

3.   The point where the graph of a linear equation crosses the *x*-axis is called the **x-intercept**.

5.   In the equation $y = 7x + 2$, *x* is called the **independent** variable.

Concepts

7.   The graph of the equation $y = 3$ is a **horizontal** line.

9a. $(0, 1)$ 9b. $(-2, 0)$ 9c.   yes

11.   Arrowheads were not drawn on both

ends of the line.

13.   Any six of the following are acceptable, (-5, -5), (-4, -4), (-3, -3), (-2, -2), (-1, -1),

(0, 0), (1, 1), (2, 2), (3, 3), (4, 4), and (5, 5).

Notation

15. $2x - 4y = 8$

$2(3) - 4y = 8$

$6 - 4y = 8$

$-4y = 2$

$y = -\dfrac{1}{2}$

Practice

17.

x	y	(x, y)
5	0	$(5, 0)$
-5	-4	$(-5, -4)$
10	2	$(10, 2)$

19.

x	y	(x, y)
3	3	$(3, 3)$
−4	−11	$(−4, −11)$
6	9	$(6, 9)$

21. *y*-intercept is $(0, 5)$

x-intercept is $(5, 0)$

23. *y*-intercept is $(0, 4)$

x-intercept is $(5, 0)$

25.

x	y
0	5
5	0
2	3

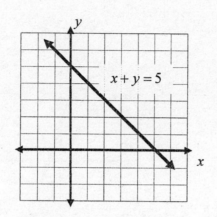

27.

x	y
0	4
5	0
1	$\frac{16}{5}$

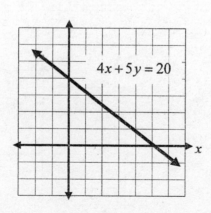

29.

x	y
0	2
−4	0
4	4

$x - 2y = -4$

31.

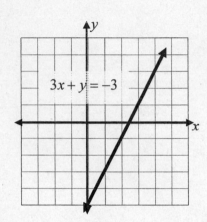

$3x + y = -3$

33.

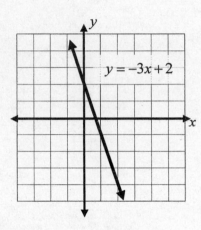

$y = -3x + 2$

35.

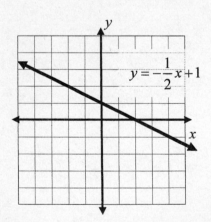

$y = -\dfrac{1}{2}x + 1$

37.

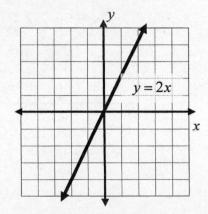

$y = 2x$

39.

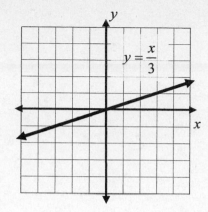

$$y = \frac{x}{3}$$

41.

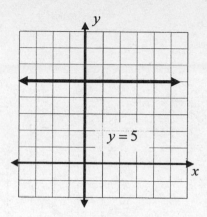

$$y = 5$$

43.

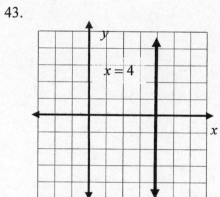

$$x = 4$$

45.

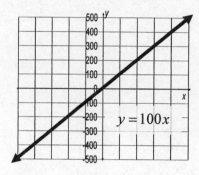

$$y = 100x$$

47.

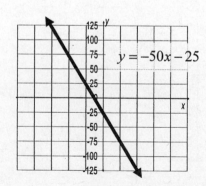

$$y = -50x - 25$$

Applications

49.   HOURLY WAGES

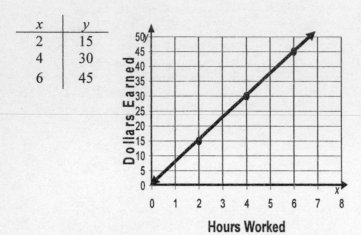

x	y
2	15
4	30
6	45

51.   DISTANCE, RATE AND TIME

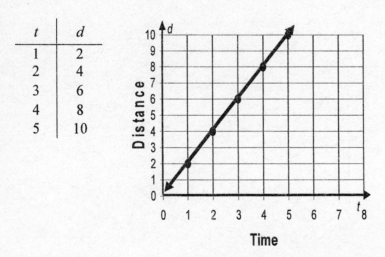

t	d
1	2
2	4
3	6
4	8
5	10

53. AIR TRAFFIC CONROL

Yes, at (2, 2) where the two lines
intersect

Writing

55. It means that when you substitute $x = -2$ and $y = -1$ into the equation $5x - 6y = -4$, you get

a true statement, which further means the point $(-2, -1)$ lies on the graph of the linear

equation $5x - 6y = -4$.

57. Both are correct, each student picked a different set of points, but every point (x, y) on each

student's list is correct.

59. For all *y*-intercepts the *x* part of the point must be zero, and when zero is substituted into

$3x + 6y = 24$ for *x*, you obtain the point (0, 4).

Review

61.

180

⎿⎾

☐2 90

⎿⎾

☐3 30

⎿⎾

☐2 15

⎿⎾

☐3 ☐5

$2^2 \bullet 3^2 \bullet 5$

63. $\dfrac{3(b-a)}{5a+7}$; $a = -2, b = 3$

$= \dfrac{3(3 - [-2])}{5(-2) + 7}$

$= \dfrac{3(3 + 2)}{-10 + 7}$

$= \dfrac{3(5)}{-3}$

$= \dfrac{15}{-3}$

$= -5$

65. LIGHTNING

$.25 = \dfrac{25}{100} = \dfrac{\cancel{25} \bullet 1}{\cancel{25} \bullet 4} = \dfrac{1}{4}$ seconds

6.3 Multiplication Rules for Exponents

Vocabulary

1.   In x^n, x is called the **base** and n is called the **exponent**.

3. $x^m \bullet x^n$ is the product of two exponential expressions with **like** bases.

5. $(2x)^n$ is a **product** raised to a power.

Concepts

7a. x^7

7b. $x^2 y^3$

7c. $3^4 a^2 b^3$

9. $x^4 \bullet x^5 = x^9$ (answers may vary)

11. $\left(5x^4\right)^3 = 125x^{12}$ (answers may vary)

13a. $x^m \bullet x^n = x^{m+n}$

13b. $\left(x^m\right)^n = x^{m \bullet n}$

13c. $(ax)^n = a^n x^n$

15a.   2

15b. -10

15c. x

17a. $x \bullet x = x^2$
$x + x = 2x$

17b. $x \bullet x^2 = x^3$
$x + x^2 = x + x^2$

17c. $x^2 \bullet x^2 = x^4$
$x^2 + x^2 = 2x^2$

19a. $4x \bullet x = 4x^2$
$4x + x = 5x$

19b. $4x \bullet 3x = 12x^2$
$4x + 3x = 7x$

19c. $4x^2 \bullet 3x = 12x^3$
$4x^2 + 3x = 4x^2 + 3x$

21. $3^{2+1} = 3^3 = 27$

Notation

23. $x^5 \bullet x^7 = x^{5+7}$
$\qquad\quad = x^{12}$

25. $\left(2x^4\right)\left(8x^3\right) = (2 \bullet 8)\left(x^4 \bullet x^3\right)$
$\qquad\qquad\qquad = 16x^{4+3}$
$\qquad\qquad\qquad = 16x^7$

27. $x^2 \bullet x^3 = x^{2+3}$

               $= x^5$

29. $x^3 x^7 = x^{3+7}$

             $= x^{10}$

31. $f^5\left(f^8\right) = f^{5+8}$

              $= f^{13}$

33. $n^{24} \bullet n^8 = n^{24+8}$

               $= n^{32}$

35. $t^4 \bullet t^5 \bullet t = t^{4+5+1}$

               $= t^{10}$

37. $x^6\left(x^3\right)x^2 = x^{6+3+2}$

               $= x^{11}$

39. $2^4 \bullet 2^8 = 2^{4+8}$

              $= 2^{12}$

41. $5^6\left(5^2\right) = 5^{6+2}$

               $= 5^8$

43. $2x^2 \bullet 4x = (2 \bullet 4)\left(x^2 \bullet x\right)$

               $= 8x^{2+1}$

               $= 8x^3$

45. $5t \bullet t^9 = (5 \bullet 1)\left(t \bullet t^9\right)$

               $= 5t^{1+9}$

               $= 5t^{10}$

47. $-6x^3\left(4x^2\right) = (-6 \bullet 4)\left(x^3 \bullet x^2\right)$

               $= -24x^{3+2}$

               $= -24x^5$

49. $-x \bullet x^3 = (-1 \bullet 1)\left(x^1 \bullet x^3\right)$

               $= -1x^{1+3}$

               $= -x^4$

51. $6y\left(2y^3\right)3y^4 = (6 \bullet 2 \bullet 3)\left(y^1 \bullet y^3 \bullet y^4\right)$

                     $= 36y^{1+3+4}$

                     $= 36y^8$

53. $-2t^3\left(-4t^2\right)\left(-5t^5\right) = (-2 \bullet -4 \bullet -5)\left(t^3 \bullet t^2 \bullet t^5\right)$

                       $= -40t^{3+2+5}$

                       $= -40t^{10}$

55.
$$xy^2 \bullet x^2 y = \left(x^1 \bullet x^2\right)\left(y^2 \bullet y^1\right)$$
$$= x^{1+2} y^{2+1}$$
$$= x^3 y^3$$

57.
$$b^3 \bullet c^2 \bullet b^5 \bullet c^6 = \left(b^3 \bullet b^5\right)\left(c^2 \bullet c^6\right)$$
$$= b^{3+5} c^{2+6}$$
$$= b^8 c^8$$

59.
$$x^4 y(xy) = \left(x^4 \bullet x^1\right)\left(y^1 \bullet y^1\right)$$
$$= x^{4+1} y^{1+1}$$
$$= x^5 y^2$$

61
$$a^2 b \bullet b^3 a^2 = \left(a^2 \bullet a^2\right)\left(b^1 \bullet b^3\right)$$
$$= a^{2+2} b^{1+3}$$
$$= a^4 b^4$$

63.
$$x^5 y \bullet y^6 = x^5 \left(y^1 \bullet y^6\right)$$
$$= x^5 y^{1+6}$$
$$= x^5 y^7$$

65.
$$3x^2 y^3 \bullet 6xy = (3 \bullet 6)\left(x^2 \bullet x^1\right)\left(y^3 \bullet y^1\right)$$
$$= 18 x^{2+1} y^{3+1}$$
$$= 18 x^3 y^4$$

67.
$$xy^2 \bullet 16 x^3 = 16\left(x^1 \bullet x^3\right) y^2$$
$$= 16 x^{1+3} y^2$$
$$= 16 x^4 y^2$$

69.
$$-6 f^2 t\left(4 f^4 t^3\right) = (-6 \bullet 4)\left(f^2 \bullet f^4\right)\left(t^1 \bullet t^3\right)$$
$$= -24 f^{2+4} t^{1+3}$$
$$= -24 f^6 t^4$$

71.
$$ab \bullet ba \bullet a^2 b$$
$$= \left(a^1 \bullet a^1 \bullet a^2\right)\left(b^1 \bullet b^1 \bullet b^1\right)$$
$$= a^{1+1+2} b^{1+1+1}$$
$$= a^4 b^3$$

73.
$$-4x^2 y\left(-3x^2 y^2\right) = (-4 \bullet -3)\left(x^2 \bullet x^2\right)\left(y^1 \bullet y^2\right)$$
$$= 12 x^{2+2} y^{1+2}$$
$$= 12 x^4 y^3$$

75.
$$\left(x^2\right)^4 = x^{2\cdot4}$$
$$= x^8$$

77.
$$\left(m^{50}\right)^{10} = m^{50\cdot10}$$
$$= m^{500}$$

79.
$$\left(2a\right)^3 = \left(2\right)^3\left(a^1\right)^3$$
$$= 8a^{1\cdot3}$$
$$= 8a^3$$

81.
$$\left(xy\right)^4 = \left(x^1\right)^4\left(y^1\right)^4$$
$$= x^{1\cdot4}y^{1\cdot4}$$
$$= x^4y^4$$

83.
$$\left(3s^2\right)^3 = \left(3\right)^3\left(s^2\right)^3$$
$$= 27s^{2\cdot3}$$
$$= 27s^6$$

85.
$$\left(2s^2t^3\right)^2 = \left(2\right)^2\left(s^2\right)^2\left(t^3\right)^2$$
$$= 4s^{2\cdot2}t^{3\cdot2}$$
$$= 4s^4t^6$$

87.
$$\left(x^2\right)^3\left(x^4\right)^2 = \left(x^{2\cdot3}\right)\left(x^{4\cdot2}\right)$$
$$= x^6 \bullet x^8$$
$$= x^{6+8}$$
$$= x^{14}$$

89.
$$\left(c^5\right)^3 \bullet \left(c^3\right)^5 = \left(c^{5\cdot3}\right)\left(c^{3\cdot5}\right)$$
$$= c^{15} \bullet c^{15}$$
$$= c^{15+15}$$
$$= c^{30}$$

91.
$$\left(2a^4\right)^2\left(3a^3\right)^2 = \left(2\right)^2\left(a^4\right)^2\left(3\right)^2\left(a^3\right)^2$$
$$= \left(4\bullet9\right)\left(a^{4\cdot2}\right)\left(a^{3\cdot2}\right)$$
$$= 36\bullet a^8 \bullet a^6$$
$$= 36a^{8+6}$$
$$= 36a^{14}$$

93.
$$\left(3a^3\right)^3\left(2a^2\right)^3 = \left(3\right)^3\left(a^3\right)^3\left(2\right)^3\left(a^2\right)^3$$
$$= \left(27\bullet8\right)\left(a^{3\cdot3}\right)\left(a^{2\cdot3}\right)$$
$$= 216\bullet a^9 \bullet a^6$$
$$= 216a^{9+6}$$
$$= 216a^{15}$$

95.
$$\left(x^2x^3\right)^{12} = \left(x^{2+3}\right)^{12}$$
$$= \left(x^5\right)^{12}$$
$$= x^{5\cdot12}$$
$$= x^{60}$$

97.
$$\left(2b^4b\right)^5 = \left(2b^{4+1}\right)^5$$
$$= \left(2b^5\right)^5$$
$$= \left(2\right)^5\left(b^5\right)^5$$
$$= 32\bullet b^{5\cdot5}$$
$$= 32b^{25}$$

Writing

99. $x^2 = x \bullet x$, while $2x = 2 \bullet x$.

101. $\left(x^5\right)^7 = x^{5\bullet7} = x^{35}$, or 35 is the product of the powers 5 and 7.(answers may vary).

Review

103.   JEWELERY

$$\frac{18}{24} = \frac{\cancel{6} \bullet 3}{\cancel{6} \bullet 4} = \frac{3}{4}$$

105. $\dfrac{-25}{-5} = 5$

107.
$$2\left(\frac{12}{-3}\right) + 3(5)$$
$$2(-4) + 3(5)$$
$$-8 + 15$$
$$7$$

109. $-x = -12$
$$\frac{-x}{-1} = \frac{-12}{-1}$$
$$x = 12$$

6.4 Introduction to Polynomials

Vocabulary

1. A polynomial with one term, such as $2x$, is called a **monomial**.

3. A polynomial with two terms, such as $x^2 - 25$, is called a **binomial**.

Concepts

5. binomial

7. monomial

9. monomial

11. trinomial

13. 3

15. 2

17. 1

19. 7

Notation

21.
$$
\begin{aligned}
3a^2 + 2a - 7 &= 3(2)^2 + 2(2) - 7 \\
&= 3(4) + 4 - 7 \\
&= 12 + 4 - 7 \\
&= 16 - 7 \\
&= 9
\end{aligned}
$$

Practice

23.
$$
\begin{aligned}
3x + 4 &= 3(3) + 4 \\
&= 9 + 4 \\
&= 13
\end{aligned}
$$

25.
$$
\begin{aligned}
2x^2 + 4 &= 2(-1)^2 + 4 \\
&= 2(1) + 4 \\
&= 2 + 4 \\
&= 6
\end{aligned}
$$

27.
$$
\begin{aligned}
0.5t^3 - 1 &= 0.5(4)^3 - 1 \\
&= 0.5(64) - 1 \\
&= 32 - 1 \\
&= 31
\end{aligned}
$$

29.

$$\frac{2}{3}b^2 - b + 1 = \frac{2}{3}(3)^2 - (3) + 1$$
$$= \frac{2}{3}(9) - 3 + 1$$
$$= 6 - 3 + 1$$
$$= 3 + 1$$
$$= 4$$

31.

$$-2s^2 - 2s + 1 = -2(-1)^2 - 2(-1) + 1$$
$$= -2(1) - 2(-1) + 1$$
$$= -2 + 2 + 1$$
$$= 0 + 1$$
$$= 1$$

33.

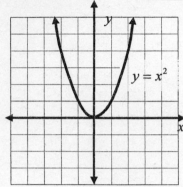

35.

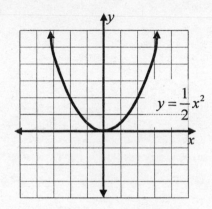

37.

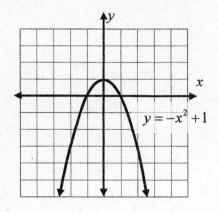

39.

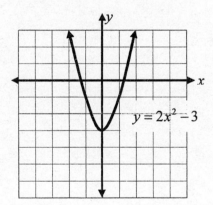

Applications

41.
$$-16t^2 + 64t = -16(0)^2 + 64(0)$$
$$= -16(0) + 64(0)$$
$$= 0 + 0$$
$$= 0\,ft$$

43.
$$-16t^2 + 64t = -16(2)^2 + 64(2)$$
$$= -16(4) + 64(2)$$
$$= -64 + 128$$
$$= 64\,ft$$

45.
$$0.04v^2 + 0.9v = 0.04(30)^2 + 0.9(30)$$
$$= 0.04(900) + 0.9(30)$$
$$= 36 + 27$$
$$= 63\,ft$$

47.
$$0.04v^2 + 0.9v = 0.04(60)^2 + 0.9(60)$$
$$= 0.04(3600) + 0.9(60)$$
$$= 144 + 54$$
$$= 198\,ft$$

49.   BRIDGES

x	0	2	4	-2	-4
y	0	1	4	1	4

Writing

51. Look for the degree of each term and the degree of the polynomial is the largest degree found; in this problem the degree is 5.

53. Substitute in -2, -1, 0, 1, 2 for x and find which values of y you obtain. Each pair forms a coordinate, then graph the coordinates.

Review

55.
$$\frac{2}{3} + \frac{4}{3} = \frac{2+4}{3}$$
$$= \frac{6}{3}$$
$$= 2$$

57.
$$\frac{36}{7} - \frac{23}{7} = \frac{36-23}{7}$$
$$= \frac{13}{7} \ or \ 1\frac{6}{7}$$

59. $$\frac{5}{12} \cdot \frac{18}{5} = \frac{\cancel{5}}{\cancel{6} \cdot 2} \cdot \frac{\cancel{6} \cdot 3}{\cancel{5}}$$

$$= \frac{3}{2} \ or \ 1\frac{1}{2}$$

61. $$x - 4 = 12$$

$$x - 4 + 4 = 12 + 4$$

$$x = 16$$

63. $$2(x - 3) = 6$$

$$2x - 6 = 6$$

$$2x - 6 + 6 = 6 + 6$$

$$2x = 12$$

$$\frac{2x}{2} = \frac{12}{2}$$

$$x = 6$$

6.5 Adding and Subtracting Polynomials

Vocabulary

1. If two algebraic terms have exactly the same variables and exponents, they are called <u>**like**</u> terms.

Concepts

3. To add two monomials, we add the <u>**coefficients**</u> and keep the same <u>**variables**</u> and exponents.

5. yes,

$$3y + 4y = (3+4)y$$
$$= 7y$$

7. no

9. yes,

$$3x^3 + 4x^3 + 6x^3 = (3+4+6)x^3$$
$$= 13x^3$$

11. yes,

$$-5x^2 + 13x^2 + 7x^2 = (-5+13+7)x^2$$
$$= 15x^2$$

Notation

13.
$$(3x^2 + 2x - 5) + (2x^2 - 7x)$$
$$= (3x^2 + 2x^2) + (2x - 7x) + (-5)$$
$$= 5x^2 + (-5x) - 5$$
$$= 5x^2 - 5x - 5$$

Practice

15. $4y + 5y = 9y$

17. $-8t^2 - 4t^2 = -8t^2 + (-4t^2)$
$$= -12t^2$$

19. $3s^2 + 4s^2 + 7s^2 = 14s^2$

21. $(3x+7) + (4x-3)$
$$= (3x+4x) + (7-3)$$
$$= 7x+4$$

23. $\left(2x^2+3\right)+\left(5x^2-10\right)$

$=\left(2x^2+5x^2\right)+(3-10)$

$=\left(2x^2+5x^2\right)+(3+[-10])$

$=7x^2-7$

25. $\left(5x^3-4.2x\right)+\left(7x^3-10.7x\right)$

$=\left(5x^3+7x^3\right)+(-4.2x-10.7x)$

$=\left(5x^3+7x^3\right)+(-4.2x+[-10.7x])$

$=12x^3+(-14.9x)$

$=12x^3-14.9x$

27. $\left(3x^2+2x-4\right)+\left(5x^2-17\right)$

$=\left(3x^2+5x^2\right)+2x+(-4-17)$

$=\left(3x^2+5x^2\right)+2x+(-4+[-17])$

$=8x^2+2x+(-21)$

$=8x^2+2x-21$

29. $\left(7y^2+5y\right)+\left(y^2-y-2\right)$

$=\left(7y^2+1y^2\right)+(5y-1y)+(-2)$

$=8y^2+4y+(-2)$

$=8y^2+4y-2$

31. $\left(3x^2-3x-2\right)+\left(3x^2+4x-3\right)$

$=\left(3x^2+3x^2\right)+(-3x+4x)+(-2-3)$

$=\left(3x^2+3x^2\right)+(-3x+4x)+(-2+[-3])$

$=6x^2+1x+(-5)$

$=6x^2+x-5$

33. $\left(3n^2-5.8n+7\right)+\left(-n^2+5.8n-2\right)$

$=\left(3n^2-1n^2\right)+(-5.8n+5.8n)+(7-2)$

$=2n^2+0n+5$

$=2n^2+5$

35. $3x^2+4x+5$

$+\ \underline{\ 2x^2-3x+6}$

$\ \ \ 5x^2+1x+11$ or $5x^2+x+11$

37. $-3x^2\ \ \ \ \ \ \ -7$

$+\ \underline{\ -4x^2-5x+6}$

$\ \ \ -7x^2-5x-1$

39. $-3x^2+4x+25.4$

$+\ \underline{\ \ \ 5x^2-3x-12.5}$

$\ \ \ 2x^2+1x+12.9$ or $2x^2+x+12.9$

41. $32u^3-16u^3=16u^3$

43. $18x^5-11x^5=7x^5$

45.

$$(4.5a + 3.7) - (2.9a - 4.3)$$
$$= (4.5a + 3.7) + (-2.9a + 4.3)$$
$$= (4.5a - 2.9a) + (3.7 + 4.3)$$
$$= 1.6a + 8$$

47.

$$(-8x^2 - 4) - (11x^2 + 1)$$
$$= (-8x^2 - 4) + (-11x^2 - 1)$$
$$= (-8x^2 - 11x^2) + (-4 - 1)$$
$$= (-8x^2 + [-11x^2]) + (-4 + [-1])$$
$$= -19x^2 + (-5)$$
$$= -19x^2 - 5$$

49.

$$(3x^2 - 2x - 1) - (-4x^2 + 4)$$
$$= (3x^2 - 2x - 1) + (4x^2 - 4)$$
$$= (3x^2 + 4x^2) + (-2x) + (-1 - 4)$$
$$= (3x^2 + 4x^2) + (-2x) + (-1 + [-4])$$
$$= 7x^2 + (-2x) + (-5)$$
$$= 7x^2 - 2x - 5$$

51.

$$(3.7y^2 - 5) - (2y^2 - 3.1y + 4)$$
$$= (3.7y^2 - 5) + (-2y^2 + 3.1y - 4)$$
$$= (3.7y^2 - 2y^2) + (3.1y) + (-5 - 4)$$
$$= (3.7y^2 - 2y^2) + (3.1y) + (-5 + [-4])$$
$$= 1.7y^2 + 3.1y + (-9)$$
$$= 1.7y^2 + 3.1y - 9$$

53.

$$(2b^2 + 3b - 5) - (2b^2 - 4b - 9)$$
$$= (2b^2 + 3b - 5) + (-2b^2 + 4b + 9)$$
$$= (2b^2 - 2b^2) + (3b + 4b) + (-5 + 9)$$
$$= 0b^2 + 7b + 4$$
$$= 7b + 4$$

55.

$$(5p^2 - p + 7.1) - (4p^2 + p + 7.1)$$
$$= (5p^2 - p + 7.1) + (-4p^2 - p - 7.1)$$
$$= (5p^2 - 4p^2) + (-p - p) + (7.1 - 7.1)$$
$$= (5p^2 - 4p^2) + (-1p + [-1p]) + (7.1 - 7.1)$$
$$= 1p^2 + (-2p) + 0$$
$$= p^2 - 2p$$

57.
$$3x^2 + 4x - 5 \qquad 3x^2 + 4x - 5$$
$$\underline{-\left(-2x^2 - 2x + 3\right)} \quad +\underline{\left(2x^2 + 2x - 3\right)}$$
$$5x^2 + 6x - 8$$

59.
$$-2x^2 - 4x + 12 \qquad -2x^2 - 4x + 12$$
$$\underline{-\left(10x^2 + 9x - 24\right)} \quad +\underline{\left(-10x^2 - 9x + 24\right)}$$
$$-12x^2 - 13x + 36$$

61.
$$4x^3 - 3x + 10 \qquad 4x^3 - 3x + 10$$
$$\underline{-\left(5x^3 - 4x - 4\right)} \quad +\underline{\left(-5x^3 + 4x + 4\right)}$$
$$-1x^3 + 1x + 14$$
$$-x^3 + x + 14$$

Applications

63.   VALUE OF A HOUSE
$$y = 700(10) + 85,000$$
$$y = 7,000 + 85,000$$
$$y = 92,000$$
Value is \$92,000

65.   VALUE OF A HOUSE
$$y = 900(12) + 102,000$$
$$y = 10,800 + 102,000$$
$$y = 112,800$$
Value is \$112,800

67a.   VALUE OF TWO HOUSES

$y = 700(15) + 85,000$

$y = 10,500 + 85,000$

$y = 95,500$

$y = 900(15) + 102,000$

$y = 13,500 + 102,000$

$y = 115,500$

Value is $95,500 + 115,500 = \$211,000$

67b. $y = \left(700[15] + 85,000\right) + \left(900[15] + 102,000\right)$

$y = \left(10,500 + 85,000\right) + \left(13,500 + 102,000\right)$

$y = 95,500 + 115,500$

$y = 211,000$

Value is \$211,000

69.   VALUE OF A CAR

$y = -800x + 8,500$ *or*

$y = 8,500 - 800x$

71.   VALUE OF TWO CARS

$y = \left(-800x + 8,500\right) + \left(-1,100x + 10,200\right)$

$y = \left(-800x - 1,100x\right) + \left(8,500 + 10,200\right)$

$y = \left(-800x + [-1,100x]\right) + \left(8,500 + 10,200\right)$

$y = -1,900x + 18,700$

Writing

73.   Like terms have to meet 2 criteria: first, they must be the same variable or variables, second, the exponent on each variable of the two terms must match.

75.   First, rewrite the subtraction as addition by distributing the subtraction sign into the polynomial following it, then add the polynomials.

Review

77.   BASKETBALL SHOES

$14.6 - 13.8 = 0.8$

The Air Garnetts are 0.8oz lighter.

79.   THE PANAMA CANAL

x is the number of feet ship must be lowered

$x + 31 = 85$

$x + 31 - 31 = 85 - 31$

$x = 54$

The ship must be lowered 54 feet.

6.6 Multiplying Polynomials

Vocabulary

1. A polynomial with exactly one term is called a **monomial**.

3. A polynomial with exactly **3** terms is called a trinomial.

Concepts

5. To multiply two monomials, multiply the **numerical** factors and then multiply the variable **factors**.

7. To multiply two binomials, multiply each **term** of one binomial by each term of the other binomial and combine **like** terms.

Notation

9.
$$3x(2x-5) = 3x(2x) - 3x(5)$$
$$= 6x^2 - 15x$$

Practice

11.
$$\left(3x^2\right)\left(4x^3\right)$$
$$= (3 \bullet 4)\left(x^2 \bullet x^3\right)$$
$$= 12x^5$$

13.
$$\left(3b^2\right)(-2b)$$
$$= (3 \bullet -2)\left(b^2 \bullet b^1\right)$$
$$= -6b^3$$

15.
$$\left(-2x^2\right)\left(3x^3\right)$$
$$= (-2 \bullet 3)\left(x^2 \bullet x^3\right)$$
$$= -6x^5$$

17.
$$\left(-\frac{2}{3}y^5\right)\left(\frac{3}{4}y^2\right)$$
$$= \left(-\frac{2}{3} \bullet \frac{3}{4}\right)\left(y^5 \bullet y^2\right)$$
$$= \left(-\frac{\cancel{2}}{\cancel{3}} \bullet \frac{\cancel{3}}{\cancel{2} \bullet 2}\right)\left(y^5 \bullet y^2\right)$$
$$= -\frac{1}{2}y^7$$

19. $3(x+4) = 3x + 12$

21. $-4(t+7) = -4t - 28$

23.
$$3x(x-2)$$
$$=3x(x)-3x(2)$$
$$=3x^2-6x$$

25.
$$-2x^2(3x^2-x)$$
$$=-2x^2(3x^2)-(-2x^2)(x)$$
$$=-6x^4+2x^3$$

27.
$$2x(3x^2+4x-7)$$
$$=2x(3x^2)+2x(4x)-2x(7)$$
$$=6x^3+8x^2-14x$$

29.
$$-p(2p^2-3p+2)$$
$$=-p(2p^2)-(-p)(3p)+(-p)(2)$$
$$=-2p^3-(-3p^2)+(-2p)$$
$$=-2p^3+3p^2-2p$$

31.
$$3q^2(q^2-2q+7)$$
$$=3q^2(q^2)-3q^2(2q)+3q^2(7)$$
$$=3q^4-6q^3+21q^2$$

33.
$$(a+4)(a+5)$$
$$=(a+4)a+(a+4)5$$
$$=a(a+4)+5(a+4)$$
$$=a^2+4a+5a+20$$
$$=a^2+9a+20$$

35.
$$(3x-2)(x+4)$$
$$=(3x-2)x+(3x-2)4$$
$$=x(3x-2)+4(3x-2)$$
$$=3x^2-2x+12x-8$$
$$=3x^2+10x-8$$

37.
$$(2a+4)(3a-5)$$
$$=(2a+4)3a-(2a+4)5$$
$$=3a(2a+4)-5(2a+4)$$
$$=6a^2+12a-10a-20$$
$$=6a^2+2a-20$$

39.
$$(2x+3)^2=(2x+3)(2x+3)$$
$$=(2x+3)2x+(2x+3)3$$
$$=2x(2x+3)+3(2x+3)$$
$$=4x^2+6x+6x+9$$
$$=4x^2+12x+9$$

41.
$$(2x-3)^2=(2x-3)(2x-3)$$
$$=(2x-3)2x-(2x-3)3$$
$$=2x(2x-3)-3(2x-3)$$
$$=4x^2-6x-6x+9$$
$$=4x^2-12x+9$$

43.
$$(5t+1)^2 = (5t+1)(5t+1)$$
$$= (5t+1)5t + (5t+1)1$$
$$= 5t(5t+1) + 1(5t+1)$$
$$= 25t^2 + 5t + 5t + 1$$
$$= 25t^2 + 10t + 1$$

45.
$$(2x+1)(3x^2 - 2x + 1)$$
$$= (2x+1)3x^2 - (2x+1)2x + (2x+1)1$$
$$= 3x^2(2x+1) - 2x(2x+1) + 1(2x+1)$$
$$= 6x^3 + 3x^2 - 4x^2 - 2x + 2x + 1$$
$$= 6x^3 - 1x^2 + 0x + 1$$
$$= 6x^3 - x^2 + 1$$

47.
$$(x-1)(x^2 + x + 1)$$
$$= (x-1)x^2 + (x-1)x + (x-1)1$$
$$= x^2(x-1) + x(x-1) + 1(x-1)$$
$$= x^3 - x^2 + x^2 - x + x - 1$$
$$= x^3 + 0x^2 + 0x - 1$$
$$= x^3 - 1$$

49.
$$(x+2)(x^2 - 3x + 1)$$
$$= (x+2)x^2 - (x+2)3x + (x+2)1$$
$$= x^2(x+2) - 3x(x+2) + 1(x+2)$$
$$= x^3 + 2x^2 - 3x^2 - 6x + 1x + 2$$
$$= x^3 - 1x^2 - 5x + 2$$
$$= x^3 - x^2 - 5x + 2$$

51.
$$\begin{array}{r} 4x+3 \\ \times\ \underline{x+2} \\ 4x^2+3x \\ \underline{+8x+6} \\ 4x^2+11x+6 \end{array}$$

53.
$$\begin{array}{r} 4x-2 \\ \times\ \underline{3x+5} \\ 12x^2-6x \\ \underline{+20x-10} \\ 12x^2+14x-10 \end{array}$$

55.
$$\begin{array}{r} x^2-x+1 \\ \times\ \underline{x+1} \\ x^3-1x^2+1x \\ \underline{+1x^2-1x+1} \\ x^3+0x^2+0x+1 \\ x^3+1 \end{array}$$

Applications

57.   GEOMETRY

$A = (x+2)(x-2)$

$A = (x+2)x - (x+2)2$

$A = x(x+2) - 2(x+2)$

$A = x^2 + 2x - 2x - 4$

$A = x^2 + 0x - 4$

$A = (x^2 - 4) \text{ ft}^2$

59.   ECONOMICS

$R = (\text{price})(\text{number sold})$

$R = \left(-\dfrac{x}{100} + 30\right)x$

$R = x\left(-\dfrac{x}{100} + 30\right)$

$R = x\left(-\dfrac{x}{100}\right) + x(30)$

$R = -\dfrac{x^2}{100} + 30x$

Writing

61.   The FOIL method could be used, or you can distribute the first binomial into the second binomial and then work out the distributions that result.

63. $(x+1)^2 = x^2 + 2x + 1$ when multiplied which is not the same as $x^2 + 1^2$ and thus, they are not equal.

Review

65.   THE EARTH

four and ninety-one thousandths

67.
$$
\begin{array}{r}
0.109375 \\
64\overline{)7.000000} \\
\underline{64} \\
60 \\
\underline{0} \\
600 \\
\underline{576} \\
240 \\
\underline{192} \\
480 \\
\underline{448} \\
320 \\
\underline{320} \\
0
\end{array}
$$

69.
$$
\begin{array}{r}
56.090 \\
78.000 \\
+ 0.567 \\
\hline
134.657
\end{array}
$$

71.
$$
\begin{aligned}
\sqrt{16} + \sqrt{36} &= 4 + 6 \\
&= 10
\end{aligned}
$$

Key Concept Graphing

1.   see graph below

2. $P = (3, 2)$

3.   see graph below

4.   The origin is $(0, 0)$

5.   see graph below for labeling.
quadrant II has a negative *x*-coordinate and a positive *y*-coordinate.

6.   see graph below for points

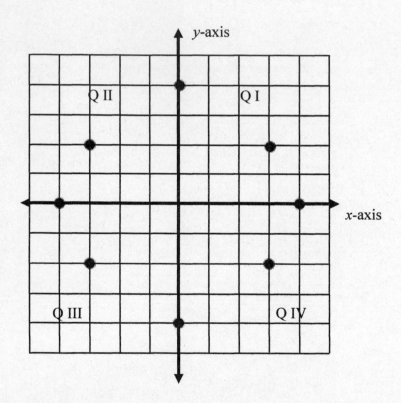

7.

x	y	(x, y)
0	-2	$(0, -2)$
4	0	$(4, 0)$
2	-1	$(2, -1)$

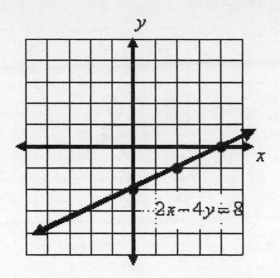

8.

x	y	(x, y)
-2	5	$(-2, 5)$
0	1	$(0, 1)$
2	-3	$(2, -3)$

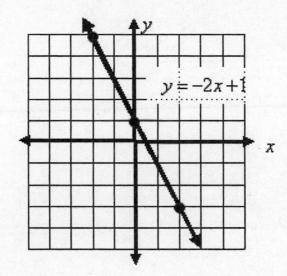

9.

x	y	(x, y)
-2	5	$(-2, 5)$
-1	2	$(-1, 2)$
0	1	$(0, 1)$
1	2	$(1, 2)$
2	5	$(2, 5)$

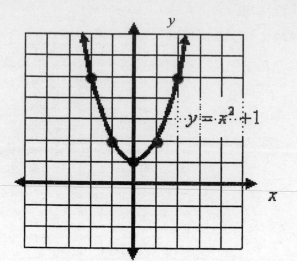

$y = x^2 + 1$

Chapter Six Review

Section 6.1 The Rectangular Coordinate System

1. $2(2)+5(-3)=-11$
    $4+(-15)=-11$
    $-11=-11$

    yes

2. $3(-3)-5(2)=19$
    $-9-10=19$
    $-19 \neq 19$

    no

3. $(0,-3)$ and $(-4,-6)$

4.

x	y	(x,y)
1	-5	$(1,-5)$
3	-11	$(3,-11)$
-2	4	$(-2,4)$

5.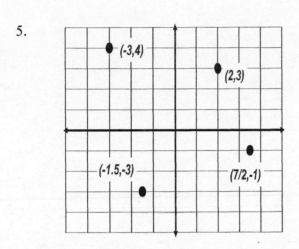

6.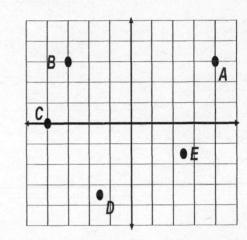

7. Quadrant III

8.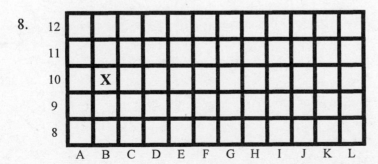

6.2 Graphing Linear Equations

9.

x	y
0	-5
$\frac{5}{3}$	0
1	-2

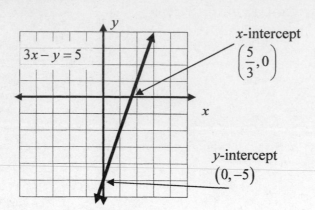

$3x - y = 5$

x-intercept $\left(\frac{5}{3}, 0\right)$

y-intercept $(0, -5)$

10.

x	y
2	0
0	-1
-4	-3

NOTE,
points may
vary but
should still
be on the line

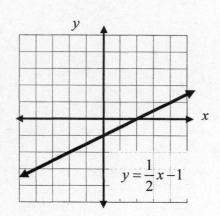

$y = \frac{1}{2}x - 1$

11.

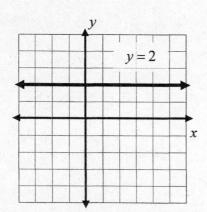

$y = 2$

12.

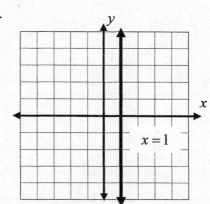

$x = 1$

13.   The line is made up of infinitely many points. Every point on the line is a solution of the equation, and every solution of the equation is on the line.

6.3 Multiplication Rules for Exponents

14. $4h \bullet 4h \bullet 4h$

15. $5^2 d^3 m^4$

16. $h^6 h^4 = h^{6+4}$
$= h^{10}$

17. $t^3 \left(t^5\right) = t^{3+5}$
$= t^8$

18. $w^2 \bullet w \bullet w^4 = w^{2+1+4}$
$= w^7$

19. $4^7 \bullet 4^5 = 4^{7+5}$
$= 4^{12}$

20. $2b^2 \bullet 4b^5 = (2 \bullet 4)\left(b^2 \bullet b^5\right)$
$= 8b^{2+5}$
$= 8b^7$

21. $-6x^3 (4x) = (-6 \bullet 4)\left(x^3 \bullet x^1\right)$
$= -24x^{3+1}$
$= -24x^4$

22. $-2f^2 (-4f)\left(3f^4\right)$
$= (-2 \bullet [-4] \bullet 3)\left(f^2 \bullet f^1 \bullet f^4\right)$
$= 24f^{2+1+4}$
$= 24f^7$

23. $-ab \bullet b \bullet a = (-a \bullet a)(b \bullet b)$
$= -a^{1+1}b^{1+1}$
$= -a^2 b^2$

24. $xy^4 \bullet xy^2 = (x \bullet x)\left(y^4 \bullet y^2\right)$
$= x^{1+1}y^{4+2}$
$= x^2 y^6$

25. $(mn)(mn) = (m \bullet m)(n \bullet n)$
$= m^{1+1}n^{1+1}$
$= m^2 n^2$

26. $3z^3 \bullet 9m^3 z^4 = (3 \bullet 9)\left(m^3\right)\left(z^3 \bullet z^4\right)$
$= 27m^3 z^{3+4}$
$= 27m^3 z^7$

27. $-5cd\left(4c^2 d^5\right) = (-5 \bullet 4)\left(c \bullet c^2\right)\left(d \bullet d^5\right)$
$= -20c^{1+2}d^{1+5}$
$= -20c^3 d^6$

28. $\left(v^3\right)^4 = v^{3 \bullet 4}$
$= v^{12}$

29. $(3y)^3 = (3)^3 \left(y^1\right)^3$
$= 27y^{1 \bullet 3}$
$= 27y^3$

30.
$$\left(5t^4\right)^2 = (5)^2\left(t^4\right)^2$$
$$= 25t^{4\bullet2}$$
$$= 25t^8$$

31.
$$\left(2a^4b^5\right)^3 = (2)^3\left(a^4\right)^3\left(b^5\right)^3$$
$$= 8a^{4\bullet3}b^{5\bullet3}$$
$$= 8a^{12}b^{15}$$

32.
$$\left(c^4\right)^5\left(c^2\right)^3 = \left(c^{4\bullet5}\right)\left(c^{2\bullet3}\right)$$
$$= c^{20}\bullet c^6$$
$$= c^{20+6}$$
$$= c^{26}$$

33.
$$\left(3s^2\right)^3\left(2s^3\right)^2 = (3)^3\left(s^2\right)^3(2)^2\left(s^3\right)^2$$
$$= (27\bullet4)\left(s^{2\bullet3}\right)\left(s^{3\bullet2}\right)$$
$$= 108\bullet s^6 \bullet s^6$$
$$= 108s^{6+6}$$
$$= 108s^{12}$$

34.
$$\left(c^4c^3\right)^2 = \left(c^{4+3}\right)^2$$
$$= \left(c^7\right)^2$$
$$= c^{7\bullet2}$$
$$= c^{14}$$

35.
$$\left(2xx^2\right)^3 = \left(2x^{1+2}\right)^3$$
$$= \left(2x^3\right)^3$$
$$= (2)^3\left(x^3\right)^3$$
$$= 8x^{3\bullet3}$$
$$= 8x^9$$

6.4 Introduction to Polynomials

36. trinomial

37. monomial

38. binomial

39. 3

40. 4

41. 5

42.
$$3x^2 - 2x - 1;\ x = 2$$
$$3(2)^2 - 2(2) - 1 = 3(4) - 2(2) - 1$$
$$= 12 - 4 - 1$$
$$= 7$$

43.
$$2t^2 + t - 2;\ t = -3$$
$$2(-3)^2 + (-3) - 2 = 2(9) + (-3) - 2$$
$$= 18 - 3 - 2$$
$$= 13$$

44.

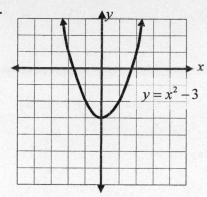

$$y = x^2 - 3$$

45.

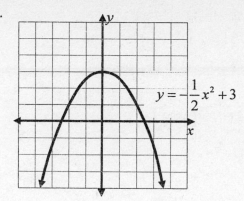

$$y = -\frac{1}{2}x^2 + 3$$

6.5 Adding and Subtracting Polynomials

46. $3x^3 + 2x^3 = 5x^3$

47. $\dfrac{1}{2}p^2 + \dfrac{5}{2}p^2 + \dfrac{7}{2}p^2 = \dfrac{1+5+7}{2}p^2$

$$= \dfrac{13}{2}p^2$$

48. $(3x - 1) + (6x + 5)$

$= (3x + 6x) + (-1 + 5)$

$= 9x + 4$

49. $(3x^2 - 2x + 4) + (-x^2 - 1)$

$= (3x^2 - 1x^2) - 2x + (4 - 1)$

$= 2x^2 - 2x + 3$

50. $\quad 5x - 2$

$\underline{+\ 3x + 5}$

$\quad 8x + 3$

51. $\quad 3x^2 - 2x + 7$

$\underline{+\ -5x^2 + 3x - 5}$

$\quad -2x^2 + 1x + 2$

$or -2x^2 + x + 2$

52. $16p^3 - 9p^3 = 7p^3$

53. $4y^2 - 9y^2 = 4y^2 + (-9y^2)$

$$= -5y^2$$

54. $(2.5x+4)-(1.4x+12)$

$=(2.5x+4)+(-1.4x-12)$

$=(2.5x-1.4x)+(4-12)$

$=(2.5x-1.4x)+(4+[-12])$

$=1.1x+(-8)$

$=1.1x-8$

55. $(3z^2-z+4)-(2z^2+3z-2)$

$=(3z^2-z+4)+(-2z^2-3z+2)$

$=(3z^2-2z^2)+(-z-3z)+(4+2)$

$=(3z^2-2z^2)+(-1z+[-3z])+(4+2)$

$=1z^2+(-4z)+6$

$=z^2-4z+6$

56.
$$\begin{array}{cc} 5x-2 & 5x-2 \\ \underline{-(3x+5)} & \underline{+(-3x-5)} \\ & 2x-7 \end{array}$$

57.
$$\begin{array}{cc} 3x^2-2x+7 & 3x^2-2x+7 \\ \underline{-(-5x^2+3x-5)} & \underline{+(5x^2-3x+5)} \\ & 8x^2-5x+12 \end{array}$$

6.6 Multiplying Polynomials

58. $3x^2 \bullet 5x^3 = (3\bullet 5)(x^2 \bullet x^3)$

$\qquad = 15x^5$

59. $(3z^2)(-2z^2) = (3\bullet -2)(z^2 \bullet z^2)$

$\qquad = -6z^4$

60. $2x^2(3x+2)$

$=2x^2(3x)+2x^2(2)$

$=6x^3+4x^2$

61. $-5t^3(7t^2-6t-2)$

$=-5t^3(7t^2)+(-5t^3)(-6t)+(-5t^3)(-2)$

$=-35t^5+30t^4+10t^3$

62. $(2x-1)(3x+2)$

$=(2x-1)3x+(2x-1)2$

$=3x(2x-1)+2(2x-1)$

$=6x^2-3x+4x-2$

$=6x^2+1x-2$

$=6x^2+x-2$

63. $(5t+4)(7t-6)$

$=(5t+4)7t-(5t+4)6$

$=7t(5t+4)-6(5t+4)$

$=35t^2+28t-30t-24$

$=35t^2-2t-24$

64.
$$\begin{array}{r} 5x - 2 \\ \times\ \underline{3x + 5} \\ 15x^2 - 6x \\ \underline{+\ 25x - 10} \\ 15x^2 + 19x - 10 \end{array}$$

65.
$$\begin{array}{r} 3x + 2 \\ \times\ \underline{5x - 5} \\ 15x^2 + 10x \\ \underline{-15x - 10} \\ 15x^2 - 5x - 10 \end{array}$$

66.
$$\begin{aligned}
(3x+2)\left(2x^2 - x + 1\right) \\
= (3x+2)\,2x^2 - (3x+2)\,x + (3x+2)\,1 \\
= 2x^2(3x+2) - x(3x+2) + 1(3x+2) \\
= 6x^3 + 4x^2 - 3x^2 - 2x + 3x + 2 \\
= 6x^3 + 1x^2 + 1x + 2 \\
= 6x^3 + x^2 + x + 2
\end{aligned}$$

67.
$$\begin{aligned}
(2r-3)\left(3r^2 + 2r - 3\right) \\
= (2r-3)\,3r^2 + (2r-3)\,2r - (2r-3)\,3 \\
= 3r^2(2r-3) + 2r(2r-3) - 3(2r-3) \\
= 6r^3 - 9r^2 + 4r^2 - 6r - 6r + 9 \\
= 6r^3 - 5r^2 - 12r + 9
\end{aligned}$$

68.
$$\begin{array}{r} 5x^2 - 2x + 3 \\ \times\ \underline{3x + 5} \\ 15x^3 - 6x^2 + 9x \\ \underline{+\ 25x^2 - 10x + 15} \\ 15x^3 + 19x^2 - 1x + 15 \\ 15x^3 + 19x^2 - x + 15 \end{array}$$

69.
$$\begin{array}{r} 3x^2 - 2x - 1 \\ \times\ \underline{5x - 2} \\ 15x^3 - 10x^2 - 5x \\ \underline{-\ 6x^2 + 4x + 2} \\ 15x^3 - 16x^2 - 1x + 2 \\ 15x^3 - 16x^2 - x + 2 \end{array}$$

70.
$$\begin{aligned}
(x+2)^2 &= (x+2)(x+2) \\
&= x^2 + 4x + 4 \neq x^2 + 4
\end{aligned}$$

Chapter Six Test

1. $4(-1)+5(2)=6$

    $-4+10=6$

    $6=6$

    yes

2. $3(3)-2(-2)=-13$

    $9-(-4)=-13$

    $9+4=-13$

    $13 \neq -13$

    no

3. $(0,-2), (4,0)$ *and* $(2,-1)$

4.

x	y	(x,y)
0	1	(0,1)
3	2	(3,2)
−3	0	(−3,0)

5. PANTS SALE

    (30,32), (30,34), (31,34), (38,30)

6.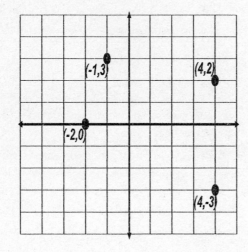

7. A(0,0), B(2.5,3.5), C(-3,-2), D(0,-2)

8.

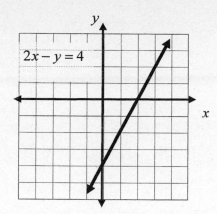

9a. (2,0)

9b. (0,−4)

10.

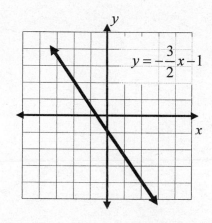

11.

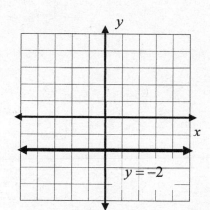

12.

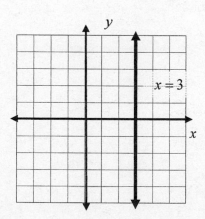

13a. $h^2 h^4 = h^{2+4} = h^6$

13b. $-7x^3\left(4x^2\right) = (-7 \bullet 4)\left(x^3 \bullet x^2\right)$

$= -28x^{2+3}$

$= -28x^5$

13c. $b^2 \bullet b \bullet b^5 = b^{2+1+5} = b^8$

13d. $-3g^2 k^3\left(-8g^3 k^{10}\right)$

$= (-3 \bullet -8)\left(g^2 \bullet g^3\right)\left(k^3 \bullet k^{10}\right)$

$= 24g^{2+3} k^{3+10}$

$= 24g^5 k^{13}$

14a. $\left(f^3\right)^5 = f^{3 \bullet 5} = f^{15}$

14b. $\left(2a^2 b\right)^2 = (2)^2 \left(a^2\right)^2 (b)^2$

$= 4a^{2 \bullet 2} b^{1 \bullet 2}$

$= 4a^4 b^2$

14c. $\left(x^2\right)^3 \left(x^3\right)^3 = \left(x^{2 \bullet 3}\right)\left(x^{3 \bullet 3}\right)$

$= x^6 \bullet x^9$

$= x^{6+9}$

$= x^{15}$

14d. $\left(x^2 x^3\right)^3 = \left(x^{2+3}\right)^3$

$= \left(x^5\right)^3$

$= x^{5 \bullet 3}$

$= x^{15}$

15.   binomial

16.   trinomial

17.   6

18.   7

19. $3x^2 - 2x + 4; \ x = 3$

$= 3(3)^2 - 2(3) + 4$

$= 3(9) - 2(3) + 4$

$= 27 - 6 + 4$

$= 21 + 4$

$= 25$

20. $-2r^2 - r + 3; \ r = -1$

$= -2(-1)^2 - (-1) + 3$

$= -2(1) - (-1) + 3$

$= -2 + 1 + 3$

$= -1 + 3$

$= 2$

21.

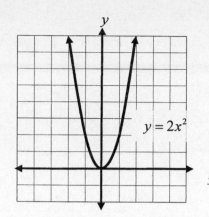

$y = 2x^2$

22.

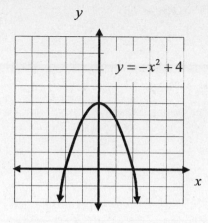

$y = -x^2 + 4$

23.
$$\left(3x^2 + 2x\right) + \left(2x^2 - 5x + 4\right)$$
$$= \left(3x^2 + 2x^2\right) + \left(2x - 5x\right) + 4$$
$$= \left(3x^2 + 2x^2\right) + \left(2x + \left[-5x\right]\right) + 4$$
$$= 5x^2 + \left(-3x\right) + 4$$
$$= 5x^2 - 3x + 4$$

24.
$$4x^2 - 5x + 5$$
$$\underline{3x^2 + 7x - 7}$$
$$7x^2 + 2x - 2$$

25.
$$\left(2.1p^2 - 2p - 2\right) - \left(3.3p^2 - 5p - 2\right)$$
$$= \left(2.1p^2 - 2p - 2\right) + \left(-3.3p^2 + 5p + 2\right)$$
$$= \left(2.1p^2 - 3.3p^2\right) + \left(-2p + 5p\right) + \left(-2 + 2\right)$$
$$= \left(2.1p^2 + \left[-3.3p^2\right]\right) + \left(-2p + 5p\right) + \left(-2 + 2\right)$$
$$= -1.2p^2 + 3p + 0$$
$$= -1.2p^2 + 3p$$

26.
$$\begin{array}{cc} 3d^2 - 3d + 7.2 & 3d^2 - 3d + 7.2 \\ \underline{-\left(-5d^2 + 6d - 5.3\right)} & +\underline{\left(5d^2 - 6d + 5.3\right)} \\ & 8d^2 - 9d + 12.5 \end{array}$$

27.
$$\left(-2x^3\right)\left(4x^2\right)$$
$$= \left(-2 \bullet 4\right)\left(x^3 \bullet x^2\right)$$
$$= -8x^5$$

28.
$$3y^2\left(y^2 - 2y + 3\right)$$
$$= 3y^2\left(y^2\right) - 3y^2\left(2y\right) + 3y^2(3)$$
$$= 3y^4 - 6y^3 + 9y^2$$

29. $(2x-5)(3x+4)$

$=(2x-5)3x+(2x-5)4$

$=3x(2x-5)+4(2x-5)$

$=6x^2-15x+8x-20$

$=6x^2-7x-20$

30. $(2x-3)(x^2-2x+4)$

$=(2x-3)x^2-(2x-3)2x+(2x-3)4$

$=x^2(2x-3)-2x(2x-3)+4(2x-3)$

$=2x^3-3x^2-4x^2+6x+8x-12$

$=2x^3-3x^2+(-4x^2)+6x+8x-12$

$=2x^3-7x^2+14x-12$

31. No; (1,-2) lies in quadrant IV while (-2,1) lies in quadrant II.

32. There are infinitely many ordered pairs (x, y) that satisfy this equation.

Chapters 1-6 Cumulative Review Exercises

1.   6,246,000

2.   6,000,000

3. $P = 8 + 3 + 8 + 3$
       $P = 22m$

4. $P = 13 + 13 + 13 + 13$
       $P = 52in$

5.   PARKING

Type	Length (ft)	Width (ft)	Area (ft^2)
Standard Space	20	9	180
Standard space adjacent to a wall	20	10	200
Parallel space	25	10	250
Compact space	17	8	136

6.   HEALTH

$$\frac{80}{45} ; \frac{120}{80} ; \frac{140}{85}$$

7.   120
               ⬋ ⬊
            |2| 60
               ⬋ ⬊
            |2| 30
               ⬋ ⬊
            |2| 15
               ⬋ ⬊
            |3| |5|

       $2^3 \bullet 3 \bullet 5$

8.   525
               ⬋ ⬊
            |5| 105
               ⬋ ⬊
            |5| 21
               ⬋ ⬊
            |3| |7|

       $3 \bullet 5^2 \bullet 7$

9.   LAKE TAHOE

       The visibility has decreased or is 39

       feet less.

10. $\dfrac{6x + x^3}{|x|}$ *for* $x = -2$

$$= \dfrac{6(-2) + (-2)^3}{|-2|}$$

$$= \dfrac{6(-2) + (-8)}{2}$$

$$= \dfrac{-12 + (-8)}{2}$$

$$= \dfrac{-20}{2} = -10$$

11. $12 - 2\big[1 - (-8 + 2)\big] = 12 - 2\big[1 - (-6)\big]$

$$= 12 - 2\big[1 + 6\big]$$

$$= 12 - 2\big[7\big]$$

$$= 12 - 14$$

$$= 12 + (-14)$$

$$= -2$$

12. $-3^2 = -(3)(3) = -9$

NOTE:

$(-3)^2 = (-3)(-3) = 9$,

These two types of problems are often confused and should be looked at more closely if there are difficulties.

13. $5x - 11x = 5x + (-11x)$

$$= -6x$$

14. $-4(x - 3y) + 5x - 2y$

$$= -4x + 12y + 5x - 2y$$

$$= (-4x + 5x) + (12y - 2y)$$

$$= 1x + 10y$$

$$= x + 10y$$

15.
$$4x + 3 = 11$$
$$4x + 3 - 3 = 11 - 3$$
$$4x = 8$$
$$\frac{4x}{4} = \frac{8}{4}$$
$$x = 2$$

Check:
$$4(2) + 3 = 11$$
$$8 + 3 = 11$$
$$11 = 11$$

16.
$$2z + 12 = 6z - 4$$
$$2z - 2z + 12 = 6z - 2z - 4$$
$$12 = 4z - 4$$
$$12 + 4 = 4z - 4 + 4$$
$$16 = 4z$$
$$\frac{16}{4} = \frac{4z}{4}$$
$$4 = z \ \text{ or } \ z = 4$$

Check:
$$2(4) + 12 = 6(4) - 4$$
$$8 + 12 = 24 - 4$$
$$20 = 20$$

17.
$$\frac{t}{3} + 2 = -4$$
$$\frac{t}{3} + 2 - 2 = -4 - 2$$
$$\frac{t}{3} = -6$$
$$\cancel{3}\left(\frac{t}{\cancel{3}}\right) = (-6)3$$
$$t = -18$$

Check:
$$\frac{-18}{3} + 2 = -4$$
$$-6 + 2 = -4$$
$$-4 = -4$$

18.
$$2y + 7 = 2 - (4y + 7)$$
$$2y + 7 = 2 - 4y - 7$$
$$2y + 7 = -4y - 5$$
$$2y + 4y + 7 = -4y + 4y - 5$$
$$6y + 7 = -5$$
$$6y + 7 - 7 = -5 - 7$$
$$6y = -12$$
$$\frac{6y}{6} = \frac{-12}{6}$$
$$y = -2$$

Check:
$$2(-2) + 7 = 2 - \left(4[-2] + 7\right)$$
$$-4 + 7 = 2 - (-8 + 7)$$
$$3 = 2 - (-1)$$
$$3 = 2 + 1$$
$$3 = 3$$

19.

$$\frac{5}{10b^3} \cdot 2b^2$$

$$= \frac{\cancel{5}}{\cancel{5} \cdot \cancel{2} \cdot \cancel{b} \cdot \cancel{b} \cdot b} \cdot \frac{\cancel{2} \cdot \cancel{b} \cdot \cancel{b}}{1}$$

$$= \frac{1}{b}$$

20.

$$-4\frac{1}{4} \div 4\frac{1}{2} = -\frac{17}{4} \div \frac{9}{2}$$

$$= -\frac{17}{4} \cdot \frac{2}{9}$$

$$= -\frac{17}{2 \cdot \cancel{2}} \cdot \frac{\cancel{2}}{9}$$

$$= -\frac{17}{18}$$

21.

$$\begin{array}{c} 34\frac{1}{9} = 34\frac{1 \cdot 2}{9 \cdot 2} = 34\frac{2}{18} \\ \underline{-13\frac{5}{6} = -13\frac{5}{6 \cdot 3} = -13\frac{15}{18}} \\ 33\frac{2}{18} + \frac{18}{18} = 33\frac{20}{18} \\ \underline{-13\frac{15}{18} \qquad = -13\frac{15}{18}} \\ 20\frac{5}{18} \end{array}$$

22.

$$\frac{5}{m} - \frac{n}{5} = \frac{(5)5}{(5)m} - \frac{(m)n}{(m)5}$$

$$= \frac{25}{5m} - \frac{mn}{5m}$$

$$= \frac{25 - mn}{5m}$$

23.
$$\frac{7}{8}t = -28$$

$$\left(\frac{8}{7}\right)\frac{7}{8}t = -\frac{28}{1}\left(\frac{8}{7}\right)$$

$$t = -\frac{4 \bullet \cancel{7}}{1}\left(\frac{8}{\cancel{7}}\right)$$

$$t = -32$$

Check:

$$\frac{7}{8}(-32) = -28$$

$$\frac{7}{\cancel{8}}(-4 \bullet \cancel{8}) = -28$$

$$-28 = -28$$

24.
$$\frac{4}{5}x = \frac{3}{4}x + \frac{1}{2}$$

$$(20)\left(\frac{4}{5}x\right) = (20)\left(\frac{3}{4}x + \frac{1}{2}\right)$$

$$16x = 15x + 10$$

$$16x - 15x = 15x - 15x + 10$$

$$x = 10$$

Check:

$$\frac{4}{5}\left(\frac{10}{1}\right) = \frac{3}{4}\left(\frac{10}{1}\right) + \frac{1}{2}$$

$$\frac{4}{\cancel{5}}\left(\frac{\cancel{5} \bullet 2}{1}\right) = \frac{3}{2 \bullet \cancel{2}}\left(\frac{\cancel{2} \bullet 5}{1}\right) + \frac{1}{2}$$

$$8 = \frac{15}{2} + \frac{1}{2}$$

$$8 = \frac{16}{2}$$

$$8 = 8$$

25. PAPER SHREDDERS

$n = $ number of strips

$$n = 8\frac{1}{2} \div \frac{1}{4}$$

$$n = \frac{17}{2} \div \frac{1}{4}$$

$$n = \frac{17}{2} \bullet \frac{4}{1}$$

$$n = \frac{17}{\cancel{2}} \bullet \frac{\cancel{2} \bullet 2}{1} = \frac{34}{1} = 34$$

The paper is cut into 34 strips.

26. To divide a number by a fraction, we multiply the number by the reciprocal of the fraction, and 4 is the reciprocal of $\frac{1}{4}$.

27.   57.57

28.   29.703
      +321.350
       351.053

29.   287.23
      −179.97
       107.26

30.   7.89
      × 0.27
       5523
      15780
       2.1303

31.

$$3.8\overline{)17.746} = 38\overline{)177.46}^{\,4.67}$$

  152
  254
  228
  266
  266
    0

32. $\dfrac{35}{99}$

$$= 99\overline{)35.0000}^{\,.3535}$$

  297
  530
  495
  350
  297
  530
  495
   35

$.3535 = .\overline{35}$

33. $5\dfrac{5}{8}$

$$\dfrac{5}{8} = 8\overline{)5.00}^{\,.62} \approx .6$$

  48
  20
  16
   4

$5\dfrac{5}{8} \approx 5.6$

34. $-4\dfrac{7}{9}$

$$\dfrac{7}{9} = 9\overline{)7.00}^{\,.77} \approx .8$$

  63
  70
  63
   7

$-4\dfrac{7}{9} \approx -4.8$

35.
$$3.2x = 74.46 - 1.9x$$
$$3.2x + 1.9x = 74.46 - 1.9x + 1.9x$$
$$5.1x = 74.46$$
$$\frac{5.1x}{5.1} = \frac{74.46}{5.1}$$
$$x = 14.6$$

Check:
$$3.2(14.6) = 74.46 - 1.9(14.6)$$
$$46.72 = 74.46 - 27.74$$
$$46.72 = 46.72$$

36.
$$-5.2x = 108 - 6.1x$$
$$-5.2x + 6.1x = 108 - 6.1x + 6.1x$$
$$0.9x = 108$$
$$\frac{0.9x}{0.9} = \frac{108}{0.9}$$
$$x = 120$$

Check:
$$-5.2(120) = 108 - 6.1(120)$$
$$-624 = 108 - 732$$
$$-624 = 108 + (-732)$$
$$-624 = -624$$

37.
$$-2(x - 2.1) = -2.4$$
$$-2x + 4.2 = -2.4$$
$$-2x + 4.2 - 4.2 = -2.4 - 4.2$$
$$-2x = -6.6$$
$$\frac{-2x}{-2} = \frac{-6.6}{-2}$$
$$x = 3.3$$

Check:
$$-2(3.3 - 2.1) = -2.4$$
$$-2(1.2) = -2.4$$
$$-2.4 = -2.4$$

38.
$$\frac{1}{5}x - 2.5 = -17.2 \quad NOTE : \left(\frac{1}{5} = 0.2\right)$$
$$0.2x - 2.5 = -17.2$$
$$0.2x - 2.5 + 2.5 = -17.2 + 2.5$$
$$0.2x = -14.7$$
$$\frac{0.2x}{0.2} = \frac{-14.7}{0.2}$$
$$x = -73.5$$

Check
$$\frac{1}{5}(-73.5) - 2.5 = -17.2$$
$$0.2(-73.5) - 2.5 = -17.2$$
$$-14.7 - 2.5 = -17.2$$
$$-14.7 + (-2.5) = -17.2$$

39. EARTHQUAKES
$$\text{mean} = \frac{9.2 + 8.8 + 8.7 + 8.3 + 8.3 + 8.2 + 8.2 + 8.0 + 7.9 + 7.9 + 7.9 + 7.9 + 7.9 + 7.8 + 7.8}{15}$$

$$\text{mean} = \frac{122.8}{15} \approx 8.2$$
$$\text{median} = 8.0$$
$$\text{mode} = 7.9$$
$$\text{range} = 9.2 - 7.8 = 1.4$$

40.   DRIVING

d = distance traveled in 3.5 hours

$d = (3.5)(55) = 192.5$

r = final odometer reading

$r = 192.5 + 41,618.9$

$r = 41,811.4$

The odometer will read 41,811.4 miles.

41.   PETITION DRIVE

x = the number of signatures

Note: 5 cents=$0.05

$0.05x + 20 = 60$

$0.05x + 20 - 20 = 60 - 20$

$0.05x = 40$

$\dfrac{0.05x}{0.05} = \dfrac{40}{0.05}$

$x = 800$

She must get 800 signatures.

42.   CONCERT TICKETS

x = total number of tickets

$\dfrac{7}{8}x + 200 = x$

$8\left(\dfrac{7}{8}x + 200\right) = 8(x)$

$7x + 1,600 = 8x$

$7x + 1,600 - 7x = 8x - 7x$

$1600 = x \ \ or \ \ x = 1,600$

There were 1,600 tickets sold.

43. $\sqrt{121} = 11$

44. $\sqrt{\dfrac{81}{4}} = \dfrac{9}{2}$

45. $\sqrt{0.25} = 0.5$

46. $3\sqrt{144} - \sqrt{49} = 3(12) - 7$

$= 36 - 7$

$= 29$

47.

$4x - 5y = -23 \ ; x = -2, y = 3$

$4(-2) - 5(3) = -23$

$-8 - 15 = -23$ yes

$-8 + (-15) = -23$

$-23 = -23$

48.

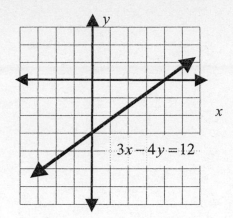

$3x - 4y = 12$

49. $p^4 p^5 = p^{4+5}$
    $= p^9$

50. $\left(2q^2\right)\left(-5q^6\right) = \left(2 \bullet -5\right)\left(q^2 \bullet q^6\right)$
    $= -10q^{2+6}$
    $= -10q^8$

51. $\left(p^3 q^2\right)\left(p^3 q^4\right) = \left(p^3 \bullet p^3\right)\left(q^2 \bullet q^4\right)$
    $= p^{3+3} q^{2+4}$
    $= p^6 q^6$

52. $\left(3a^2\right)^3 \left(-a^3\right)^3 = \left(3\right)^3 \left(a^2\right)^3 \left(-1\right)^3 \left(a^3\right)^3$
    $= \left(27 \bullet -1\right)\left(a^6 \bullet a^9\right)$
    $= -27a^{6+9}$
    $= -27a^{15}$

53. $\left(3x^2 - 5x\right) - \left(2x^2 + x - 3\right)$
    $= \left(3x^2 - 5x\right) + \left(-2x^2 - x + 3\right)$
    $= \left(3x^2 - 2x^2\right) + \left(-5x - 1x\right) + 3$
    $= \left(3x^2 - 2x^2\right) + \left(-5x + \left[-1x\right]\right) + 3$
    $= 1x^2 + \left(-6x\right) + 3$
    $= x^2 - 6x + 3$

54. $\left(2x + 3\right)\left(3x - 1\right)$
    $= \left(2x + 3\right)3x - \left(2x + 3\right)1$
    $= 3x\left(2x + 3\right) - 1\left(2x + 3\right)$
    $= 6x^2 + 9x - 2x - 3$
    $= 6x^2 + 7x - 3$

Chapter 7 Percents

7.1 Percents, Decimals, and Fractions

Vocabulary

1. **Percent** means parts per one hundred.

Concepts

3. To write a percent as a fraction, drop the percent symbol and write the given number over **100**, then simplify the fraction if possible.

5. To change a decimal to a percent, multiply the decimal by 100 by moving the decimal point two places to the **right**, and then insert a % symbol.

Notation

7a. $0.84, 84\%, \dfrac{84}{100} = \dfrac{21 \bullet \cancel{4}}{25 \bullet \cancel{4}} = \dfrac{21}{25}$

7b. 16%

Practice

9. $17\% = \dfrac{17}{100}$

11. $5\% = \dfrac{5}{100} = \dfrac{1 \bullet \cancel{5}}{20 \bullet \cancel{5}} = \dfrac{1}{20}$

13. $60\% = \dfrac{60}{100} = \dfrac{3 \bullet \cancel{20}}{5 \bullet \cancel{20}} = \dfrac{3}{5}$

15. $125\% = \dfrac{125}{100} = \dfrac{5 \bullet \cancel{25}}{4 \bullet \cancel{25}} = \dfrac{5}{4}$

17. $\dfrac{2}{3}\% = \dfrac{\frac{2}{3}}{100} = \dfrac{3\left(\frac{2}{3}\right)}{3(100)} = \dfrac{2}{300} = \dfrac{1 \bullet \cancel{2}}{150 \bullet \cancel{2}} = \dfrac{1}{150}$

19. $5\dfrac{1}{4}\% = \dfrac{21}{4}\% = \dfrac{\frac{21}{4}}{100} = \dfrac{4\left(\frac{21}{4}\right)}{4(100)} = \dfrac{21}{400}$

21. $0.6\% = \dfrac{0.6}{100} = \dfrac{10(0.6)}{10(100)} = \dfrac{6}{1,000} = \dfrac{3 \bullet \cancel{2}}{500 \bullet \cancel{2}} = \dfrac{3}{500}$

23. $1.9\% = \dfrac{1.9}{100} = \dfrac{10(1.9)}{10(100)} = \dfrac{19}{1,000}$

25. $19\% = 0.19$

27. $6\% = 0.06$

29. $40.8\% = 0.408$

31. $250\% = 2.5$

33. $0.79\% = 0.0079$

35. $\dfrac{1}{4}\% = 0.25\% = 0.0025$

37. $0.93 = 93\%$

39. $0.612 = 61.2\%$

41. $0.0314 = 3.14\%$

43. $8.43 = 843\%$

45. $50 = 5,000\%$

47. $9.1 = 910\%$

49. $\dfrac{17}{100} = 0.17 = 17\%$

51. $\dfrac{4}{25} = 0.16 = 16\%$

53. $\dfrac{2}{5} = 0.4 = 40\%$

55. $\dfrac{21}{20} = 1.05 = 105\%$

57. $\dfrac{5}{8} = 0.625 = 62.5\%$

59. $\dfrac{3}{16} = 0.1875 = 18.75\%$

61. $\dfrac{2}{3} = 0.6666\ldots = 66.66\ldots\% = 66\dfrac{2}{3}\%$

63. $\dfrac{1}{12} = 0.08333\ldots = 8.333\ldots\% = 8\dfrac{1}{3}\%$

65. $\dfrac{1}{9} \approx 0.1111 = 11.11\%$

67. $\dfrac{5}{9} \approx 0.5556 = 55.56\%$

Applications

69a. U.N. SECURITY COUNCIL

$\dfrac{15}{188}$

69b. $\dfrac{15}{188} \approx .0798 = 7.98\% \approx 8\%$

71a. PIANO KEYS

$\dfrac{36}{88} = \dfrac{9 \bullet \cancel{4}}{22 \bullet \cancel{4}} = \dfrac{9}{22}$

71b. $\dfrac{9}{22} \approx 0.409 = 40.9\% \approx 41\%$

73a. THE HUMAN SPINE

29 bones total, 5 are lumbar

$\dfrac{5}{29}$

73b. $\dfrac{5}{29} \approx 0.172 = 17.2\% \approx 17\%$

73c. 29 bones total, 7 are cervical

$\dfrac{7}{29} \approx 0.241 = 24.1\% \approx 24\%$

75. STEEP GRADE

The road rises 5 ft for every 100-foot run.

77. SOAP

$99\dfrac{44}{100}\% = 99.44\% = 0.9944$

79. BASKETBALL

They are given as a decimal

$0.896 = 89.6\%$

81.   SKIN

t = missing percent for torso

$t = 100 - (8 + 2.5 + 4 + 4 + 3 + 3 + 3 + 3 + 10.5 + 10.5 + 7 + 7 + 3.5 + 3.5)$

$t = 100 - 72.5$

$t = 27.5\%$

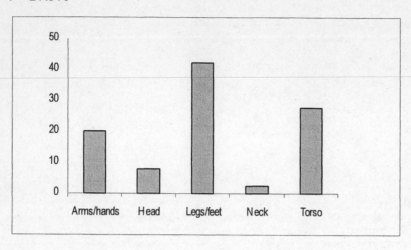

83.   CHARITIES

92 cents on the dollar

implies $\dfrac{92}{100} = 0.92 = 92\%$

85.   BIRTHDAYS

$\dfrac{1}{365} \approx 0.00274 = 0.274\% \approx 0.27\%$

Writing

87.   Answers may vary.

89.   To change a fraction to a percent you must first convert the fraction to a decimal, then change the decimal to a percent.

91.   CHAMPIONS

No, it means that the fraction that is made when victories are compared to total fights fought comes to 0.92, which would be 92%.

Review

93.
$$-\frac{2}{3}x = -6$$

$$-\frac{3}{2}\left(-\frac{2}{3}x\right) = -\frac{3}{2}(-6)$$

$$x = -\frac{3}{\cancel{2}}\left(-\frac{3 \bullet \cancel{2}}{1}\right)$$

$$x = 9$$

95.
$$y = 2(2) + 3$$
$$y = 4 + 3$$
$$y = 7 \quad \text{which gives } (2, 7)$$
The other ordered pairs

are $(4, 11), (0, 3)$

97.
$$(x+1)(x+2) = x(x+2) + 1(x+2)$$
$$= x^2 + 2x + 1x + 2$$
$$= x^2 + 3x + 2$$

7.2 Solving Percent Problems

Vocabulary

1. In a circle **graph**, pie-shaped wedges are used to show the division of a whole quantity into its component parts.

Concepts

3. What number is 10% of 50? $x = 0.10 \bullet 50$

5. 48 is what percent of 47? $48 = x \bullet 47$

7a. $12\% = 0.12$

7b. $5.6\% = 0.056$

7c. $125\% = 1.25$

7d. $\dfrac{1}{4}\% = 0.25\% = 0.0025$

9.   more since $120\% > 100\%$

11a.   25

11b.   100%

11c.   87

13.   E-MAIL

   s = percent of mail that is spam

   $s = 100 - (31 + 16 + 5 + 4)$

   $s = 100 - 56$

   $s = 44$

   44% of e-mail is spam.

Notation

15a.   multiply

15b.   equals

15c. x (as a variable)

Practice

17. $x = 0.36 \bullet 250$
$x = 90$

19. $16 = x \bullet 20$
$16 = 20x$
$\dfrac{16}{20} = \dfrac{20x}{20}$
$0.8 = x$
$x = 80\%$

21. $7.8 = 0.12 \bullet x$
$\dfrac{7.8}{0.12} = \dfrac{0.12x}{0.12}$
$65 = x$ or $x = 65$

23. $x = 0.008 \bullet 12$
$x = 0.096$

25. $0.5 = x \bullet 40,000$
$0.5 = 40,000x$
$\dfrac{0.5}{40,000} = \dfrac{40,000x}{40,000}$
$0.0000125 = x$
$x = 0.00125\%$

27. $3.3 = 0.075 \bullet x$
$\dfrac{3.3}{0.075} = \dfrac{0.075x}{0.075}$
$44 = x$ or $x = 44$

29. $x = .0725 \bullet 600$
$x = 43.5$

31. $1.02 \bullet 105 = x$
$107.1 = x$ or $x = 107.1$

33. $33\dfrac{1}{3}\% = \dfrac{1}{3}$
$\dfrac{1}{3}x = 33$
$3\left(\dfrac{1}{3}x\right) = 3(33)$
$x = 99$

35.
$$9\frac{1}{2}\% = 0.095$$
$$0.095x = 5.7$$
$$\frac{0.095x}{0.095} = \frac{5.7}{0.095}$$
$$x = 60$$

37.
$$x \bullet 8,000 = 2,500$$
$$8,000x = 2,500$$
$$\frac{8,000x}{8,000} = \frac{2,500}{8,000}$$
$$x = 0.3125$$
$$x = 31.25\%$$

39.   Note: For problem 39, the percentages indicate the number of tick marks used by each slice of pie. In other words, 8% means that the 8% slice of pie is 8 tick marks wide.

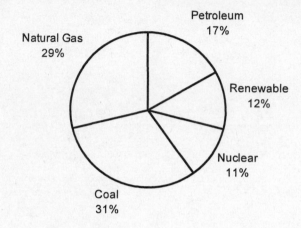

Natural Gas
29%

Petroleum
17%

Renewable
12%

Nuclear
11%

Coal
31%

Applications

41.   CHILD CARE

70% of what number is 84?
$$0.70x = 84$$
$$\frac{0.7x}{0.7} = \frac{84}{0.7}$$
$$x = 120$$

120 is the maximum.

43.   GOVERNMENT SPENDING

37% of 1,800 billion is what?
$$0.37 \bullet 1,800 = x$$
$$666 = x \text{ or } x = 666$$

$666 billion was spent on Social Security, Medicare, and other retirement programs.

45.   THE INTERNET

If 24% is downloaded, that must leave

76% left to download.

76% of 50,000 is what?

$0.76 \bullet 50,000 = x$

$38,000 = x$

There are 38,000 bytes (or 38K) left.

47.   PRODUCT PROMOTIONS

25% of what number is 6?

$0.25 \bullet x = 6$

$\dfrac{0.25x}{0.25} = \dfrac{6}{0.25}$

$x = 24$

The large bottle contains 24 oz.

49.   DRIVER'S LICENSE

$\dfrac{28}{40} = 0.70 = 70\%$

Yes, he passed.

51.   MIXTURES

	Gallons of solution in tank	% sulfuric acid	Gallons of sulfuric acid in tank
Tank 1	60	50%	30
Tank 2	40	30%	12

53.   MAKING COPIES

What is 180% of 1.5?

$x = 1.8 \bullet 1.5$

$x = 2.7$

The resulting height is 2.7 in.

55.   INSURANCE

200 is what percent of 4,000?

$$200 = x \bullet 4,000$$

$$200 = 4,000x$$

$$\frac{200}{4,000} = \frac{4,000x}{4,000}$$

$$.05 = x$$

$$x = 5\%$$

The driver paid 5% of the cost.

57.   MAJORITIES

There are a total of 17,177 votes.

50% of that would be 8588.5. Since the most votes received were 8,501, then yes, there must be a runoff held.

Writing

59.   Amount is equal to percent time the base.

61.   100% of a number is that number, so anything larger than 100% is more than the number.

Review

63.   2.78
          6
          9.09
        +0.3
        ―――
        18.17

65.   5.001

67. $34.5464 \bullet 1,000 = 34,546.4$

69. $$0.4x + 1.2 = -7.8$$
         $$0.4 + 1.2 - 1.2 = -7.8 - 1.2$$
         $$0.4x = -9$$
         $$\frac{0.4x}{0.4} = -\frac{9}{0.4}$$
         $$x = -22.5$$

7.3 Applications of Percent

Vocabulary

1. Some salespeople are paid on <u>**commission**</u>. It is based on a percent of the total dollar amount of the goods or services they sell.

3. The difference between the original price and the sale price of an item is called the <u>**discount**</u>.

Concepts

5. To find the percent decrease, <u>**subtract**</u> the smaller number from the larger number to find the amount of decrease. Then find what percent that difference is to the <u>**original**</u> amount.

Applications

7. SALES TAXES

What number is 4.45% of 900?

t = sales tax

$t = 0.0475 \bullet 900$

$t = 42.75$

The tax is $42.75.

9. ROOM TAXES

10.32 is what percent of 129?

x = the room tax rate

$10.32 = x \bullet 129$

$10.32 = 129x$

$\dfrac{10.32}{129} = \dfrac{129x}{129}$

$0.08 = x$

$x = 8\%$

The room tax is 8%.

11. SALES RECEIPTS

subtotal $= 8.97 + 9.87 + 28.50 = \$47.34$

sales tax $= (0.06)(47.34) = \$2.84$

total $= 47.34 + 2.84 = \$50.18$

13. SALES TAX

What is 1% of 15,000?

$x = 0.01 \bullet 15,000$

$x = 150$

$150 more would be collected.

15.   PAYCHECKS

Each problem has the same basic question, what percent of 360 is the amount of tax?

f = Fed Tax	w = Work Comp	m = Medicare	s = Social Security
$f \bullet 360 = 28.80$	$w \bullet 360 = 4.32$	$m \bullet 360 = 5.22$	$s \bullet 360 = 22.32$
$360f = 28.80$	$360w = 4.32$	$360m = 5.22$	$360s = 22.32$
$\dfrac{360f}{360} = \dfrac{28.80}{360}$	$\dfrac{360w}{360} = \dfrac{4.32}{360}$	$\dfrac{360m}{360} = \dfrac{5.22}{360}$	$\dfrac{360s}{360} = \dfrac{22.32}{360}$
$f = 0.08$	$w = 0.012$	$m = 0.0145$	$s = 0.062$
$f = 8\%$	$w = 1.2\%$	$m = 1.45\%$	$s = 6.2\%$

17.   OVERTIME

What is 25% of 480?

$x = 0.25 \bullet 480$

$x = 120$

Now we must subtract that from the original amount of 480.

target number $= 480 - 120 = 360$

The target number is 360 hours.

19.   REDUCED CALORIES

What is 36% of 150?

$c = 0.36 \bullet 150$

$c = 54$

Now we must subtract that from the original amount of 150.

calories in new chips $= 150 - 54 = 96$

There are 96 calories in the new chips.

21.   ENDANGERED SPECIES

The decrease occurred in 1995-1996.

The amount of decrease was 76.

What percent of 1599 is 76?

$x \bullet 1,599 = 76$

$1,599x = 76$

$\dfrac{1,599x}{1,599} = \dfrac{76}{1,599}$

$x \approx 0.0475$

$x \approx 5\%$

The percent decline was about 5%.

23.   CAR INSURANCE

The amount of decrease was $40.

What percent of 400 is 40?

$x \bullet 400 = 40$

$400x = 40$

$\dfrac{400x}{400} = \dfrac{40}{400}$

$x = 0.1$

$x = 10\%$

The decrease was 10%.

25. **LAKE SHORELINES**

   Amount of increase was 1.8 miles.

   What percent of 5.8 is 1.8?

   $x \bullet 5.8 = 1.8$

   $5.8x = 1.8$

   $\dfrac{5.8x}{5.8} = \dfrac{1.8}{5.8}$

   $x \approx 0.3103$

   $x \approx 31\%$

   The increase was about 31%.

27a. **EARTH MOVING**

   Amount of increase is 0.25 cu yd.

   What percent of 1 is 0.25?

   $x \bullet 1 = 0.25$

   $x = 0.25$

   $x = 25\%$

   Step one gives a 25% increase.

27b. Amount of decrease is 0.45 cu yd.

   What percent of 1.25 is 0.45?

   $x \bullet 1.25 = 0.45$

   $1.25x = 0.45$

   $\dfrac{1.25x}{1.25} = \dfrac{0.45}{1.25}$

   $x = 0.36$

   $x = 36\%$

   Step two gives a 36% decrease.

29. **REAL ESTATE**

   What is 6% of 98,500?

   $x = 0.06 \bullet 98,500$

   $x = 5,910$

   Each person will get half of the

   commission, $\dfrac{5,910}{2} = \$2,955$.

31. **SPORTS AGENTS**

   What percent of 2,500,000 is 37,500?

   $x \bullet 2,500,000 = 37,500$

   $2,500,000x = 37,500$

   $\dfrac{2,500,000x}{2,500,000} = \dfrac{37,500}{2,500,000}$

   $x = 0.015$

   $x = 1.5\%$

   The agent charged 1.5%.

33.   **CONCERT PARKING**

total parking money $= 6 \bullet 6,000$

total parking money $= \$36,000$

What is $33\frac{1}{3}\%$ of 36,000?

Remember, $33\frac{1}{3}\% = \frac{1}{3}$.

$x = \frac{1}{3}(36,000)$

$x = \frac{1}{\cancel{3}}\left(\dfrac{12,000 \bullet \cancel{3}}{1}\right)$

$x = 12,000$

The promoter makes $12,000 from parking.

35.   **WATCHES**

regular price $= 29.95 + 10$

regular price $= \$39.95$

What percent of 39.95 is 10?

$x \bullet 39.95 = 10$

$39.95x = 10$

$\dfrac{39.95x}{39.95} = \dfrac{10}{39.95}$

$x \approx 0.2503$

$x \approx 25\%$

The discount is about 25%.

37.   **RINGS**

80% of what number is 149.99?

$0.80 \bullet x = 149.99$

$\dfrac{0.8x}{0.8} = \dfrac{149.99}{0.8}$

$x \approx 187.488$

$x = 187.49$

The regular price is $187.49.

39.   **VCRS**

sale price $= 399.97 - 50$

sale price $= \$349.97$

What percent of 399.97 is 50?

$x \bullet 399.97 = 50$

$399.97x = 50$

$\dfrac{399.97x}{399.97} = \dfrac{50}{399.97}$

$x \approx 0.12501$

$x \approx 13\%$

The discount is about 13%.

41. REBATES

The discount is $3.60.

The reduced price is

$15.48 - 3.60 = \$11.88$.

What percent of 15.48 is 3.60?

$x \bullet 15.48 = 3.60$

$15.48x = 3.60$

$\dfrac{15.48x}{15.48} = \dfrac{3.60}{15.48}$

$x \approx 0.2326$

$x \approx 23\%$

The discount is about 23%.

43. SHOPPING

What is 55% of 170?

$x = 0.55 \bullet 170$

$x = 93.5$

The discount is $93.50.

HSN price $= 170 - 93.50$

HSN price $= \$76.50$

Writing

45. Answers will vary.

47. A tax is a flat fee you may pay in addition to the cost of a good or service, while a tax rate is a percentage that you pay in addition to the cost of a good or service.

Review

49. $-5(-5)(-2) = -50$

51. $\begin{aligned} 2(-2)^2(-2 + [-1]) &= 2(-2)^2(-3) \\ &= 2(4)(-3) \\ &= -24 \end{aligned}$

53. $12d$

55. $\begin{aligned} |-5 - 8| &= |-5 + (-8)| \\ &= |-13| \\ &= 13 \end{aligned}$

57.

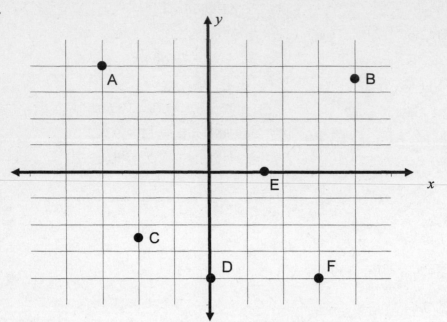

Estimation

1. COLLEGE COURSES
    about 164

3. DISCOUNTS
    about $60

5. FIRE DAMAGE
    about $54,000

7. WEIGHTLIFTING
    about 320 lb

9. TRAFFIC STUDIES
    about 130 motorists

11. NO-SHOWS
    about 21 people

13. INTERNET SURVEY
    about 18,000 people

15. VOTING
    about 3,100 volunteers

7.4 Interest

Vocabulary

1. In banking, the original amount of money borrowed or deposited is known as the **principal**.

3. The percent that is used to calculate the amount of interest to be paid is called the **interest** rate.

5. Interest computed only on the original principal is called **simple** interest.

Concepts

7a. $7\% = 0.07$ 7b. $9.8\% = 0.098$ 7c. $6\dfrac{1}{4}\% = 6.25\% = 0.0625$

9.

Principal	Rate	Time	Interest Earned
$10,000	6%	3 yr	$1,800

11a.   compound interest   11b. $1,000

11c.   4 times   11d. $50   11e.   1 year

Notation

13.   multiplication

Applications

15.   RETIREMENT INCOME   17.   REMODELING

$I = 5,000 \bullet 0.06 \bullet 1$ $I = 8,000 \bullet 0.092 \bullet 2$

$I = 300$ $I = 1,472$

Balance $= 5,000 + 300$ $1,472 interest would be owed.

Balance $= \$5,300$

The account is worth $5,300.

19.   MEETING PAYROLLS

$$30 \text{ days} = \frac{30}{365} = \frac{6 \cdot \cancel{5}}{73 \cdot \cancel{5}} = \frac{6}{73} \text{ year}$$

$$I = 4,200 \cdot 0.18 \cdot \frac{6}{73}$$

$$I \approx 62.137$$

$$I = 62.14$$

total repayment $= 4,200 + 62.14$

total repayment $= \$4,262.14$

They would repay $4,262.14 total.

21.   SAVINGS ACCOUNTS

P	r	t	I
$10,000	0.0725	2 yr	$1,450

23.   LOAN APPLICATIONS

1. Amount of loan (principal) <u>$1,200.00</u>

2. Length of loan (time)   <u>2 YEARS</u>

3. Annual percentage rate <u>8%</u>

4. Interest charged   <u>$192</u>

5. Total amount to be repaid <u>$1,392</u>

6. Monthly payments is checked

Borrower agrees to by <u>24</u> equal payments of <u>$58</u> to repay loan.

25.   LOW-INTEREST LOANS

$$I = 18 \cdot 0.023 \cdot 2$$

$$I = 0.828$$

total repayment $= 18 + 0.828$

total repayment $= \$18.828$ million

The country would pay back $18.828 million.

27.   **COMPOUNDING ANNUALLY**

$$A = 600\left(1 + \frac{0.08}{1}\right)^{1 \bullet 3}$$

$$A = 600\left(1 + 0.08\right)^{3}$$

$$A = 600\left(1.08\right)^{3}$$

$$A = 600\left(1.259712\right)$$

$$A \approx 755.827$$

$$A = 755.83$$

The account is worth \$755.83.

29.   **COLLEGE FUNDS**

$$A = 1,000\left(1 + \frac{0.06}{365}\right)^{365 \bullet 4}$$

$$A = 1,000\left(1 + 0.000164384\right)^{1460}$$

$$A = 1,000\left(1.000164384\right)^{1460}$$

$$A = 1,000\left(1.27122\right)$$

$$A \approx 1,271.224$$

$$A = 1,271.22$$

The account would be worth \$1,271.22.

31.   **TAX REFUNDS**

$$A = 545\left(1 + \frac{0.046}{365}\right)^{365 \bullet 1}$$

$$A = 545\left(1 + 0.000126027\right)^{365}$$

$$A = 545\left(1.000126027\right)^{365}$$

$$A = 545\left(1.0470714\right)$$

$$A \approx 570.653$$

$$A = 570.65$$

The account will be worth \$570.65.

33.   **LOTTERIES**

$$A = 500,000\left(1 + \frac{0.06}{365}\right)^{365 \bullet 1}$$

$$A = 500,000\left(1 + 0.000164384\right)^{365}$$

$$A = 500,000\left(1.000164384\right)^{365}$$

$$A = 500,000\left(1.061831311\right)$$

$$A \approx 530,915.655$$

$$A = 530,915.66$$

The interest is $A - P$.

$$\text{Interest} = 530,915.66 - 500,000$$

$$\text{Interest} = \$30,915.66$$

The interest would be \$30,915.66 per year.

Writing

35.   Simple interest is interest earned on the original principal, while compound interest is earned on the accumulated amount.

37.   Banks would charge a penalty for early withdrawal to dissuade customers from taking out their money to make sure that the bank can continue to earn the maximum profits and to recoup any lost money as a result of an early withdrawal by a customer.

Review

39.
$$\sqrt{\frac{1}{4}} = \frac{1}{2}$$

41.
$$-3 = 2(2) - 10$$
$$-3 = 4 - 10$$
$$-3 \neq -6$$
no

43.
$$\frac{2}{3}x = -2$$
$$\frac{3}{2}\left(\frac{2}{3}x\right) = \frac{3}{\cancel{2}}\left(-\frac{\cancel{2}}{1}\right)$$
$$x = -3$$

45.   3

47.   Quadrant III

Key Concept Equivalent Expressions

1. $\dfrac{10}{24} = \dfrac{5 \bullet \cancel{2}}{12 \bullet \cancel{2}} = \dfrac{5}{12}$

simplifying a fraction

3. $2x + 3x = 5x$

combining like terms

5. $x^2 \bullet x^3 = x^{2+3} = x^5$

when multiplying like bases, add the exponents

7. $\dfrac{2}{3} = 0.\overline{6} = 66\dfrac{2}{3}\%$

divide the numerator by the denominator, move the decimal point two places to the right, and insert a % symbol

9. $4x^2 + 1 - 2x^2 = 2x^2 + 1$

combining like terms

11. $\dfrac{6}{6} = 1$

a number divided by itself is 1

13. $(-5)(-6) = 30$

the product of two negative numbers is positive

15. $\dfrac{2x}{2} = x$

2 times x divided by 2 is x

17. $2 + 3 \bullet 5 = 2 + 15 = 17$

order of operations, do multiplications before additions

19. $2(x + 5) = 2x + 10$

the distributive property

21. $\dfrac{2}{3} \bullet \dfrac{3}{2} = 1$

the product of a number and its reciprocal is 1

Chapter Seven Review

Section 7.1 Percents, Decimals, and Fractions

1. $39\%, 0.39, \dfrac{39}{100}$

2. $111\%, 1.11, \dfrac{111}{100}$

3. $100 - 39 = 61\%$

4. $15\% = \dfrac{15}{100} = \dfrac{3 \bullet \cancel{5}}{20 \bullet \cancel{5}} = \dfrac{3}{20}$

5. $120\% = \dfrac{120}{100} = \dfrac{6 \bullet \cancel{20}}{5 \bullet \cancel{20}} = \dfrac{6}{5}$

6. $9\dfrac{1}{4}\% = \dfrac{37}{4}\% = \dfrac{\frac{37}{4}}{100} = \dfrac{4\left(\frac{37}{4}\right)}{4(100)} = \dfrac{37}{400}$

7. $0.1\% = \dfrac{0.1}{100} = \dfrac{10(0.1)}{10(100)} = \dfrac{1}{1,000}$

8. $27\% = 0.27$

9. $8\% = 0.08$

10. $155\% = 1.55$

11. $1\dfrac{4}{5}\% = 1.8\% = 0.018$

12. $0.83 = 83\%$

13. $0.625 = 62.5\%$

14. $0.051 = 5.1\%$

15. $6 = 600\%$

16. $\dfrac{1}{2} = 0.5 = 50\%$

17. $\dfrac{4}{5} = 0.8 = 80\%$

18. $\dfrac{7}{8} = 0.875 = 87.5\%$

19. $\dfrac{1}{16} = 0.0625 = 6.25\%$

20. $\dfrac{1}{3} = 0.\overline{3} = 33\dfrac{1}{3}\%$

21. $\dfrac{5}{6} = 0.8\overline{3} = 83\dfrac{1}{3}\%$

22. $\dfrac{5}{9} \approx 0.5556 \approx 55.56\%$

23. $\dfrac{8}{3} \approx 2.6667 \approx 266.67\%$

24. BILL OF RIGHTS

27-10 leaves 17 amendments

$\dfrac{17}{27} \approx 0.629 \approx 63\%$

About 63% of the Amendments were adopted after the Bill of Rights.

25. From problem 7 of this review,

$0.1\% = \dfrac{1}{1,000}$

Section 7.2 Solving Percent Problems

26. amount: 15, base: 45, percent: $33\dfrac{1}{3}\%$

27. $x = 32\% \bullet 96$

28. $x = 0.40 \bullet 500$
$x = 200$

29. $0.16 \bullet x = 20$
$\dfrac{0.16x}{0.16} = \dfrac{20}{0.16}$
$x = 125$

30.
$$1.4 = x \cdot 80$$
$$\frac{1.4}{80} = \frac{80x}{80}$$
$$0.0175 = x$$
$$x = 1.75\%$$

31.
Remember, $66\frac{2}{3}\% = \frac{2}{3}$.

$$66\frac{2}{3}\% \cdot 3{,}150 = x$$
$$\frac{2}{3} \cdot 3{,}150 = x$$
$$\frac{2}{\cancel{3}} \cdot \frac{1{,}050 \cdot \cancel{3}}{1} = x$$
$$2{,}100 = x \ \text{ or } \ x = 2{,}100$$

32.
$$x = 2.20 \cdot 55$$
$$x = 121$$

33.
$$x = 0.0005 \cdot 60{,}000$$
$$x = 30$$

34. RACING

Both amounts answer the question,

What is some percent of 15?

$n =$ amount of nitro

$$n = 0.96 \cdot 15$$
$$n = 14.4$$
$m =$ amount of methane
$$m = 0.04 \cdot 15$$
$$m = 0.6$$

14.4 gallons of nitro, and

0.6 gallons of methane are needed.

35. HOME SALES

51 is 75% of what number?

$$51 = 0.75 \cdot x$$
$$\frac{51}{0.75} = \frac{0.75x}{0.75}$$
$$68 = x \ \text{ or } \ x = 68$$

68 homes were originally for sale.

36. HURRICANES

96 is what percent of 110?

$$96 = x \cdot 110$$
$$96 = 110x$$
$$\frac{96}{110} = \frac{110x}{110}$$
$$0.873 \approx x$$
$$x \approx 87\%$$

87% of the homes were damaged.

37. TIPPING

What is 15% of 36.20?

$$x = 0.15 \cdot 36.20$$
$$x = 5.43$$

They should tip $5.43.

38. AIR POLLUTION

   Note: For problem 38, the percentages indicate the number of tick marks used by each slice of pie. In other words, 8% means that the 8% slice of pie is 8 tick marks wide.

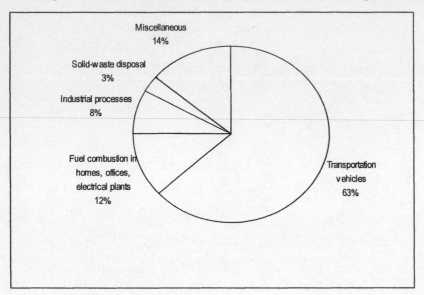

39. EARTH'S SURFACE

   What is 70.9% of 196,800,000?

   $x = 0.709 \bullet 196,800,000$

   $x = 139,531,200$

   139,531,200 mi^2 of the Earth's surface is covered by water.

Section 7.3 Applications of Percent

40. SALES RECEIPT

   SUBTOTAL = $59.99

   SALES TAX = $(59.99)(0.055)$ = $3.30

   TOTAL = $59.99 + 3.30$ = $63.29

41. TAX RATES

   492 is what percent of 12,300?

   $x \bullet 12,300 = 492$

   $12,300x = 492$

   $\dfrac{12,300x}{12,300} = \dfrac{492}{12,300}$

   $x = 0.04$

   $x = 4\%$

   The tax rate is 4%.

42.   COMMISSIONS

total sales $= 369.97 + 299.97 = \$669.94$

commission $= 0.06 \bullet 669.94$

commission $= 40.20$

The commission is $40.20.

43.   Always find the percent of increase or decrease of a quantity with respect to the **original** amount.

44.   TROOP SIZE

The amount of increase is 2,500.

What percent of 10,000 is 2,500?

$x \bullet 10,000 = 2,500$

$10,000x = 2,500$

$\dfrac{10,000x}{10,000} = \dfrac{2,500}{10,000}$

$x = 0.25$

$x = 25\%$

The increase was 25%.

45.   GAS MILEAGE

The amount of decrease was 1.8.

What percent of 18.8 is 1.8?

$x \bullet 18.8 = 1.8$

$18.8x = 1.8$

$\dfrac{18.8x}{18.8} = \dfrac{1.8}{18.8}$

$x \approx 0.096$

$x \approx 9.6\%$

It is about a 9.6% decrease.

46.   TOOL CHESTS

discount $= \$50$

original price $= 139.99 + 50 = \$189.99$

What percent of 189.99 is 50?

$x \bullet 189.99 = 50$

$189.99x = 50$

$\dfrac{189.99x}{189.99} = \dfrac{50}{189.99}$

$x \approx 0.263$

$x \approx 26\%$

The tool chest is discounted about 26%.

Section 7.4 Interest

47.

P	r	t	I
$6,000	8%	2 yrs	$960

48. CODE VIOLATIONS

$$\frac{90}{365} = \frac{18 \bullet \cancel{5}}{73 \bullet \cancel{5}} = \frac{18}{73}$$

$$I = 10,000 \bullet 0.125 \bullet \frac{18}{73}$$

$$I \approx 308.219$$

$$I = 308.22$$

total payback $= 10,000 + 308.22$

total payback $= \$10,308.22$

The total payback is $10,308.22.

49. MONTHLY PAYMENTS

$$7\frac{3}{4}\% = 7.75\% = 0.0775$$

$$I = 1,500 \bullet 0.0775 \bullet 1$$

$$I = 116.25$$

total payback $= 1,500 + 116.25$

total payback $= \$1,616.25$

Dividing this by 12 gives monthly

monthly payment $= 1,616.25 \div 12$

monthly payment ≈ 134.687

monthly payment $= \$134.69$

The monthly payment is $134.69.

50.

$$A = 2,000\left(1 + \frac{0.07}{2}\right)^{2 \bullet 1}$$

$$A = 2,000(1 + 0.035)^2$$

$$A = 2,000(1.035)^2$$

$$A = 2,000(1.071225)$$

$$A = \$2,142.45$$

51.

$$A = 5,000\left(1 + \frac{0.065}{365}\right)^{365\bullet 3}$$

$$A = 5,000\left(1 + 0.000178082\right)^{1095}$$

$$A = 5,000\left(1.000178082\right)^{1095}$$

$$A = 5,000\left(1.21528989\right)$$

$$A \approx 6,076.449$$

$$A = \$6,076.45$$

52. CASH GRANTS

$$A = 500,000\left(1 + \frac{0.083}{365}\right)^{365\bullet 1}$$

$$A = 500,000\left(1 + 0.000227397\right)^{365}$$

$$A = 500,000\left(1.000227397\right)^{365}$$

$$A = 500,000\left(1.08653156\right)$$

$$A \approx 543,265.778$$

$$A = 543,265.78$$

The interest is $A - P$.

$$\text{Interest} = 543,265.78 - 500,000$$

$$\text{Interest} = \$43,265.78$$

The grant will be $43,265.78.

Chapter Seven Test

1. $61\%, \dfrac{61}{100}, \; 0.61$

2. $199\%, \dfrac{199}{100}, 1.99$

3a. $67\% = 0.67$

3b. $12.3\% = 0.123$

3c. $9\dfrac{3}{4}\% = 9.75\% = 0.0975$

4a. $\dfrac{1}{4} = 0.25 = 25\%$

4b. $\dfrac{5}{8} = 0.625 = 62.5\%$

4c. $\dfrac{3}{25} = 0.12 = 12\%$

5a. $0.19 = 19\%$

5b. $3.47 = 347\%$

5c. $0.0005 = 0.5\%$

6a $55\% = \dfrac{55}{100} = \dfrac{11 \bullet \cancel{5}}{20 \bullet \cancel{5}} = \dfrac{11}{20}$

6b. $0.01\% = \dfrac{0.01}{100} = \dfrac{100(0.01)}{100(100)} = \dfrac{1}{10,000}$

6c. $125\% = \dfrac{125}{100} = \dfrac{5 \bullet \cancel{25}}{4 \bullet \cancel{25}} = \dfrac{5}{4}$

7. $\dfrac{7}{30} \approx 0.23333 \approx 23.33\%$

8.   WEATHER REPORTS

If there is a 40% of rain, there is a

$(100 - 40)\%$, or 60% chance of no rain.

9. $\dfrac{2}{3} = 66.\overline{6} = 66\dfrac{2}{3}\%$

10. $\dfrac{1}{4} = 0.25 = 25\%$

11a.   SHRINKAGE

l = length lost due to shrinkage

$l = 34 \bullet 0.03$

$l = 1.02$

1.02 in. of length will be lost.

11b. r = resulting length

$r = 34 - 1.02$

$r = 32.98$

The resulting length is 32.98 in.

12.
$$65 = x \bullet 1,000$$
$$65 = 1,000x$$
$$\frac{65}{1,000} = \frac{1,000x}{1,000}$$
$$0.065 = x$$
$$x = 6.5\%$$

13. TIPPING

What is 15% of 25.40?

$$x = 0.15 \bullet 25.40$$
$$x = 3.81$$

The tip would be $3.81.

14. FUGITIVES

What percent of 450 is 479?

$$x \bullet 479 = 450$$
$$479x = 450$$
$$\frac{479x}{479} = \frac{450}{479}$$
$$x \approx 0.9394$$
$$x \approx 93.9\%$$

93.9% of the fugitives were caught.

15. SWIMMING WORKOUTS

20% of what number is 18?

$$0.20 \bullet x = 18$$
$$\frac{0.2x}{0.2} = \frac{18}{0.2}$$
$$x = 90$$

He normally completes 90 laps during a workout.

16. COLLEGE EMPLOYEES

What number is 3% of 700?

$$x = 0.03 \bullet 700$$
$$x = 21$$

21 employees are in administration.

17.
$$x = 0.24 \bullet 600$$
$$x = 144$$

18. HAIRCUTS

The amount of change is 17 minutes.

What percent of 63 is 17?

$$x \bullet 63 = 17$$
$$63x = 17$$
$$\frac{63x}{63} = \frac{17}{63}$$
$$x \approx 0.2698$$
$$x \approx 27\%$$

A decrease of about 27%.

19. INSURANCE

What is 4% of 898?

$$x = 0.04 \bullet 898$$
$$x = 35.92$$

The commission is $35.92.

20.   COST-OF-LIVING INCREASES

Find the amount of increase, then add that to the original salary.

What is 3.6% of 40,000?

$x = 0.036 \bullet 40,000$

$x = 1,440$

new salary $= 40,000 + 1,440$

new salary $= 41,440$

Her new salary is $41,440.

21.   CAR WAX SALE

sale price $= 14.95 - 3 = \$11.95$

discount $= \$3$

What percent of 14.95 is 3?

$x \bullet 14.95 = 3$

$14.95x = 3$

$$\frac{14.95x}{14.95} = \frac{3}{14.95}$$

$x \approx 0.2007$

$x \approx 20\%$

The discount is about 20%.

22.   POPULATION INCREASES

The amount of increase was 2757.

What percent of 12,808 is 2757?

$x \bullet 12,808 = 2,757$

$12,808x = 2,757$

$$\frac{12,808x}{12,808} = \frac{2,757}{12,808}$$

$x \approx 0.215$

$x \approx 22\%$

The increase was about 22%.

23. $I = 3,000 \bullet 0.05 \bullet 1$

$I = 150$

The interest is $150.

24.

$$A = 24,000\left(1 + \frac{0.064}{365}\right)^{365 \cdot 3}$$

$$A = 24,000\left(1 + 0.000175342\right)^{1095}$$

$$A = 24,000\left(1.000175342\right)^{1095}$$

$$A = 24,000\left(1.211650124\right)$$

$$A \approx 29,079.602$$

$$A = 29,079.60$$

The interest is $A - P$, or

$$29,079.60 - 24,000 = 5,079.60,$$

of the interest earned was $5,079.60.

25. The phrase "bringing crime down to 37%" is unclear. The question that arises is 37% of what?

26. Interest is money that is paid for the use of money.

Chapter 1-7 Cumulative Review Exercises

1.   SHAQUILLE O'NEAL

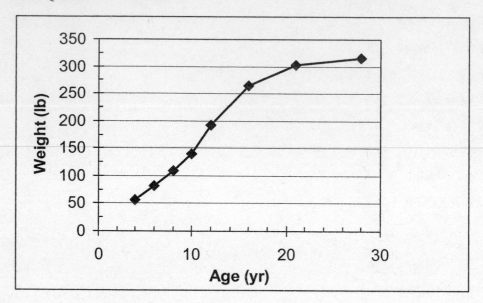

2.   If a and b represent numbers, then

$ab = ba$.

3a.   1, 2, 4, 5, 8, 10, 20, 40

3b.   40

4.   AUTO INSURANCE

$$\text{mean} = \frac{2,672 + 1,680 + 2,485 + 1,370 + 2,737 + 1,692}{6}$$

$$\text{mean} = \frac{12,635}{6} \approx 2,105.83$$

$$\text{mean} \approx \$2,106$$

5. PAINTING

$A = s^2$

$A = (8)^2$

$A = 64$

The tarp covers 64 ft^2.

6. $-12 - (-5) = -12 + 5$

$\qquad = -7$

7. $12 - 2\left[-8 - 2^4(-1)\right]$

$= 12 - 2\left[-8 - 16(-1)\right]$

$= 12 - 2\left[-8 - (-16)\right]$

$= 12 - 2\left[-8 + 16\right]$

$= 12 - 2\left[8\right]$

$= 12 - 16$

$= 12 + (-16)$

$= -4$

8. $\left|-55\right| = 55$

9. $\qquad 6 = 2 - 2x$

$6 - 2 = 2 - 2x - 2$

$\qquad 4 = -2x$

$\dfrac{4}{-2} = \dfrac{-2x}{-2}$

$-2 = x \ \text{ or } \ x = -2$

10. $2t - 16$

11. FRUIT STORAGE

$C = \dfrac{5(59 - 32)}{9}$

$C = \dfrac{5(27)}{9}$

$C = \dfrac{135}{9} = 15°C$

12. $\qquad -(3x - 3) = 6(2x - 7)$

$-3x + 3 = 12x - 42$

$-3x + 3 + 3x = 12x - 42 + 3x$

$3 = 15x - 42$

$3 + 42 = 15x - 42 + 42$

$45 = 15x$

$\dfrac{45}{15} = \dfrac{15x}{15}$

$3 = x \ \text{ or } \ x = 3$

13.   SPELLING

4 vowels out of 11 total letters, $\dfrac{4}{11}$

14. $\dfrac{10y}{15y} = \dfrac{2 \bullet \cancel{5} \bullet \cancel{y}}{3 \bullet \cancel{5} \bullet \cancel{y}} = \dfrac{2}{3}$

15. $-\dfrac{16a}{35} \bullet \dfrac{25}{48a^2} = -\dfrac{\cancel{16} \bullet \cancel{a}}{7 \bullet \cancel{5}} \bullet \dfrac{5 \bullet \cancel{5}}{3 \bullet \cancel{16} \bullet a \bullet \cancel{a}}$

$= -\dfrac{5}{21a}$

16. $4\dfrac{2}{5} \div 11 = \dfrac{22}{5} \bullet \dfrac{1}{11}$

$= \dfrac{2 \bullet \cancel{11}}{5} \bullet \dfrac{1}{\cancel{11}}$

$= \dfrac{2}{5}$

17. $\dfrac{4}{m} + \dfrac{2}{7} = \dfrac{4 \bullet 7}{m \bullet 7} + \dfrac{2 \bullet m}{7 \bullet m}$

$= \dfrac{28}{7m} + \dfrac{2m}{7m}$

$= \dfrac{2m + 28}{7m}$

18. $34\dfrac{1}{9} = \ 34\dfrac{1 \bullet 2}{9 \bullet 2} = \ \ 34\dfrac{2}{18}$

$\underline{-13\dfrac{5}{6} = -13\dfrac{5 \bullet 3}{6 \bullet 3} = \ \ -13\dfrac{15}{18}}$

$33\dfrac{2}{18} + \dfrac{18}{18} = \ \ 33\dfrac{20}{18}$

$\underline{-13\dfrac{15}{18} \qquad = -13\dfrac{15}{18}}$

$= 20\dfrac{5}{18}$

19. $\dfrac{5}{6}y = -25$

$\dfrac{6}{5}\left(\dfrac{5}{6}y\right) = \dfrac{6}{5}\left(-\dfrac{25}{1}\right)$

$y = \dfrac{6}{\cancel{5}}\left(-\dfrac{5 \bullet \cancel{5}}{1}\right)$

$y = -30$

20. $\dfrac{y}{6} = \dfrac{y}{12} + \dfrac{2}{3}$

$12\left(\dfrac{y}{6}\right) = 12\left(\dfrac{y}{12} + \dfrac{2}{3}\right)$

$2y = y + 8$

$2y - y = y + 8 - y$

$y = 8$

21. $78.1 - 7.81 = 70.29$

22. $2.13(-4.05) = -8.6265$

23. $0.752(1,000) = 752$

24. $\dfrac{241.86}{2.9} = 83.4$

25. $\dfrac{3.6-(-1.5)}{0.5(-1.5)-0.4(3.6)} = \dfrac{3.6+1.5}{-0.75-1.44}$

$$= \dfrac{5.1}{-0.75+(-1.44)}$$

$$= \dfrac{5.1}{-2.19}$$

$$\approx -2.328$$

$$= -2.33$$

26. 452.030

27. $\dfrac{11}{15} = 0.7\overline{3}$

28. $\dfrac{y}{2.22} = -5$

$$2.22\left(\dfrac{y}{2.22}\right) = 2.22(-5)$$

$$y = -11.1$$

29. $3\sqrt{81} - 8\sqrt{49} = 3(9) - 8(7)$

$$= 27 - 56$$

$$= -29$$

30. **LABOR COSTS**

$h =$ hours of labor

$$35h + 175 = 297.50$$

$$35h + 175 - 175 = 297.50 - 175$$

$$35h = 122.5$$

$$\dfrac{35h}{35} = \dfrac{122.5}{35}$$

$$h = 3.5$$

It took 3.5 hours of labor on the car.

31. $(m^2 - m - 5) - (3m^2 + 2m - 8)$

$$= (m^2 - m - 5) + (-3m^2 - 2m + 8)$$

$$= (m^2 + [-3m^2]) + (-m + [-2m]) + (-5 + 8)$$

$$= -2m^2 + (-3m) + 3$$

$$= -2m^2 - 3m + 3$$

32. $(3x-2)(x+4)$
$=(3x-2)x+(3x-2)4$
$=x(3x-2)+4(3x-2)$
$=3x^2-2x+12x-8$
$=3x^2+10x-8$

33. $(2y-5)^2=(2y-5)(2y-5)$
$=(2y-5)2y-(2y-5)5$
$=2y(2y-5)-5(2y-5)$
$=4y^2-10y-10y+25$
$=4y^2-20y+25$

34. $y^2y^5=y^{2+5}$
$=y^7$

35. $\left(h^4\right)^5=h^{4\cdot5}$
$=h^{20}$

36. $\left(2a^3b^6\right)^3=(2)^3\left(a^3\right)^3\left(b^6\right)^3$
$=8a^{3\cdot3}b^{6\cdot3}$
$=8a^9b^{18}$

37. $-7g^5\left(8g^4\right)=(-7\cdot8)\left(g^5\cdot g^4\right)$
$=-56g^{5+4}$
$=-56g^9$

38.

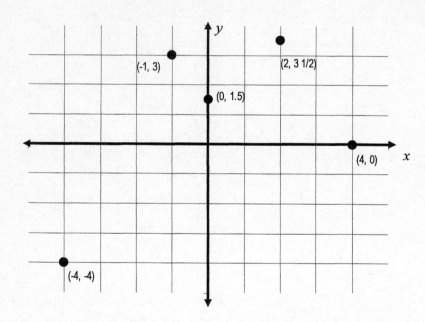

39.

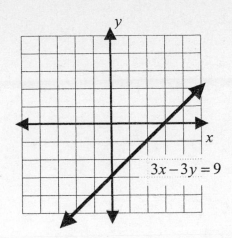

$3x - 3y = 9$

40.

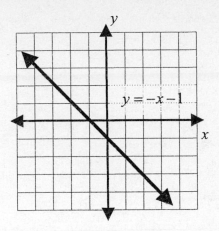

$y = -x - 1$

41.

Percent	Decimal	Fraction
29%	0.29	$\frac{29}{100}$
47.3%	0.473	$\frac{473}{1,000}$
87.5%	0.875	$\frac{7}{8}$

42.

$0.16 \cdot x = 20$

$\dfrac{0.16x}{0.16} = \dfrac{20}{0.16}$

$x = 125$

43.   TIPPING

Gratuity $= 0.15 \cdot 75.18 = 11.277$

rounded up to nearest dollar $= \$12$

Amount $75.18

Gratuity $12.00

Total $87.18

44.   GENEALOGY

$\dfrac{180}{10,000,000} = 0.000018 = 0.0018\%$

45.   SAVINGS ACCOUNT

$I = 10,000 \cdot 0.0725 \cdot 2$

$I = 1,450$

The interest is $1,450.

Chapter 8 Ratio, Proportion, and Measurement

8.1 Ratio

Vocabulary

1. A **ratio** is a quotient of two numbers or a quotient of two quantities with the same unit.

3. When the price of candy is advertised as \$1.75 per pound, we are told its unit **cost**.

Concepts

5. common factor of 3

7. multiply by 10

9.
$$\frac{11 \text{ minutes}}{1 \text{ hour}} = \frac{11 \text{ minutes}}{60 \text{ minutes}} = \frac{11}{60}$$

Notation

11.
$\dfrac{13}{9}, 13 \text{ to } 9, 13:9$

Practice

13.
$5 \text{ to } 7 \text{ is } \dfrac{5}{7}$

15.
$17 \text{ to } 34 \text{ is } \dfrac{17}{34} = \dfrac{17 \cdot 1}{17 \cdot 2} = \dfrac{1}{2}$

17.
$22:33 \text{ is } \dfrac{22}{33} = \dfrac{2 \cdot 11}{3 \cdot 11} = \dfrac{2}{3}$

19.
$1.5:2.4 \text{ is } \dfrac{1.5}{2.4} = \dfrac{10(1.5)}{10(2.4)} = \dfrac{15}{24} = \dfrac{5 \cdot 3}{8 \cdot 3} = \dfrac{5}{8}$

21.
$7 \text{ to } 24.5 \text{ is } \dfrac{7}{24.5} = \dfrac{10(7)}{10(24.5)} = \dfrac{70}{245} = \dfrac{35 \cdot 2}{35 \cdot 7} = \dfrac{2}{7}$

23.
$4 \text{ ounces to } 12 \text{ ounces is } \dfrac{4 \text{ ounces}}{12 \text{ ounces}} = \dfrac{4 \cdot 1}{4 \cdot 3} = \dfrac{1}{3}$

25.

12 minutes to 1 hour is $\dfrac{12 \ \cancel{\text{minutes}}}{60 \ \cancel{\text{minutes}}} = \dfrac{\cancel{12} \cdot 1}{\cancel{12} \cdot 5} = \dfrac{1}{5}$

27.

3 days to 1 week is $\dfrac{3 \ \cancel{\text{days}}}{7 \ \cancel{\text{days}}} = \dfrac{3}{7}$

29.

18 months to 2 years is $\dfrac{18 \ \cancel{\text{months}}}{24 \ \cancel{\text{months}}} = \dfrac{\cancel{6} \cdot 3}{\cancel{6} \cdot 4} = \dfrac{3}{4}$

31.   total budget $= 800 + 600 + 180 + 100 + 120$
total budget $= \$1,800$

33. $\dfrac{600}{1,800} = \dfrac{1 \cdot \cancel{600}}{3 \cdot \cancel{600}} = \dfrac{1}{3}$

35.   total $= 875 + 1,250 + 1,750 + 4,375 + 500$
total $= \$8,750$

37. $\dfrac{1,750}{8,750} = \dfrac{1 \cdot \cancel{1,750}}{5 \cdot \cancel{1,750}} = \dfrac{1}{5}$

39. $\dfrac{64 \text{ ft}}{6 \text{ sec}} = \dfrac{32 \cdot \cancel{2} \text{ ft}}{3 \cdot \cancel{2} \text{ sec}} = \dfrac{32 \text{ ft}}{3 \text{ sec}}$

41. $\dfrac{84 \text{ made}}{100 \text{ attempts}} = \dfrac{21 \cdot \cancel{4} \text{ made}}{25 \cdot \cancel{4} \text{ attempts}} = \dfrac{21 \text{ made}}{25 \text{ attempts}}$

43. $\dfrac{3,000 \text{ students}}{16 \text{ years}} = \dfrac{375 \cdot \cancel{8} \text{ students}}{2 \cdot \cancel{8} \text{ years}} = \dfrac{375 \text{ students}}{2 \text{ years}}$

45. $\dfrac{18 \text{ beats}}{12 \text{ measures}} = \dfrac{3 \cdot \cancel{6} \text{ beats}}{2 \cdot \cancel{6} \text{ measures}} = \dfrac{3 \text{ beats}}{2 \text{ measures}}$

47. $\dfrac{60 \text{ revolutions}}{5 \text{ min}} = 12$ revolutions per min

49. $\dfrac{12 \text{ errors}}{8 \text{ hours}} = 1.5$ errors per hour

51. $\dfrac{245 \text{ presents}}{35 \text{ children}} = 7$ presents per child

53. $\dfrac{4{,}000{,}000 \text{ people}}{12{,}500 \text{ square miles}}$

 $= 320$ people per square mile

55. $\dfrac{\$3.50}{50 \text{ ft}} = \$0.07$ per foot

57. $\dfrac{78 \text{ cents}}{65 \text{ ounces}} = 1.2$ cents per ounce

59. $\dfrac{\$272}{4 \text{ people}} = \$68$ per person

61. $\dfrac{\$4 \text{ billion}}{5 \text{ months}} = \$0.8$ billion per month

Applications

63. ART HISTORY

 Since all sides are the same, the proportion will reduce to $\dfrac{1}{1}$.

65. GEAR RATIOS

 18 teeth on large wheel and 12 teeth on the small wheel gives

 $$\dfrac{18}{12} = \dfrac{3 \bullet \cancel{6}}{2 \bullet \cancel{6}} = \dfrac{3}{2}$$

67. COOKING

 $$\dfrac{\dfrac{2}{3}}{3\dfrac{1}{2}}$$

69. SOFTBALL

 $$\dfrac{12 \text{ hits}}{22 \text{ at bats}} = \dfrac{6 \bullet \cancel{2} \text{ hits}}{11 \bullet \cancel{2} \text{ at bats}} = \dfrac{6 \text{ hits}}{11 \text{ at bats}}$$

71. CPR

 $$\dfrac{125 \text{ compressions}}{50 \text{ breaths}} = \dfrac{5 \bullet \cancel{25} \text{ compressions}}{2 \bullet \cancel{25} \text{ breaths}} = \dfrac{5 \text{ compressions}}{2 \text{ breaths}}$$

73. AIRLINE COMPLAINTS

 $$\dfrac{3.29 \text{ complaints}}{1{,}000 \text{ passengers}} = \dfrac{100\left(3.29 \text{ complaints}\right)}{100\left(1{,}000 \text{ passengers}\right)} = \dfrac{329 \text{ complaints}}{100{,}000 \text{ passengers}}$$

75.　FACULTY-STUDENTS RATIOS

$$\frac{125 \text{ faculty members}}{2,000 \text{ students}} = \frac{1 \cdot \cancel{125} \text{ faculty members}}{16 \cdot \cancel{125} \text{ students}} = \frac{1 \text{ faculty member}}{16 \text{ students}}$$

77.　UNIT COSTS

$$\frac{\$32.13}{20 \text{ gal}} = \$1.89 \text{ per gallon}$$

79.　UNIT COSTS

$$\frac{84 \text{ cents}}{12 \text{ oz}} = 7 \text{ cents per oz}$$

81.　COMPARISON SHOPPING

$$\frac{\$0.89}{6 \text{ oz}} \approx \$0.1483 \text{ per ounce}$$

$$\frac{\$1.19}{8 \text{ oz}} \approx \$0.1488 \text{ per ounce}$$

The 6 oz can is a better buy.

83.　COMPARISON SHOPPING

$$\frac{\$4.29}{20 \text{ tablets}} \approx \$0.2145 \text{ per tablet}$$

$$\frac{\$9.59}{50 \text{ tablets}} \approx \$0.1918 \text{ per tablet}$$

The 50 tablet box is a better buy.

85.　COMPARING SPEEDS

$$\frac{345 \text{ miles}}{6 \text{ hr}} = 57.5 \text{ mph for the car}$$

$$\frac{376 \text{ miles}}{6.2 \text{ hr}} \approx 60.65 \text{ mph for the truck}$$

The truck is going faster.

87.　EMPTYING TANKS

$$\frac{11,880 \text{ gal}}{27 \text{ min}} = 440 \text{ gal per min}$$

89.　AUTO TRAVEL

The car traveled

$$35,071 - 34,746 = 325 \text{ miles}$$

$$\frac{325 \text{ miles}}{5 \text{ hr}} = 65 \text{ mph}$$

The car's speed was 65 mph.

91.　GAS MILEAGE

$$\frac{1,235 \text{ miles}}{51.3 \text{ gal}} \approx 24.07 \text{ mpg for first car}$$

$$\frac{1,456 \text{ miles}}{55.78 \text{ gal}} \approx 26.10 \text{ mpg for second car}$$

The second car had better mileage.

Writing

93. No, 3 to 1 is $\dfrac{3}{1}$, while 1 to 3 is $\dfrac{1}{3}$.

95. You can use unit cost to compare different sized items and select the best deal.

Review

97.
$$
\begin{array}{r}
3.05 \\
17.17 \\
+25.317 \\
\hline
45.537
\end{array}
$$

99.
$$
\begin{aligned}
13.2 + 25.07 \cdot 7.16 &= 13.2 + 179.5012 \\
&= 192.7012
\end{aligned}
$$

101.
$$
5 - 3\frac{1}{4} = \frac{5}{1} - \frac{13}{4} = \frac{5 \cdot 4}{1 \cdot 4} - \frac{13}{4} = \frac{20}{4} - \frac{13}{4}
$$
$$
= \frac{7}{4} = 1\frac{3}{4}
$$

8.2 Proportions

Vocabulary

1.   A **proportion** is a statement that two ratios or rates are equal.

3. $3 \bullet 20 = 60, \ 4 \bullet 15 = 60$ are the two cross products.

Concepts

5.   The equation $\dfrac{a}{b} = \dfrac{c}{d}$ will be a proportion if the product **_ad_** is equal to the product **_bc_**.

7a. $\dfrac{5}{8} = \dfrac{15}{24}$

7b. $\dfrac{3 \text{ teacher's aides}}{25 \text{ children}} = \dfrac{12 \text{ teacher's aides}}{100 \text{ children}}$

9.   i and iv

Notation

11.
$$\frac{12}{18} = \frac{x}{24}$$
$$12 \bullet 24 = 18x$$
$$288 = 18x$$
$$\frac{288}{18} = \frac{18x}{18}$$
$$16 = x$$

Practice

13. $9 \bullet 70 = 630, \ 7 \bullet 81 = 567$

no

15. $7 \bullet 6 = 42, \ 3 \bullet 14 = 42$

yes

17. $9 \bullet 80 = 720, \ 19 \bullet 38 = 722$

no

19. $10.4 \bullet 14.4 = 149.76,$
$3.6 \bullet 41.6 = 149.76$

yes

21. $\dfrac{2}{3} \cdot \dfrac{9}{16} = \dfrac{\cancel{2}}{\cancel{3}} \cdot \dfrac{3 \cdot \cancel{3}}{8 \cdot \cancel{2}} = \dfrac{3}{8}$,

$\dfrac{5}{8} \cdot \dfrac{4}{5} = \dfrac{\cancel{5}}{4 \cdot \cancel{2}} \cdot \dfrac{2 \cdot \cancel{2}}{\cancel{5}} = \dfrac{2}{4} = \dfrac{1}{2}$

no

23. $\dfrac{25}{6} \cdot \dfrac{9}{10} = \dfrac{5 \cdot \cancel{5}}{2 \cdot \cancel{3}} \cdot \dfrac{3 \cdot \cancel{3}}{2 \cdot \cancel{5}} = \dfrac{15}{4}$,

$\dfrac{12}{7} \cdot \dfrac{35}{16} = \dfrac{3 \cdot \cancel{4}}{\cancel{7}} \cdot \dfrac{5 \cdot \cancel{7}}{4 \cdot \cancel{4}} = \dfrac{15}{4}$

yes

25. $\dfrac{2}{3} = \dfrac{x}{6}$

$12 = 3x$

$\dfrac{12}{3} = \dfrac{3x}{3}$

$4 = x \ \text{ or } \ x = 4$

Check:

$\dfrac{2}{3} = \dfrac{4}{6}$

$2 \cdot 6 = 12$

$3 \cdot 4 = 12$

27. $\dfrac{5}{10} = \dfrac{3}{c}$

$5c = 30$

$\dfrac{5c}{5} = \dfrac{30}{5}$

$c = 6$

Check:

$\dfrac{5}{10} = \dfrac{3}{6}$

$5 \cdot 6 = 30$

$10 \cdot 3 = 30$

29. $\dfrac{6}{x} = \dfrac{8}{4}$

$24 = 8x$

$\dfrac{24}{8} = \dfrac{8x}{8}$

$3 = x \ \text{ or } \ x = 3$

Check:

$\dfrac{6}{3} = \dfrac{8}{4}$

$6 \cdot 4 = 24$

$3 \cdot 8 = 24$

31. $\dfrac{x}{8} = \dfrac{9}{2}$

$2x = 72$

$\dfrac{2x}{2} = \dfrac{72}{2}$

$x = 36$

Check:

$\dfrac{36}{8} = \dfrac{9}{2}$

$36 \cdot 2 = 72$

$8 \cdot 9 = 72$

33.
$$\frac{x+1}{5} = \frac{3}{15}$$

$$15(x+1) = 15$$

$$15x + 15 = 15$$

$$15x + 15 - 15 = 15 - 15$$

$$15x = 0$$

$$\frac{15x}{15} = \frac{0}{15}$$

$$x = 0$$

Check:

$$\frac{0+1}{5} = \frac{3}{15}$$

$$\frac{1}{5} = \frac{3}{15}$$

$$1 \bullet 15 = 15$$

$$5 \bullet 3 = 15$$

35.
$$\frac{x+3}{12} = \frac{-7}{6}$$

$$6(x+3) = -84$$

$$6x + 18 = -84$$

$$6x + 18 - 18 = -84 - 18$$

$$6x = -102$$

$$\frac{6x}{6} = \frac{-102}{6}$$

$$x = -17$$

Check:

$$\frac{-17+3}{12} = \frac{-7}{6}$$

$$\frac{-14}{12} = \frac{-7}{6}$$

$$-14 \bullet 6 = -84$$

$$12 \bullet (-7) = -84$$

37.
$$\frac{4-x}{13}=\frac{11}{26}$$
$$26(4-x)=143$$
$$104-26x=143$$
$$104-26x-104=143-104$$
$$-26x=39$$
$$\frac{-26x}{-26}=\frac{39}{-26}$$
$$x=-\frac{39}{26}$$
$$x=-\frac{3\bullet\cancel{13}}{2\bullet\cancel{13}}=-\frac{3}{2}$$

Check:
$$\frac{4-\left(-\dfrac{3}{2}\right)}{13}=\frac{11}{26}$$
$$\frac{2\left(4+\dfrac{3}{2}\right)}{2(13)}=\frac{11}{26}$$
$$\frac{8+3}{26}=\frac{11}{26}$$
$$\frac{11}{26}=\frac{11}{26}$$

39.
$$\frac{2x+1}{18}=\frac{14}{3}$$
$$3(2x+1)=252$$
$$6x+3=252$$
$$6x+3-3=252-3$$
$$6x=249$$
$$\frac{6x}{6}=\frac{249}{6}$$
$$x=\frac{249}{6}$$
$$x=\frac{83\bullet\cancel{3}}{2\bullet\cancel{3}}=\frac{83}{2}$$

Check:
$$\frac{2\left(\dfrac{83}{2}\right)+1}{18}=\frac{14}{3}$$
$$\frac{83+1}{18}=\frac{14}{3}$$
$$\frac{84}{18}=\frac{14}{3}$$
$$3\bullet84=252$$
$$18\bullet14=252$$

41.
$$\frac{4,000}{x} = \frac{3.2}{2.8}$$
$$11,200 = 3.2x$$
$$\frac{11,200}{3.2} = \frac{3.2x}{3.2}$$
$$3,500 = x \ \text{ or } \ x = 3,500$$

Check:
$$\frac{4,000}{3,500} = \frac{3.2}{2.8}$$
$$4,000 \bullet 2.8 = 3,500 \bullet 3.2$$
$$11,200 = 11,200$$

43.
$$\frac{\frac{1}{2}}{\frac{1}{5}} = \frac{x}{2\frac{1}{4}}$$
$$\frac{0.5}{0.2} = \frac{x}{2.25}$$
$$0.2x = 1.125$$
$$\frac{0.2x}{0.2} = \frac{1.125}{0.2}$$
$$x = 5.625 \ \text{ or } \ x = 5\frac{5}{8}$$

Check:
$$\frac{0.5}{0.2} = \frac{5.625}{2.25}$$
$$0.5 \bullet 2.25 = 0.2 \bullet 5.625$$
$$1.125 = 1.125$$

Applications

45.   SCHOOL LUNCHES
$$x = \text{cost of 750 cups}$$
$$\frac{6 \text{ cups}}{\$1.75} = \frac{750 \text{ cups}}{\$x}$$
$$6x = 1,312.5$$
$$\frac{6x}{6} = \frac{1,312.5}{6}$$
$$x = 218.75$$

It will cost $218.75.

47.   GARDENING
$$3 \text{ dozen} = 36 \text{ packets}$$
$$x = \text{cost of 36 packets}$$
$$\frac{3 \text{ packets}}{\$0.98} = \frac{36 \text{ packets}}{\$x}$$
$$3x = 35.28$$
$$\frac{3x}{3} = \frac{35.28}{3}$$
$$x = 11.76$$

It will cost $11.76.

49. **BUSINESS PERFORMANCE**

For 2003, $\dfrac{5}{10}$.

For 2004, $\dfrac{10}{20}$.

Since the cross products give

$$\frac{5}{10} = \frac{10}{20}$$

$100 = 100$, they are the same.

51. **MIXING PERFUMES**

$x =$ drops of pure essence

$$\frac{3 \text{ drops pure essence}}{7 \text{ drops alcohol}} = \frac{x \text{ drops pure essence}}{56 \text{ drops alcohol}}$$

$$7x = 168$$

$$\frac{7x}{7} = \frac{168}{7}$$

$$x = 24$$

24 drops of pure essence will be needed.

53. **LAB WORK**

$x =$ red blood cells on whole grid

$$\frac{5 \text{ squares}}{195 \text{ cells}} = \frac{25 \text{ squares}}{x \text{ cells}}$$

$$5x = 4,875$$

$$\frac{5x}{5} = \frac{4,875}{5}$$

$$x = 975$$

The grid contains 975 cells total.

55. **MAKING COOKIES**

$x =$ cups of flour needed

$$\frac{1\frac{1}{4} \text{ cups of flour}}{3\frac{1}{2} \text{ dozen}} = \frac{x \text{ cups of flour}}{12 \text{ dozen}}$$

$$\frac{1.25 \text{ cups of flour}}{3.5 \text{ dozen}} = \frac{x \text{ cups of flour}}{12 \text{ dozen}}$$

$$3.5x = 15$$

$$\frac{3.5x}{3.5} = \frac{15}{3.5}$$

$$x \approx 4.286$$

It will take about $4\frac{1}{4}$ cups of flour.

57. COMPUTER SPEED

$x =$ time for 100 calculations

$$\frac{15 \text{ calculations}}{2.85 \text{ sec}} = \frac{100 \text{ calculations}}{x \text{ sec}}$$

$$15x = 285$$

$$\frac{15x}{15} = \frac{285}{15}$$

$$x = 19$$

It would take 19 seconds.

59. FUEL CONSUMPTION

$x =$ miles Hummer can go on 17 gal

$$\frac{325 \text{ miles}}{25 \text{ gal}} = \frac{x \text{ miles}}{17 \text{ gal}}$$

$$25x = 5,525$$

$$\frac{25x}{25} = \frac{5,525}{25}$$

$$x = 221$$

The Hummer can go 221 miles on 17 gallons of diesel.

61. PAYCHECKS

$x =$ pay for 30 hour week

$$\frac{40 \text{ hours}}{\$412} = \frac{30 \text{ hours}}{\$x}$$

$$40x = 12,360$$

$$\frac{40x}{40} = \frac{12,360}{40}$$

$$x = 309$$

He will make $309 for 30 hours.

63. BLUEPRINTS

$x =$ actual kitchen length

$$\frac{\frac{1}{4} \text{ in}}{1 \text{ ft}} = \frac{2\frac{1}{2} \text{ in}}{x \text{ ft}}$$

$$\frac{0.25 \text{ in}}{1 \text{ ft}} = \frac{2.5 \text{ in}}{x \text{ ft}}$$

$$0.25x = 2.5$$

$$\frac{0.25x}{0.25} = \frac{2.5}{0.25}$$

$$x = 10$$

The kitchen is 10 feet long.

65. MODEL RAILROADING

$x =$ length of real engine

$$\frac{87 \text{ ft}}{12 \text{ in}} = \frac{x \text{ ft}}{9 \text{ in}}$$

$$12x = 783$$

$$\frac{12x}{12} = \frac{783}{12}$$

$$x = 65.25$$

The real engine is 65.25 ft long, or 65 ft 3 in.

67. MINIATURES

$35 \text{ ft} = 420 \text{ in}$

$x = $ width of model

$$\frac{160}{1} = \frac{420}{x}$$

$160x = 420$

$$\frac{160x}{160} = \frac{420}{160}$$

$x = 2.625 \text{ or } x = 2\frac{5}{8}$

The width of the model is 2.625 in or

$2\frac{5}{8}$ in.

Writing

69. A ratio is a quotient of two numbers, while a proportion is a statement that 2 ratios are equal.

71. It means that for every 1 foot of measurement of the actual house or item, the miniature would measure 1 inch per foot.

Review

73. $\frac{9}{10} = 0.9 = 90\%$

75.

$$33\frac{1}{3}\% = \frac{33\frac{1}{3}}{100} = \frac{\frac{100}{3}}{100} = \frac{3\left(\frac{100}{3}\right)}{3(100)} = \frac{100}{300}$$

$$= \frac{1 \cdot \cancel{100}}{3 \cdot \cancel{100}} = \frac{1}{3}$$

77. $x = 0.005 \cdot 520$

$x = 2.6$

8.3 American Units of Measurement

Vocabulary

1. Inches, feet, and miles are examples of American units of **length**.

3. The value of any unit conversion factor is **1**.

5. Some examples of American units of **capacity** are cups, pints, quarts, and gallons.

Concepts

7.   12 in. = 1 ft   9.   1 mi = 5,280 ft

11.   16 ounces = 1 pound   13.   1 cup = 8 fluid ounces

15.   2 pints = 1 quart   17.   1 day = 24 hours

19.
left arrow is $\dfrac{5}{8}$ in.

middle arrow is $1\dfrac{3}{4}$ in.

right arrow is $2\dfrac{5}{16}$ in.

21a. $\dfrac{1 \text{ ton}}{2,000 \text{ lb}}$ 21b. $\dfrac{2 \text{ pt}}{1 \text{ qt}}$

23a.   iv   23b.   i

23c.   ii   23d.   iii

25a.   iii   25b.   iv

25c.   i   25d.   ii

Notation

27.
$$12 \text{ yd} = 12 \cancel{\text{ yd}} \cdot \frac{36 \text{ in.}}{1 \cancel{\text{ yd}}}$$
$$= 12 \cdot 36 \text{ in.}$$
$$= 432 \text{ in.}$$

29.
$$12 \text{ pt} = 12 \cancel{\text{ pt}} \cdot \frac{1 \cancel{\text{ qt}}}{2 \cancel{\text{ pt}}} \cdot \frac{1 \text{ gal}}{4 \cancel{\text{ qt}}}$$
$$= 12 \cdot \frac{1}{2} \cdot \frac{1}{4} \text{ gal}$$
$$= 1.5 \text{ gal}$$

Practice

31.
$$2\frac{5}{8} \text{ in.}$$

33.
$$10\frac{3}{4} \text{ in.}$$

35.
$$4 \text{ ft} = 4 \cancel{\text{ ft}} \cdot \frac{12 \text{ in.}}{1 \cancel{\text{ ft}}}$$
$$= 4 \cdot 12 \text{ in.}$$
$$= 48 \text{ in.}$$

37.
$$3\frac{1}{2} \text{ ft} = \frac{7}{2} \cancel{\text{ ft}} \cdot \frac{12 \text{ in.}}{1 \cancel{\text{ ft}}}$$
$$= \frac{7}{\cancel{2}} \cdot \frac{6 \cdot \cancel{2}}{1} \text{ in.}$$
$$= 7 \cdot 6 \text{ in.}$$
$$= 42 \text{ in.}$$

39.
$$24 \text{ in.} = 24 \cancel{\text{ in.}} \cdot \frac{1 \text{ ft}}{12 \cancel{\text{ in.}}}$$
$$= \frac{24}{12} \text{ ft}$$
$$= 2 \text{ ft}$$

41.
$$8 \text{ yd} = 8 \cancel{\text{ yd}} \cdot \frac{36 \text{ in.}}{1 \cancel{\text{ yd}}}$$
$$= 8 \cdot 36 \text{ in.}$$
$$= 288 \text{ in.}$$

43.
$$90 \text{ in.} = 90 \cancel{\text{ in.}} \cdot \frac{1 \text{ yd}}{36 \cancel{\text{ in.}}}$$
$$= \frac{90}{36} \text{ yd}$$
$$= 2.5 \text{ yd}$$

45.
$$56 \text{ in.} = 56 \cancel{\text{ in.}} \cdot \frac{1 \text{ ft}}{12 \cancel{\text{ in.}}}$$
$$= \frac{56}{12} \text{ ft}$$
$$= 4.\overline{6} \text{ ft} = 4\frac{2}{3} \text{ ft}$$

47.
$$5 \text{ yd} = 5 \text{ yd} \cdot \frac{3 \text{ ft}}{1 \text{ yd}}$$
$$= 5 \cdot 3 \text{ ft}$$
$$= 15 \text{ ft}$$

49.
$$7 \text{ ft} = 7 \text{ ft} \cdot \frac{1 \text{ yd}}{3 \text{ ft}}$$
$$= \frac{7}{3} \text{ yd}$$
$$= 2.\overline{3} \text{ yd} = 2\frac{1}{3} \text{ yd}$$

51.
$$15{,}840 \text{ ft} = 15{,}840 \text{ ft} \cdot \frac{1 \text{ mi}}{5{,}280 \text{ ft}}$$
$$= \frac{15{,}840}{5{,}280} \text{ mi}$$
$$= 3 \text{ mi}$$

53.
$$\frac{1}{2} \text{ mi} = \frac{1}{2} \text{ mi} \cdot \frac{5{,}280 \text{ ft}}{1 \text{ mi}}$$
$$= \frac{1}{2} \cdot \frac{5{,}280}{1} \text{ ft}$$
$$= \frac{1}{2} \cdot \frac{2{,}640 \cdot 2}{1} \text{ ft}$$
$$= 2{,}640 \text{ ft}$$

55.
$$80 \text{ oz} = 80 \text{ oz} \cdot \frac{1 \text{ lb}}{16 \text{ oz}}$$
$$= \frac{80}{16} \text{ lb}$$
$$= 5 \text{ lb}$$

57.
$$7{,}000 \text{ lb} = 7{,}000 \text{ lb} \cdot \frac{1 \text{ ton}}{2{,}000 \text{ lb}}$$
$$= \frac{7{,}000}{2{,}000} \text{ ton}$$
$$= \frac{7 \cdot 1{,}000}{2 \cdot 1{,}000} \text{ ton}$$
$$= 3.5 \text{ ton} = 3\frac{1}{2} \text{ ton}$$

59.
$$12.4 \text{ ton} = 12.4 \text{ ton} \cdot \frac{2{,}000 \text{ lb}}{1 \text{ ton}}$$
$$= 12.4 \cdot 2{,}000 \text{ lb}$$
$$= 24{,}800 \text{ lb}$$

61.
$$3 \text{ qt} = 3 \text{ qt} \cdot \frac{2 \text{ pt}}{1 \text{ qt}}$$
$$= 3 \cdot 2 \text{ pt}$$
$$= 6 \text{ pt}$$

63.
$$16 \text{ pt} = 16 \text{ pt} \cdot \frac{1 \text{ qt}}{2 \text{ pt}} \cdot \frac{1 \text{ gal}}{4 \text{ qt}}$$
$$= \frac{16}{2 \cdot 4} \text{ gal}$$
$$= 2 \text{ gal}$$

65.
$$32 \text{ fl oz} = 32 \text{ fl oz} \cdot \frac{1 \text{ cup}}{8 \text{ fl oz}} \cdot \frac{1 \text{ pt}}{2 \text{ cups}}$$
$$= \frac{32}{8 \cdot 2} \text{ pt}$$
$$= 2 \text{ pt}$$

67.
$$240 \text{ min} = 240 \cancel{\text{ min}} \cdot \frac{1 \text{ hr}}{60 \cancel{\text{ min}}}$$
$$= \frac{240}{60} \text{ hr}$$
$$= 4 \text{ hr}$$

69.
$$7,200 \text{ min} = 7,200 \cancel{\text{ min}} \cdot \frac{1 \cancel{\text{ hr}}}{60 \cancel{\text{ min}}} \cdot \frac{1 \text{ day}}{24 \cancel{\text{ hr}}}$$
$$= \frac{7,200}{60 \cdot 24} \text{ days}$$
$$= 5 \text{ days}$$

Applications

71.   THE GREAT PYRAMID

$$450 \text{ ft} = 450 \cancel{\text{ ft}} \cdot \frac{1 \text{ yd}}{3 \cancel{\text{ ft}}}$$
$$= \frac{450}{3} \text{ yd}$$
$$= 150 \text{ yd}$$

73.   THE GREAT SPHINX

$$240 \text{ ft} = 240 \cancel{\text{ ft}} \cdot \frac{12 \text{ in.}}{1 \cancel{\text{ ft}}}$$
$$= 240 \cdot 12 \text{ in.}$$
$$= 2,880 \text{ in.}$$

75.   THE SEARS TOWER

$$1,454 \text{ ft} = 1,454 \cancel{\text{ ft}} \cdot \frac{1 \text{ mi}}{5,280 \cancel{\text{ ft}}}$$
$$= \frac{1,454}{5,280} \text{ mi}$$
$$\approx 0.275 \text{ mi}$$
$$\approx 0.28 \text{ mi}$$

77.   NFL RECORDS

$$35 \text{ mi} = 35 \cancel{\text{ mi}} \cdot \frac{5,280 \cancel{\text{ ft}}}{1 \cancel{\text{ mi}}} \cdot \frac{1 \text{ yd}}{3 \cancel{\text{ ft}}}$$
$$= \frac{35 \cdot 5,280}{3} \text{ yd}$$
$$= 61,600 \text{ yd}$$

79.   WEIGHT OF WATER

$$8 \text{ lb} = 8 \text{ \cancel{lb}} \cdot \frac{16 \text{ oz}}{1 \text{ \cancel{lb}}}$$

$$= 8 \cdot 16 \text{ oz}$$

$$= 128 \text{ oz}$$

81.   HIPPOS

$$9{,}900 \text{ lb} = 9{,}900 \text{ \cancel{lb}} \cdot \frac{1 \text{ ton}}{2{,}000 \text{ \cancel{lb}}}$$

$$= \frac{9{,}900}{2{,}000} \text{ ton}$$

$$= 4.95 \text{ tons}$$

83.   BUYING PAINT

$$17 \text{ gal} = 17 \text{ \cancel{gal}} \cdot \frac{4 \text{ qt}}{1 \text{ \cancel{gal}}}$$

$$= 17 \cdot 4 \text{ qt}$$

$$= 68 \text{ qt}$$

85.   SCHOOL LUNCHES

$$575 \text{ pt} = 575 \text{ \cancel{pt}} \cdot \frac{1 \text{ \cancel{qt}}}{2 \text{ \cancel{pt}}} \cdot \frac{1 \text{ gal}}{4 \text{ \cancel{qt}}}$$

$$= \frac{575}{2 \cdot 4} \text{ gal}$$

$$= 71.875 \text{ gal} = 71\frac{7}{8} \text{ gal}$$

87.   CAMPING

$$2\frac{1}{2} \text{ gal} = 2.5 \text{ \cancel{gal}} \cdot \frac{4 \text{ \cancel{qt}}}{1 \text{ \cancel{gal}}} \cdot \frac{2 \text{ \cancel{pt}}}{1 \text{ \cancel{qt}}} \cdot \frac{2 \text{ \cancel{cups}}}{1 \text{ \cancel{pt}}} \cdot \frac{8 \text{ fl oz}}{1 \text{ \cancel{cup}}}$$

$$= 2.5 \cdot 4 \cdot 2 \cdot 2 \cdot 8 \text{ fl oz}$$

$$= 320 \text{ fl oz}$$

89.   SPACE TRAVEL

$$147 \text{ hr} = 147 \text{ \cancel{hr}} \cdot \frac{1 \text{ day}}{24 \text{ \cancel{hr}}}$$

$$= \frac{147}{24} \text{ days}$$

$$= 6.125 \text{ days} = 6\frac{1}{8} \text{ days}$$

Writing

91. To convert feet to inches, make sure that feet is in the denominator of the conversion factor so that they will cancel out and leave inches. The factor is $\dfrac{12 \text{ in.}}{1 \text{ ft}}$.

Review

93. 3,700

95. 3,673.26

97. 0.101

99. 0.1

8.4 Metric Units of Measurement

Vocabulary

1. *Deka* means <u>**tens**</u>.

3. *Kilo* means <u>**thousands**</u>.

5. *Centi* means <u>**hundredths**</u>.

7. Meters, grams, and liters are units of measurement in the <u>**metric**</u> system.

Concepts

9.   left arrow is 1 cm

    middle arrow is 3 cm

    right arrow is 6 cm

11a. $\dfrac{1 \text{ km}}{1,000 \text{ m}}$ 11b. $\dfrac{100 \text{ cg}}{1 \text{ g}}$ 11c. $\dfrac{1,000 \text{ milliliters}}{1 \text{ liter}}$

13a.   iii   13b.   i   13c.   ii

15a.   ii   15b.   iii   15c.   i

17.   1 dekameter = 10 meters   19. $1 \text{ centimeter} = \dfrac{1}{100} \text{ meter}$

21. $1 \text{ millimeter} = \dfrac{1}{1,000} \text{ meter}$ 23.   1 gram = 1,000 milligrams

25.   1 kilogram = 1,000 grams   27.   1 liter = 1,000 cubic centimeters

29. $1 \text{ centiliter} = \dfrac{1}{100} \text{ liter}$ 31.   100 liters = 1 hectoliter

Notation

33.
$$20 \text{ cm} = 20 \text{ cm} \cdot \frac{1 \text{ m}}{100 \text{ cm}}$$
$$= \frac{20}{100} \text{ m}$$
$$= 0.2 \text{ m}$$

35.
$$2 \text{ km} = 2 \text{ km} \cdot \frac{1{,}000 \text{ m}}{1 \text{ km}} \cdot \frac{10 \text{ dm}}{1 \text{ m}}$$
$$= 2 \cdot 1{,}000 \cdot 10 \text{ dm}$$
$$= 20{,}000 \text{ dm}$$

Practice

For questions 41 through 88, we use the following chart for conversions. Substitute grams or liters instead of meters when needed. For 87 and 88, remember 1 cc = 1 mL.

km   hm   dam   m   dm   cm   mm

37. 156 mm

39. 28 cm

41. 3 m = 300 cm

43. 5.7 m = 570 cm

45. 0.31dm = 3.1 cm

47. 76.8 hm = 7,680,000 mm

49. 4.72 cm = 0.472 dm

51. 453.2 cm = 4.532 m

53. 0.325 dm = 0.0325 m

55. 3.75 cm = 37.5 mm

57. 0.125 m = 125 mm

59. 675 dam = 675,000 cm

61. 638.3 m = 6.383 hm

63. 6.3 mm = 0.63 cm

65. 695 dm = 69.5 m

67. 5,689 m = 5.689 km

69. 576.2 mm = 5.762 dm

71. 6.45 dm = 0.00068 km

73.   658.23 m = 0.65823 km   75.   3g = 3,000 mg

77.   2 kg = 2,000 g   79.   1,000 kg = 1,000,000 g

81.   500 mg = 0.5 g   83.   3 kL = 3,000 L

85.   500 cL = 5,000 mL   87.   10 mL = 10 cc

Applications

89.   SPEED SKATING   91.   HEALTH CARE

       500 m = 0.5 km   120 mm = 12 cm

       1,000 m = 1 km   80 mm = 8 cm

       1,500 m = 1.5 km

       5,000 m = 5 km

       10,000 m = 10 km

93.   WEIGHT OF A BABY   95.   CONTAINERS

       4 kg = 400,000 cg $2 \bullet 2L = 4L$

                                          4L = 40 dL

97.   BUYING OLIVES   99.   MEDICINE

       1 kg = 1,000 g. Since each bottle $50 \text{ mg} \bullet 60 = 3,000 \text{ mg total}$

       weighs 284 g, you will need 4 bottles   3,000 mg = 3g

       to have at least 1 kg.

Writing

101.   Since the metric system is a base 10 system and since kilometers are larger than meters, we
        can move the decimal 3 units to the right to convert km to m.

103.   .century is 100 years, cent is 1 hundredth of a dollar, etc...

Review

105. $x = 0.07 \bullet 342.72$

   $x \approx 23.9904$

   $x = \$23.99$

107. $32.16 = 0.08 \bullet x$

   $\dfrac{32.16}{0.08} = \dfrac{0.08x}{0.08}$

   $402 = x$ or $x = 402$

109.

$$3\frac{1}{7} + 2\frac{1}{2} \bullet 3\frac{1}{3} = 3\frac{1}{7} + \frac{5}{2} \bullet \frac{10}{3}$$

$$= 3\frac{1}{7} + \frac{5}{\cancel{2}} \bullet \frac{5 \bullet \cancel{2}}{3}$$

$$= \frac{22}{7} + \frac{25}{3}$$

$$= \frac{22 \bullet 3}{7 \bullet 3} + \frac{25 \bullet 7}{3 \bullet 7}$$

$$= \frac{66}{21} + \frac{175}{21}$$

$$= \frac{241}{21} = 11\frac{10}{21}$$

8.5 Converting between American and Metric Units

Vocabulary

1. In the American system, temperatures are measured in degrees **Fahrenheit**. In the metric system, temperatures are measured in degrees **Celsius**.

Concepts

3a.   a meter is longer   3b.   a meter is longer

3c.   an inch is longer   3d.   a mile is longer

5a.   a liter has more capacity   5b.   a liter has more capacity

5c.   a gallon has more capacity

Notation

7.
$$4,500 \text{ ft} = 4,500 \ (0.3048 \text{ m})$$
$$= 4,500(0.3048)\text{m}$$
$$= 1371.6 \text{ m}$$
$$= 1.3716 \text{ km}$$

9.
$$8 \text{ L} = 8 \ (0.264 \text{ gal})$$
$$= 8(0.264) \text{ gal}$$
$$= 2.112 \text{ gal}$$

Practice

11.
$$3 \text{ ft} = 3(0.3048 \text{ m})$$
$$= 0.9144 \text{ m}$$
$$= 91.4 \text{ cm}$$

13.
$$3.75 \text{ m} = 3.75(3.2808 \text{ ft})$$
$$= 12.303 \ \cancel{\text{ft}} \cdot \frac{12 \text{ in.}}{1 \ \cancel{\text{ft}}}$$
$$= 12.303 \cdot 12 \text{ in.}$$
$$= 147.636 \text{ in.}$$
$$\approx 147.6 \text{ in.}$$

15. $12 \text{ km} = 12{,}000 \text{ m}$
$= 12{,}000(3.2808 \text{ ft})$
$= 39{,}369.6 \text{ ft}$
$\approx 39{,}370 \text{ ft}$

17. $5{,}000 \text{ in.} = 5{,}000(2.54 \text{ cm})$
$= 12{,}700 \text{ cm}$
$= 127 \text{ m}$

19. $37 \text{ oz} = 37(28.35 \text{ g})$
$= 1{,}048.95 \text{ g}$
$\approx 1.049 \text{ kg}$
$\approx 1 \text{ kg}$

21. $25 \text{ lb} = 25(0.454 \text{ kg})$
$= 11.35 \text{ kg}$
$= 11{,}350 \text{ g}$

23. $0.5 \text{ kg} = 0.5(2.2 \text{ lb})$
$= 1.1 \, \cancel{\text{lb}} \bullet \dfrac{16 \text{ oz}}{1 \, \cancel{\text{lb}}}$
$= 1.1 \bullet 16 \text{ oz}$
$= 17.6 \text{ oz}$

25. $17 \text{ g} = 17(0.035 \text{ oz})$
$= 0.595 \text{ oz}$
$\approx 0.6 \text{ oz}$

27. $3 \text{ fl oz} = 3(0.030 \text{ L})$
$= 0.09 \text{ L}$
$\approx 0.1 \text{ L}$

29. $7.2 \text{ L} = 7.2(33.8 \text{ fl oz})$
$= 243.36 \text{ fl oz}$
$\approx 243.4 \text{ fl oz}$

31. $0.75 \text{ qt} = 0.75(0.946 \text{ L})$
$= 0.7095 \text{ L}$
$= 709.5 \text{ mL}$
$\approx 710 \text{ mL}$

33. $500 \text{ mL} = 0.5 \text{ L}$
$= 0.5(1.06 \text{ qt})$
$= 0.53 \text{ qt}$
$\approx 0.5 \text{ qt}$

35. $C = \dfrac{5(F-32)}{9}$, when $F = 50$
$C = \dfrac{5(50-32)}{9}$
$C = \dfrac{5(18)}{9}$
$C = \dfrac{90}{9} = 10° \text{ C}$

37. $F = \dfrac{9}{5}C + 32$, when $C = 50$
$F = \dfrac{9}{5}(50) + 32$
$F = 90 + 32$
$F = 122° \text{ F}$

39.
$$F = \frac{9}{5}C + 32, \text{ when } C = -10$$
$$F = \frac{9}{5}(-10) + 32$$
$$F = -18 + 32$$
$$F = 14° \text{ F}$$

41.
$$C = \frac{5(F - 32)}{9}, \text{ when } F = -5$$
$$C = \frac{5(-5 - 32)}{9}$$
$$C = \frac{5(-37)}{9}$$
$$C = \frac{-185}{9} \approx -20.556$$
$$C \approx -20.6° \text{ C}$$

Applications

43.   THE MIDDLE EAST
$$8 \text{ km} = 8(0.6214 \text{ mi})$$
$$= 4.9712 \text{ mi}$$
$$\approx 5 \text{ mi}$$

45.   CHEETAHS
$$112 \text{ km per hour} = 112(0.6214 \text{ mi})$$
$$= 69.5968 \text{ mi}$$
$$= 70 \text{ mph}$$

47.   MOUNT WASHINGTON
$$6,288 \text{ ft} = 6,288(0.3048 \text{ m})$$
$$= 1916.5824 \text{ m}$$
$$\approx 1.9 \text{ km}$$

49.   HAIR GROWTH
$$\frac{3}{4} \text{ in} = 0.75 \text{ in}$$
$$= 0.75(2.54 \text{ cm})$$
$$= 1.905 \text{ cm}$$
$$\approx 1.9 \text{ cm}$$

51.   WEIGHTLIFING
$$187 \text{ kg} = 187(2.2 \text{ lb})$$
$$= 411.4 \text{ lb}$$
$$\approx 411 \text{ lb}$$
$$338 \text{ kg} = 338(2.2 \text{ lb})$$
$$= 743.6 \text{ lb}$$
$$\approx 744 \text{ lb}$$

53a.   OUNCES AND FLUID OUNCES

$8 \text{ oz} = 8(28.35 \text{ g})$

$\quad = 226.8 \text{ g}$

53b. $8 \text{ fl oz} = 8(0.030 \text{ L})$

$\quad = 0.24 \text{ L}$

55.   POSTAL REGULATIONS

$32 \text{ kg} = 32(2.2 \text{ lb})$

$\quad = 70.4 \text{ lb}$

You cannot mail a 32 kg package since it is over the 70 pound limit.

57.   COMPARISON SHOPPING

$3 \text{ qt} = 3(0.946 \text{ L})$

$\quad = 2.838 \text{ L}$

$\dfrac{\$4.50}{2.838 \text{ L}} = \$1.586 \text{ per liter}$

$\dfrac{\$3.60}{2 \text{ L}} = \$1.80 \text{ per liter}$

The 3 quarts is the better buy.

59.   HOT SPRINGS

$C = \dfrac{5(F-32)}{9}, \text{ when } F = 143$

$C = \dfrac{5(143-32)}{9}$

$C = \dfrac{5(111)}{9}$

$C = \dfrac{555}{9} \approx 61.667$

$C \approx 62° \text{ C}$

61.   TAKING A SHOWER

$F = \dfrac{9}{5}C + 32, \text{ when } C = 15 \quad F = \dfrac{9}{5}C + 32, \text{ when } C = 28 \quad F = \dfrac{9}{5}C + 32, \text{ when } C = 50$

$F = \dfrac{9}{5}(15) + 32 \qquad\qquad F = \dfrac{9}{5}(28) + 32 \qquad\qquad F = \dfrac{9}{5}(50) + 32$

$F = 27 + 32 \qquad\qquad\qquad F = 50.4 + 32 \qquad\qquad\quad F = 90 + 32$

$F = 59° \text{ F} \qquad\qquad\qquad F = 82.4° \text{ F} \qquad\qquad\quad F = 122° \text{ F}$

28° C seems like a reasonable shower temperature.

63.   SNOWY WEATHER

$F = \dfrac{9}{5}C + 32$, when $C = -5$ $F = \dfrac{9}{5}C + 32$, when $C = 0$ $F = \dfrac{9}{5}C + 32$, when $C = 10$

$F = \dfrac{9}{5}(-5) + 32$ $F = \dfrac{9}{5}(0) + 32$ $F = \dfrac{9}{5}(10) + 32$

$F = -9 + 32$ $F = 32°\ F$ $F = 18 + 32$

$F = 23°\ F$ $F = 50°\ F$

$-5°$ C and $0°$ C are both cold enough for snow to occur.

Writing

65.   Multiply the number of kilometers by 0.6214 mi.

67.   Answers will vary.

Review

69. $6y + 7 - y - 3 = 5y + 4$

71. $-3(x - 4) - 2(2x + 6) = -3x + 12 - 4x - 12$
$$= -7x$$

73. $x \cdot x \cdot x = x^3$ 75. $3b(5b) = 15b^{1+1} = 15b^2$

Key Concept Proportions

1.   **Step 1:** Let x = the number of **teacher's aides needed to supervise 75 children**.

15 children are to 2 aides as 75 children are to x aides.

Expressing this as a proportion, we have

$$\frac{15}{2} = \frac{75}{x}$$

Step 2: Solve for x:

$$\frac{15}{2} = \frac{75}{x}$$
$$15 \bullet x = 2 \bullet 75$$
$$15x = 150$$
$$\frac{15x}{15} = \frac{150}{15}$$
$$x = 10$$

Ten teachers aides are needed to supervise 75 children.

Step 3:

$$15 \bullet 10 = 150 \qquad\qquad 2 \bullet 75 = 150$$
$$\frac{15}{2} = \frac{75}{10}$$

Since the cross products are equal, 10 is a solution of the proportion.

3.   MOTION PICTURES

There are 7,200 seconds in 120 minutes.

x = feet of film that run through projector in 120 minutes

$$\frac{2}{3} = \frac{7,200}{x}$$
$$2x = 21,600$$
$$\frac{2x}{2} = \frac{21,600}{2}$$
$$x = 10,800$$

10,800 feet of film run through a projector in 120 minutes.

Chapter Eight Review

Section 8.1 Ratio

1. $\dfrac{4 \text{ inches}}{12 \text{ inches}} = \dfrac{1 \bullet \cancel{4}}{3 \bullet \cancel{4}} = \dfrac{1}{3}$

2. $\dfrac{8 \text{ ounces}}{2 \text{ lb}} = \dfrac{8 \text{ ounces}}{32 \text{ ounces}} = \dfrac{1 \bullet \cancel{8}}{4 \bullet \cancel{8}} = \dfrac{1}{4}$

3. $21 : 14$ is $\dfrac{21}{14} = \dfrac{3 \bullet \cancel{7}}{2 \bullet \cancel{7}} = \dfrac{3}{2}$

4. 24 to 36 is $\dfrac{24}{36} = \dfrac{2 \bullet \cancel{12}}{3 \bullet \cancel{12}} = \dfrac{2}{3}$

5. AIRCRAFT

    $\dfrac{185}{160} = \dfrac{37 \bullet \cancel{5}}{32 \bullet \cancel{5}} = \dfrac{37}{32}$

6. PAY SCALE

    $\dfrac{\$333.25}{43 \text{ hr}} = \$7.75 \text{ per hour}$

7. COMPARISON SHOPPING

    $\dfrac{\$4.95}{12 \text{ oz}} = \$0.4125 \text{ per ounce}$

    $\dfrac{\$3.25}{8 \text{ oz}} = \$0.40625 \text{ per ounce}$

    The 8oz can is the better buy.

Section 8.2 Proportions

8. 75

9. 15

10. $15 \bullet 204 = 3{,}060; \; 29 \bullet 105 = 3{,}045$

     no

11. $17 \bullet 84 = 1{,}428; \; 7 \bullet 204 = 1{,}428$

     yes

12. $\dfrac{5}{9} = \dfrac{20}{36}$?

     $5 \bullet 36 = 180; \; 9 \bullet 20 = 180$

     yes

13. $\dfrac{7}{13} = \dfrac{29}{54}$?

     $7 \bullet 54 = 378; \; 13 \bullet 29 = 377$

     no

14.
$$\frac{12}{18} = \frac{3}{x}$$
$$12x = 54$$
$$\frac{12x}{12} = \frac{54}{12}$$
$$x = \frac{54}{12} = \frac{9 \cdot \cancel{6}}{2 \cdot \cancel{6}}$$
$$x = \frac{9}{2} = 4.5$$

15.
$$\frac{4}{b} = \frac{2}{8}$$
$$32 = 2b$$
$$\frac{32}{2} = \frac{2b}{2}$$
$$16 = b \quad \text{or} \quad b = 16$$

16.
$$\frac{c+1}{5} = \frac{1}{5}$$
$$5(c+1) = 5$$
$$5c + 5 = 5$$
$$5c + 5 - 5 = 5 - 5$$
$$5c = 0$$
$$\frac{5c}{5} = \frac{0}{5}$$
$$c = 0$$

17.
$$\frac{3p+15}{2} = \frac{5}{3}$$
$$3(3p+15) = 10$$
$$9p + 45 = 10$$
$$9p + 45 - 45 = 10 - 45$$
$$9p = -35$$
$$\frac{9p}{9} = \frac{-35}{9}$$
$$p = -\frac{35}{9}$$

18. PICKUP TRUCKS

$x =$ miles truck can go on 11 gal
$$\frac{35 \text{ miles}}{2 \text{ gal}} = \frac{x \text{ miles}}{11 \text{ gal}}$$
$$2x = 385$$
$$\frac{2x}{2} = \frac{385}{2}$$
$$x = 192.5$$

The pickup can go 192.5 miles on 11 gallons of gas.

19. QUALITY CONTROL

$x =$ number of rejects per 1,650
$$\frac{66 \text{ parts}}{12 \text{ defective}} = \frac{1,650 \text{ parts}}{x \text{ defective}}$$
$$66x = 19,800$$
$$\frac{66x}{66} = \frac{19,800}{66}$$
$$x = 300$$

One could expect 300 defective parts.

20. SCALE DRAWING

$x =$ actual kitchen length

$$\frac{\frac{1}{8} \text{ in}}{1 \text{ ft}} = \frac{1\frac{1}{2} \text{ in}}{x \text{ ft}}$$

$$\frac{0.125 \text{ in}}{1 \text{ ft}} = \frac{1.5 \text{ in}}{x \text{ ft}}$$

$$0.125x = 1.5$$

$$\frac{0.125x}{0.125} = \frac{1.5}{0.125}$$

$$x = 12$$

The kitchen is 12 feet long.

Section 8.3 American Units of Measurement

21. $1\frac{1}{2}$ in.

22. $\dfrac{1 \text{ mi}}{5,280 \text{ ft}} = 1, \ \dfrac{5,280 \text{ ft}}{1 \text{ mi}} = 1$

23.
$$5 \text{ yd} = 5 \ \cancel{\text{yd}} \bullet \frac{3 \text{ ft}}{1 \ \cancel{\text{yd}}}$$
$$= 5 \bullet 3 \text{ ft}$$
$$= 15 \text{ ft}$$

24.
$$6 \text{ yd} = 6 \ \cancel{\text{yd}} \bullet \frac{36 \text{ in.}}{1 \ \cancel{\text{yd}}}$$
$$= 6 \bullet 36 \text{ in.}$$
$$= 216 \text{ in.}$$

25.
$$66 \text{ in.} = 66 \ \cancel{\text{in.}} \bullet \frac{1 \text{ ft}}{12 \ \cancel{\text{in.}}}$$
$$= \frac{66}{12} \text{ ft}$$
$$= 5.5 \text{ ft}$$

26.
$$25.5 \text{ ft} = 25.5 \ \cancel{\text{ft}} \bullet \frac{12 \text{ in.}}{1 \ \cancel{\text{ft}}}$$
$$= 25.5 \bullet 12 \text{ in.}$$
$$= 306 \text{ in.}$$

27.
$$9,240 \text{ ft} = 9,240 \ \cancel{\text{ft}} \bullet \frac{1 \text{ mi}}{5,280 \ \cancel{\text{ft}}}$$
$$= \frac{9,240}{5,280} \text{ mi}$$
$$= 1.75 \text{ mi}$$

28.
$$1 \text{ mi} = 1 \ \cancel{\text{mi}} \bullet \frac{5,280 \ \cancel{\text{ft}}}{1 \ \cancel{\text{mi}}} \bullet \frac{1 \text{ yd}}{3 \ \cancel{\text{ft}}}$$
$$= \frac{1 \bullet 5,280}{3} \text{ yd}$$
$$= 1,760 \text{ yd}$$

29.
$$32 \text{ oz} = 32 \ \cancel{\text{oz}} \cdot \frac{1 \text{ lb}}{16 \ \cancel{\text{oz}}}$$

$$= \frac{32}{16} \text{ lb}$$
$$= 2 \text{ lb}$$

30.
$$17.2 \text{ lb} = 17.2 \ \cancel{\text{lb}} \cdot \frac{16 \text{ oz}}{1 \ \cancel{\text{lb}}}$$

$$= 17.2 \cdot 16 \text{ oz}$$
$$= 275.2 \text{ oz}$$

31.
$$3 \text{ ton} = 3 \ \cancel{\text{ton}} \cdot \frac{2,000 \ \cancel{\text{lb}}}{1 \ \cancel{\text{ton}}} \cdot \frac{16 \text{ oz}}{1 \ \cancel{\text{lb}}}$$

$$= 3 \cdot 2,000 \cdot 16 \text{ oz}$$
$$= 96,000 \text{ oz}$$

32.
$$4,500 \text{ lb} = 4,500 \ \cancel{\text{lb}} \cdot \frac{1 \text{ ton}}{2,000 \ \cancel{\text{lb}}}$$

$$= \frac{4,500}{2,000} \text{ ton}$$
$$= 2.25 \text{ tons}$$

33.
$$5 \text{ pt} = 5 \ \cancel{\text{pt}} \cdot \frac{2 \text{ cups}}{1 \ \cancel{\text{pt}}} \cdot \frac{8 \text{fl oz}}{1 \ \cancel{\text{cup}}}$$

$$= 5 \cdot 2 \cdot 8 \text{ fl oz}$$
$$= 80 \text{ fl oz}$$

34.
$$8 \text{ cups} = 8 \ \cancel{\text{cups}} \cdot \frac{1 \ \cancel{\text{pt}}}{2 \ \cancel{\text{cups}}} \cdot \frac{1 \ \cancel{\text{qt}}}{2 \ \cancel{\text{pt}}} \cdot \frac{1 \text{ gal}}{4 \ \cancel{\text{qt}}}$$

$$= \frac{8}{2 \cdot 2 \cdot 4} \text{ gal}$$
$$= 0.5 \text{ gal}$$

35.
$$17 \text{ qt} = 17 \ \cancel{\text{qt}} \cdot \frac{2 \ \cancel{\text{pt}}}{1 \ \cancel{\text{qt}}} \cdot \frac{2 \text{ cups}}{1 \ \cancel{\text{pt}}}$$

$$= 17 \cdot 2 \cdot 2 \text{ cups}$$
$$= 68 \text{ cups}$$

36.
$$176 \text{ fl oz} = 176 \ \cancel{\text{fl oz}} \cdot \frac{1 \ \cancel{\text{cup}}}{8 \ \cancel{\text{fl oz}}} \cdot \frac{1 \ \cancel{\text{pt}}}{2 \ \cancel{\text{cups}}} \cdot \frac{1 \text{ qt}}{2 \ \cancel{\text{pt}}}$$

$$= \frac{176}{8 \cdot 2 \cdot 2} \text{ qt}$$
$$= 5.5 \text{ qt}$$

37.
$$5 \text{ gal} = 5 \ \cancel{\text{gal}} \cdot \frac{4 \ \cancel{\text{qt}}}{1 \ \cancel{\text{gal}}} \cdot \frac{2 \text{ pt}}{1 \ \cancel{\text{qt}}}$$

$$= 5 \cdot 4 \cdot 2 \text{ pt}$$
$$= 40 \text{ pt}$$

38.
$$3.5 \text{ gal} = 3.5 \ \cancel{\text{gal}} \cdot \frac{4 \ \cancel{\text{qt}}}{1 \ \cancel{\text{gal}}} \cdot \frac{2 \ \cancel{\text{pt}}}{1 \ \cancel{\text{qt}}} \cdot \frac{2 \text{ cups}}{1 \ \cancel{\text{pt}}}$$

$$= 3.5 \cdot 4 \cdot 2 \cdot 2 \text{ cups}$$
$$= 56 \text{ cups}$$

39.
$$20 \text{ min} = 20 \ \cancel{\text{min}} \cdot \frac{60 \text{ sec}}{1 \ \cancel{\text{min}}}$$

$$= 20 \cdot 60 \text{ sec}$$
$$= 1,200 \text{ sec}$$

40.
$$900 \text{ sec} = 900 \ \cancel{\text{sec}} \cdot \frac{1 \text{ min}}{60 \ \cancel{\text{sec}}}$$

$$= \frac{900}{60} \text{ min}$$
$$= 15 \text{ min}$$

41.
$$200 \text{ hr} = 200 \cancel{\text{ hr}} \bullet \frac{1 \text{ day}}{24 \cancel{\text{ hr}}}$$
$$= \frac{200}{24} \text{ days}$$
$$= 8.\overline{3} \text{ days} = 8\frac{1}{3} \text{ days}$$

42.
$$6 \text{ hr} = 6 \cancel{\text{ hr}} \bullet \frac{60 \text{ min}}{1 \cancel{\text{ hr}}}$$
$$= 6 \bullet 60 \text{ min}$$
$$= 360 \text{ min}$$

43.
$$4.5 \text{ days} = 4.5 \cancel{\text{ days}} \bullet \frac{24 \text{ hr}}{1 \cancel{\text{ day}}}$$
$$= 4.5 \bullet 24 \text{ hr}$$
$$= 108 \text{ hr}$$

44.
$$1 \text{ day} = 1 \cancel{\text{ day}} \bullet \frac{24 \cancel{\text{ hr}}}{1 \cancel{\text{ day}}} \bullet \frac{60 \cancel{\text{ min}}}{1 \cancel{\text{ hr}}} \bullet \frac{60 \text{ sec}}{1 \cancel{\text{ min}}}$$
$$= 1 \bullet 24 \bullet 60 \bullet 60 \text{ sec}$$
$$= 86,400 \text{ sec}$$

45. SKYSCRAPERS
$$1,454 \text{ ft} = 1,454 \cancel{\text{ ft}} \bullet \frac{1 \text{ yd}}{3 \cancel{\text{ ft}}}$$
$$= \frac{1,454}{3} \text{ yd}$$
$$= 484.\overline{6} \text{ yd}$$
$$= 484\frac{2}{3} \text{ yd}$$

46. BOTTLING
$$50 \text{ gal} = 50 \cancel{\text{ gal}} \bullet \frac{4 \cancel{\text{ qt}}}{1 \cancel{\text{ gal}}} \bullet \frac{1 \text{ magnum}}{2 \cancel{\text{ qt}}}$$
$$= \frac{50 \bullet 4}{2} \text{ magnum}$$
$$= 100 \text{ magnums}$$

Section 8.4 Metric Units of Measurement

47. 4 cm

48. $\dfrac{1 \text{ km}}{1,000 \text{ m}} = 1, \dfrac{1,000 \text{ m}}{1 \text{ km}} = 1$

For questions 49 through 68, we use the following chart for conversions. Substitute grams or liters instead of meters when needed. Also remember 1 cc = 1 mL.

km   hm   dam   m   dm   cm   mm

49. 475 cm = 4.75 m

50. 8 m = 8,000 mm

51. 3 dam = 0.03 km

52. 2 hm = 2,000 dm

53. 5 km = 50 hm

54. 2,500 m = 25 hm

55. 7 cg = 70 mg

56. 800 cg = 8 g

57. 5,425 g = 5.425 kg

58. 5,425 g = 5,425,000 mg

59. 7,500 mg = 7.5 g

60. 5,000 cg = 0.05 kg

61. PAIN RELIEVER

$$100 \; \cancel{\text{caplets}} \cdot \frac{500 \text{ mg}}{1 \; \cancel{\text{caplet}}} = 50,000 \text{ mg}$$

50,000 mg = 50 g

There are 50 g of Tylenol in the bottle.

62. 150 cL = 1.5 L

63. 3,250 L = 3.25 kL

64. 1 hL = 1,000 dL

65. 400 mL = 40 cL

66. 2 kL = 20 hL

67. 4 dL = 400 mL

68. SURGERY

1L = 1,000 mL

The bag holds 1,000 mL of dextrose.

Section 8.5 Converting between American and Metric Units

69. SWIMMING

$$50 \text{ m} = 50(3.2808 \text{ ft})$$
$$= 164.04 \text{ ft}$$

70. HIGH-RISE BUILDINGS

$$443 \text{ m} = 443(3.2808 \text{ ft})$$
$$= 1,453.3944 \text{ ft}$$

The Sears Tower is taller.

71. WESTERN SETTLERS

$$1{,}930 \text{ mi} = 1{,}930(1.6093 \text{ km})$$
$$= 3{,}105.949 \text{ km}$$
$$\approx 3{,}106 \text{ km}$$

72. AIR JORDAN

6 ft 6 in is 78 inches

$$78 \text{ in.} = 78(2.54 \text{ cm})$$
$$= 198.12 \text{ cm}$$

73. $30 \text{ oz} = 30(28.35 \text{ g})$
$$= 850.5 \text{ g}$$

74. $15 \text{ kg} = 15(2.2 \text{ lb})$
$$= 33 \text{ lb}$$

75. $25 \text{ lb} = 25(0.454 \text{ kg})$
$$= 11.35 \text{ kg}$$
$$= 11{,}350 \text{ g}$$
$$\approx 11{,}000 \text{ g}$$

76. $2{,}000 \text{ lb} = 2{,}000(0.454 \text{ kg})$
$$= 908 \text{ kg}$$
$$\approx 910 \text{ kg}$$

77. POLAR BEARS

$$910 \text{ g} = 910(0.035 \text{ oz})$$
$$= 31.85 \, \cancel{\text{oz}} \cdot \frac{1 \text{ lb}}{16 \, \cancel{\text{oz}}}$$
$$= \frac{31.85}{16} \text{ lb}$$
$$\approx 1.9906 \text{ lb}$$
$$\approx 2 \text{ lb}$$

78. BOTTLED WATER

$$17 \text{ fl oz} = 17(0.030 \text{ L})$$
$$= 0.51 \text{ L}$$

LaCroix contains more water.

79. COMPARISON SHOPPING

$$1 \text{ gal} = 1(3.785 \text{ L})$$
$$= 3.785 \text{ L}$$

$$\frac{\$1.39}{3.785 \text{ L}} \approx \$0.3672 \text{ per liter}$$

$$\frac{\$1.80}{5 \text{ L}} = \$0.36 \text{ per liter}$$

The 5 liter bottle is the better buy.

80. $C = \dfrac{5(F - 32)}{9}$, when $F = 77$

$$C = \frac{5(77 - 32)}{9}$$
$$C = \frac{5(45)}{9}$$
$$C = \frac{225}{9} = 25^\circ \text{ C}$$

81.

$F = \dfrac{9}{5}C + 32$, when $C = 10$ $F = \dfrac{9}{5}C + 32$, when $C = 30$

$F = \dfrac{9}{5}(10) + 32$ $F = \dfrac{9}{5}(30) + 32$

$F = 18 + 32$ $F = 54 + 32$

$F = 50° \text{ F}$ $F = 86° \text{ F}$

$F = \dfrac{9}{5}C + 32$, when $C = 50$ $F = \dfrac{9}{5}C + 32$, when $C = 70$

$F = \dfrac{9}{5}(50) + 32$ $F = \dfrac{9}{5}(70) + 32$

$F = 90 + 32$ $F = 126 + 32$

$F = 122° \text{ F}$ $F = 158° \text{ F}$

$30°$ C seems like a good temperature to swim in.

Chapter Eight Test

1. $$\frac{6 \text{ ft}}{8 \text{ ft}} = \frac{3 \bullet 2}{4 \bullet 2} = \frac{3}{4}$$

2. 3 pounds = 48 ounces

    $$\frac{8 \text{ ounces}}{48 \text{ ounces}} = \frac{1 \bullet 8}{6 \bullet 8} = \frac{1}{6}$$

3. COMPARISON SHOPPING

    $$\frac{\$3.38}{2 \text{ lb}} = \$1.69 \text{ per lb}$$

    $$\frac{\$8.50}{5 \text{ lb}} \approx \$1.70 \text{ per lb}$$

    The 2 lb can is the better buy.

4. UTILITY COSTS

    $$\frac{675 \text{ kwh}}{30 \text{ days}} = 22.5 \text{ kwh per day}$$

5. CHECKERS

    $\frac{1}{1}, 1:1, 1$ to 1

6. $25 \bullet 460 = 11,500; \ 33 \bullet 350 = 11,550$

    no

7. $2.2 \bullet 2.8 = 6.16; \ 3.5 \bullet 1.76 = 6.16$

    yes

8. $\dfrac{7}{15} = \dfrac{245}{525}$?

    $7 \bullet 525 = 3,675; \ 15 \bullet 245 = 3,675$

    yes

9. $$\frac{x}{3} = \frac{35}{7}$$
    $$7x = 105$$
    $$\frac{7x}{7} = \frac{105}{7}$$
    $$x = 15$$

10. $$\frac{15.3}{x} = \frac{3}{12.4}$$
    $$3x = 189.72$$
    $$\frac{3x}{3} = \frac{189.72}{3}$$
    $$x = 63.24$$

11.
$$\frac{2x+3}{5}=\frac{5}{1}$$
$$1(2x+3)=25$$
$$2x+3=25$$
$$2x+3-3=25-3$$
$$2x=22$$
$$\frac{2x}{2}=\frac{22}{2}$$
$$x=11$$

12.
$$\frac{3}{2z-1}=\frac{3}{5}$$
$$3(2z-1)=15$$
$$6z-3=15$$
$$6z-3+3=15+3$$
$$6z=18$$
$$\frac{6z}{6}=\frac{18}{6}$$
$$z=3$$

13. **SHOPPING**

$x=$ cost of 16 ounces

$$\frac{13\text{ ounces}}{\$2.79}=\frac{16\text{ ounces}}{\$x}$$
$$13x=44.64$$
$$\frac{13x}{13}=\frac{44.64}{13}$$
$$x\approx3.434$$
$$x=3.43$$

You would pay \$3.43 for 16 oz.

14. **COOKING**

$x=$ cups of sugar needed

$$\frac{\frac{2}{3}\text{ cups of sugar}}{2\text{ cups of flour}}=\frac{x\text{ cups of sugar}}{5\text{ cups of flour}}$$
$$2x=\frac{10}{3}$$
$$3(2x)=3\left(\frac{10}{3}\right)$$
$$6x=10$$
$$\frac{6x}{6}=\frac{10}{6}$$
$$x=\frac{10}{6}=\frac{5\cdot\cancel{2}}{3\cdot\cancel{2}}$$
$$x=\frac{5}{3}=1\frac{2}{3}$$

$1\frac{2}{3}$ cups of sugar are needed.

15.
$$180\text{ in.}=180\ \cancel{\text{in.}}\cdot\frac{1\text{ ft}}{12\ \cancel{\text{in.}}}$$
$$=\frac{180}{12}\text{ ft}$$
$$=15\text{ ft}$$

16. **TOOLS**

$$25\text{ ft}=25\ \cancel{\text{ft}}\cdot\frac{1\text{ yd}}{3\ \cancel{\text{ft}}}$$
$$=\frac{25}{3}\text{ yd}$$
$$=8.\overline{3}\text{ yd}=8\frac{1}{3}\text{ yd}$$

17.
$$10 \text{ lb} = 10 \text{ lb} \cdot \frac{16 \text{ oz}}{1 \text{ lb}}$$
$$= 10 \cdot 16 \text{ oz}$$
$$= 160 \text{ oz}$$

18.
$$1.6 \text{ tons} = 1.6 \text{ tons} \cdot \frac{2,000 \text{ lb}}{1 \text{ ton}}$$
$$= 1.6 \cdot 2,000 \text{ lb}$$
$$= 3,200 \text{ lb}$$

19.
$$1 \text{ gal} = 1 \text{ gal} \cdot \frac{4 \text{ qt}}{1 \text{ gal}} \cdot \frac{2 \text{ pt}}{1 \text{ qt}} \cdot \frac{2 \text{ cups}}{1 \text{ pt}} \cdot \frac{8 \text{fl oz}}{1 \text{ cup}}$$
$$= 1 \cdot 4 \cdot 2 \cdot 2 \cdot 8 \text{ fl oz}$$
$$= 128 \text{ fl oz}$$

20. LITERATURE
$$80 \text{ days} = 80 \text{ days} \cdot \frac{24 \text{ hr}}{1 \text{ day}} \cdot \frac{60 \text{ min}}{1 \text{ hr}}$$
$$= 80 \cdot 24 \cdot 60 \text{ min}$$
$$= 115,200 \text{ min}$$

21. The one on the left is the one-liter carton. A liter is slightly larger than a quart.

22. The blue one is the meter stick. A yard is slightly longer than a meter.

23. The gram is on the right side. An ounce is much heavier than a gram.

24. SPEED SKATING

500 m = 0.5 km

25. 5 m = 500 cm

26. 8,000 cg = 0.08 kg

27. 70 L = 70,000 mL

28. PRESCRIPTIONS

$$50 \text{ tablets} \cdot \frac{150 \text{ mg}}{1 \text{ tablet}} = 7,500 \text{ mg}$$

7,500 mg = 7.5 g

There are 7.5 g of medicine in the bottle.

29.
$$100 \text{ yd} = 100(0.9144 \text{ m})$$
$$= 91.44 \text{ m}$$
The 100-yd race is longer.

30.
$$160 \text{ lb} = 160(0.454 \text{ kg})$$
$$= 72.64 \text{ kg}$$
Jim is slightly heavier.

31.　COMPARISON SHOPPING

$$2 \text{ qt} = 2(0.946 \text{ L})$$
$$= 1.892 \text{ L}$$

$$\frac{\$1.73}{1.892 \text{ L}} \approx \$0.9144 \text{ per liter}$$

The one-liter bottle is the better buy.

32.　COOKING MEAT

$$F = \frac{9}{5}C + 32, \text{ when } C = 83$$

$$F = \frac{9}{5}(83) + 32$$

$$F = 149.4 + 32$$

$$F = 181.4° \text{ F} \approx 182° \text{ F}$$

33.　A scale is a ratio (or rate) comparing the size of a drawing and the size of an actual object. For example, 1 in to 6 feet (1 in. = 6 ft).

34.　It is easier to convert from one unit to another in the metric system, because it is based on the number 10.

Chapter 1-8 Cumulative Review Exercises

1.   6 ten thousands + 4 thousands + 5 hundreds + 2 ones

2.
$$\begin{array}{r} 20 \\ 37\overline{)743} \\ \underline{74} \\ 03 \\ \underline{0} \\ 3 \end{array} = 20\,\text{R}3$$

3.   ENLISTMENT

	Goal	Enlistments	Outcome
Active	77,000	77,587	587
Reserve	21,000	21,278	278
National Guard	56,000	49,210	$-6,790$

4a. $0+(-8)=-8$

4b. $\dfrac{-8}{0}=$ undefined

4c. $0-|-8|=0-8$
    $\quad\quad\quad\;\;=0+(-8)$
    $\quad\quad\quad\;\;=-8$

4d. $\dfrac{0}{-8}=0$

4e. $0-(-8)=0+8$
    $\quad\quad\quad\;\;\;=8$

4f. $0(-8)=0$

5.   GOLF

margin of victory $=3-(-12)$

margin of victory $=3+12$

margin of victory $=15$

The margin was 15 shots.

6. $-3^2=-3\bullet 3$
    $\quad\quad=-9$

$(-3)^2=(-3)\bullet(-3)$
    $\quad\quad\;\;=9$

7.
$$2+3\big[5(-6)-(1-10)\big]$$
$$=2+3\big[5(-6)-(1+[-10])\big]$$
$$=2+3\big[5(-6)-(-9)\big]$$
$$=2+3\big[-30-(-9)\big]$$
$$=2+3\big[-30+9\big]$$
$$=2+3\big[-21\big]$$
$$=2+(-63)$$
$$=-61$$

8. 1 is the coefficient of 1st term.

16 is the coefficient of the 2nd term.

9. $60h$

10. $3x-5-2x-2=x-7$

11.
$$7+2x=2-(4x+7)$$
$$7+2x=2-4x-7$$
$$7+2x=-4x-5$$
$$7+2x+4x=-4x-5+4x$$
$$7+6x=-5$$
$$7+6x-7=-5-7$$
$$6x=-12$$
$$\frac{6x}{6}=\frac{-12}{6}$$
$$x=-2$$

12. $d=rt$

13. PHONE BOOKS

x = number of office buildings
$$5x+105=500$$
$$5x+105-105=500-105$$
$$5x=395$$
$$\frac{5x}{5}=\frac{395}{5}$$
$$x=79$$

The driver went to 79 office buildings.

14.
$$A=\frac{1}{2}bh$$

15. $\dfrac{16}{20} = \dfrac{4 \bullet \cancel{4}}{5 \bullet \cancel{4}} = \dfrac{4}{5}$

16. $\dfrac{9}{10} = \dfrac{9 \bullet 6t}{10 \bullet 6t} = \dfrac{54t}{60t}$

17. $-\dfrac{7}{8h} \div \dfrac{7}{8} = -\dfrac{7}{8h} \bullet \dfrac{8}{7}$

$= -\dfrac{\cancel{7}}{8 \bullet h} \bullet \dfrac{\cancel{8}}{\cancel{7}} = -\dfrac{1}{h}$

18. $\dfrac{1}{2} \bullet \dfrac{1}{2} = \dfrac{1}{4}$

19. MOTORS

$1\dfrac{1}{2} - \dfrac{3}{4} = \dfrac{3}{2} - \dfrac{3}{4}$

$\qquad = \dfrac{3 \bullet 2}{2 \bullet 2} - \dfrac{3}{4}$

$\qquad = \dfrac{6}{4} - \dfrac{3}{4} = \dfrac{3}{4}$

The difference is $\dfrac{3}{4}$ hp.

20. $\dfrac{5}{8} y = \dfrac{1}{10}$

$\dfrac{8}{5}\left(\dfrac{5}{8} y\right) = \dfrac{8}{5}\left(\dfrac{1}{10}\right)$

$y = \dfrac{4 \bullet \cancel{2}}{5} \bullet \dfrac{1}{5 \bullet \cancel{2}}$

$y = \dfrac{4}{25}$

21.

$\dfrac{2}{5} y + 1 = \dfrac{1}{3} + y$

$15\left(\dfrac{2}{5} y + 1\right) = 15\left(\dfrac{1}{3} + y\right)$

$6y + 15 = 5 + 15y$

$6y + 15 - 6y = 5 + 15y - 6y$

$15 = 5 + 9y$

$15 - 5 = 5 + 9y - 5$

$10 = 9y$

$\dfrac{10}{9} = \dfrac{9y}{9}$

$\dfrac{10}{9} = y \ \text{ or } \ y = \dfrac{10}{9}$

22. GLOBAL WARMING

$x = $ previous record temp

$x + 0.4 = 55.5$

$x + 0.4 - 0.4 = 55.5 - 0.4$

$x = 55.1$

55.1° F was the previous record.

23.

$$\frac{6.7 - (-0.3)^2 + 1.6}{-(-0.3)^3} = \frac{6.7 - 0.09 + 1.6}{-(-0.027)} = \frac{8.21}{0.027} \approx 304.074 \approx 304.07$$

24.

$$\frac{1}{12} = 0.08\overline{3}$$

25.

$$6(y - 1.1) + 3.2 = -1 + 3y$$
$$6y - 6.6 + 3.2 = -1 + 3y$$
$$6y - 3.4 = -1 + 3y$$
$$6y - 3.4 - 3y = -1 + 3y - 3y$$
$$3y - 3.4 = -1$$
$$3y - 3.4 + 3.4 = -1 + 3.4$$
$$3y = 2.4$$
$$\frac{3y}{3} = \frac{2.4}{3}$$
$$y = 0.8$$

26.

$$3\sqrt{25} + 4\sqrt{4} = 3(5) + 4(2)$$
$$= 15 + 8 = 23$$

27.

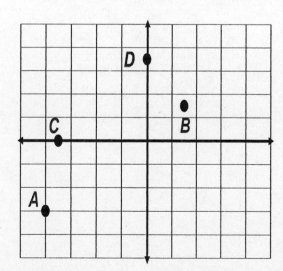

28a.

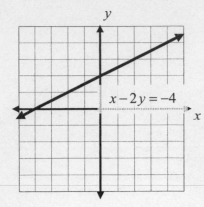

$x - 2y = -4$

28b.

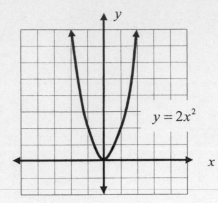

$y = 2x^2$

29.

$$\left(5x^2 - 8x + 1\right) - \left(3x^2 - 2x + 3\right)$$
$$= \left(5x^2 - 8x + 1\right) + \left(-3x^2 + 2x - 3\right)$$
$$= \left(5 + [-3]\right)x^2 + \left(-8 + 2\right)x + \left(1 + [-3]\right)$$
$$= 2x^2 + \left(-6\right)x + \left(-2\right)$$
$$= 2x^2 - 6x - 2$$

30.

$$\left(3x + 2\right)\left(2x - 5\right)$$
$$= 3x\left(2x - 5\right) + 2\left(2x - 5\right)$$
$$= 6x^2 - 15x + 4x - 10$$
$$= 6x^2 - 11x - 10$$

31a. $s^6 \bullet s^7 = s^{6+7} = s^{13}$

31b. $\left(s^6\right)^7 = s^{6 \bullet 7} = s^{42}$

31c.

$$\left(3a^2 b^4\right)^3 = \left(3\right)^3 \left(a^2\right)^3 \left(b^4\right)^3$$
$$= 27 \bullet a^{2 \bullet 3} \bullet b^{4 \bullet 3}$$
$$= 27a^6 b^{12}$$

31d. $-w^5 \left(8w^3\right) = -8 \bullet w^{5+3} = -8w^8$

32.

$$16 = x \bullet 24$$
$$16 = 24x$$
$$\frac{16}{24} = \frac{24x}{24}$$
$$0.\overline{6} = x$$
$$x = 66.\overline{6}\% \text{ or } x = 66\frac{2}{3}\%$$

33. $I = Prt$

34.

Percent	Decimal	Fraction
99%	0.99	$\dfrac{99}{100}$
1.3%	0.013	$\dfrac{13}{1,000}$
31.25%	0.3125	$\dfrac{5}{16}$

35. GUITARS

regular price $= 299.99 + 128$

regular price $= \$427.99$

What percent of 427.99 is 128?

$$x \bullet 427.99 = 128$$

$$427.99x = 128$$

$$\frac{427.99x}{427.99} = \frac{128}{427.99}$$

$$x \approx 0.299$$

$$x \approx 30\%$$

The discount is about 30%.

36.

$$\frac{3 \cancel{\text{ inches}}}{15 \cancel{\text{ inches}}} = \frac{1 \bullet \cancel{3}}{5 \bullet \cancel{3}} = \frac{1}{5}$$

37a.

$$40 \text{ days} = 40 \cancel{\text{ days}} \bullet \frac{24 \text{ hr}}{1 \cancel{\text{ day}}}$$

$$= 40 \bullet 24 \text{ hr}$$

$$= 960 \text{ hr}$$

37b.

$$3 \text{ days} = 3 \cancel{\text{ days}} \bullet \frac{24 \cancel{\text{ hr}}}{1 \cancel{\text{ day}}} \bullet \frac{60 \text{ min}}{1 \cancel{\text{ hr}}}$$

$$= 3 \bullet 24 \bullet 60 \text{ min}$$

$$= 4,320 \text{ min}$$

37c.

$$8 \text{ min} = 8 \cancel{\text{ min}} \bullet \frac{60 \text{ sec}}{1 \cancel{\text{ min}}}$$

$$= 8 \bullet 60 \text{ sec}$$

$$= 480 \text{ sec}$$

38.

$$40 \text{ oz} = 40 \cancel{\text{ oz}} \bullet \frac{1 \text{ lb}}{16 \cancel{\text{ oz}}}$$

$$= \frac{40}{16} \text{ lb}$$

$$= 2.5 \text{ lb}$$

39. $2.4 \text{ m} = 2,400 \text{ mm}$

40. $320 \text{ g} = 0.32 \text{ kg}$

41a. $2 \text{ L} = 2(0.264 \text{ gal})$
$= 0.528 \text{ gal}$

One gallon bottle holds more.

41b.   A meter stick is longer than a yardstick.

42.   BUILDING MATERIALS

$45 \text{ kg} = 45(2.2 \text{ lb})$
$= 99 \text{ lb}$

$\dfrac{\$4.48}{94 \text{ lb}} \approx \$0.0477 \text{ per lb}$

$\dfrac{\$4.56}{99 \text{ lb}} \approx \$0.0461 \text{ per lb}$

The 45 kg bag is the better buy.

Chapter 9 Introduction to Geometry

9.1 Some Basic Definitions

Vocabulary

1. A line **segment** has two endpoints.

3. A **midpoint** divides a line segment into two parts of equal length.

5. A **protractor** is used to measure angles.

7. A **right** angle measures 90°.

9. The measure of a straight angle is 180°.

11. The sum of two **supplementary** angles is 180°.

Concepts

13.	true	15.	false
17.	true	19.	true
21.	acute	23.	obtuse
25.	right	27.	straight
29.	true	31.	false
33.	yes	35.	yes
37.	no	39.	true
41.	true	43.	true
45.	true		

Notation

47.   The symbol $\angle$ means **angle**.

49.   The symbol $\overrightarrow{AB}$ is read as **"ray** *AB*.**"**

Practice

51. $\overline{AC} = 3$

53. $\overline{CE} = 3$

55. $\overline{CD} = 1$

57.   midpoint of $\overline{AD} = B$

59.   40°

61.   135°

63.
$$x + 45 = 55$$
$$x + 45 - 45 = 55 - 45$$
$$x = 10$$

65.
$$x + 22.5 = 50$$
$$x + 22.5 - 22.5 = 50 - 22.5$$
$$x = 27.5$$

67.
$$2x = x + 30$$
$$2x - x = x + 30 - x$$
$$x = 30$$

69.
$$4x + 15 = 7x - 60$$
$$4x + 15 - 4x = 7x - 60 - 4x$$
$$15 = 3x - 60$$
$$15 + 60 = 3x - 60 + 60$$
$$75 = 3x$$
$$\frac{75}{3} = \frac{3x}{3}$$
$$25 = x \text{ or } x = 25$$

71. $90 - 30 = 60°$

73. $180 - 105 = 75°$

75. $m(\angle 4) = 180 - 50 = 130°$

77. $m(\angle 1) + m(\angle 2) + m(\angle 3)$
$$= 50 + 130 + 50 = 230°$$

79. $m(\angle 1) = 100°$

81.   Since $\angle 3 \cong \angle 4$,

$$m(\angle 3) = (180 - 100) \div 2$$
$$m(\angle 3) = 80 \div 2 = 40°$$

Applications

83.   BASEBALL

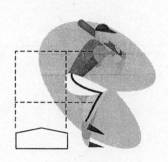

85.   SYNTHESIZERS

$$x = 180 - 115 = 65$$
$$y = 115$$

87.   GARDENING

The handle of the lawnmower makes

$180 - 150 = 30°$ angle with the ground.

Writing

89.   a. It implies that the president took a directly opposing position than he had previously.

b. It implies that the rollerblader turned completely around as she jumped.

91.   A protractor is a device used to measure angles.

93.   A 105° angle is already larger than 90°, so it cannot have a complement.

Review

95. $2^4 = 2 \bullet 2 \bullet 2 \bullet 2 = 16$

97. $\dfrac{3}{4} - \dfrac{1}{8} - \dfrac{1}{3} = \dfrac{3 \bullet 6}{4 \bullet 6} - \dfrac{1 \bullet 3}{8 \bullet 3} - \dfrac{1 \bullet 8}{3 \bullet 8}$

$$= \dfrac{18}{24} - \dfrac{3}{24} - \dfrac{8}{24}$$

$$= \dfrac{7}{24}$$

99.

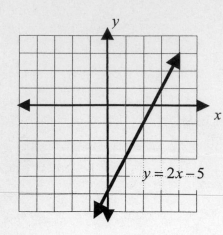

$$y = 2x - 5$$

101.

$$\frac{x+1}{18} = \frac{12.5}{45}$$

$$45(x+1) = 225$$

$$45x + 45 = 225$$

$$45x + 45 - 45 = 225 - 45$$

$$45x = 180$$

$$\frac{45x}{45} = \frac{180}{45}$$

$$x = 4$$

9.2 Parallel and Perpendicular Lines

Vocabulary

1. Two lines in the same plane are **coplanar**.

3. If two lines intersect and form right angles, they are **perpendicular**.

5. In the illustration, $\angle 4$ and $\angle 6$ are **alternate** interior angles.

Concepts

7. $\angle 4$ and $\angle 6$, $\angle 3$ and $\angle 5$

9. $\angle 3, \angle 4, \angle 5, \angle 6$

11. They are parallel.

13. The symbol $\llcorner$ indicates **a right angle**.

15. The symbol $\perp$ is read as "**is perpendicular to**."

Practice

17. $m(\angle 1)=130°, m(\angle 2)=50°,$
    $m(\angle 3)=50°, m(\angle 4)=130°,$
    $m(\angle 5)=130°, m(\angle 6)=50°,$
    $m(\angle 7)=50°, m(\angle 8)=130°$

19. $m(\angle A)=50°, m(\angle 1)=85°,$
    $m(\angle 2)=45°, m(\angle 4)3=135°$

21.
$$6x-10=5x$$
$$6x-10-5x=5x-5x$$
$$x-10=0$$
$$x-10+10=0+10$$
$$x=10$$

23.
$$2x+10+4x-10=180$$
$$6x=180$$
$$\frac{6x}{6}=\frac{180}{6}$$
$$x=30$$

25.
$$3x + 20 + x = 180$$
$$4x + 20 = 180$$
$$4x + 20 - 20 = 180 - 20$$
$$4x = 160$$
$$\frac{4x}{4} = \frac{160}{4}$$
$$x = 40$$

27.
$$9x - 38 = 6x - 2$$
$$9x - 38 - 6x = 6x - 2 - 6x$$
$$3x - 38 = -2$$
$$3x - 38 + 38 = -2 + 38$$
$$3x = 36$$
$$\frac{3x}{3} = \frac{36}{3}$$
$$x = 12$$

Applications

29.   PYRAMIDS

If the stones are level, the plumb bob string should pass through the midpoint of the crossbar of the A-frame.

31.   LOGOS

The perpendicular lines occur in the middle of the logo.

33.   WALLPAPER

When hanging wallpaper, the paper should be perpendicular to the ceiling and the floor, and each sheet should be parallel to one another.

Writing

35.   Answers will vary.

37.   They are both interior angles and they are on alternate sides of the transversal.

39.   Yes, because the alternate interior angle is congruent to the vertical angle of the angle in question of a diagram.

Review

41. $x = 0.60 \bullet 120$

    $x = 72$

43. $x \bullet 500 = 225$

    $500x = 225$

    $\dfrac{500x}{500} = \dfrac{225}{500}$

    $x = 0.45 = 45\%$

45. yes

47. $\dfrac{4 \; \cancel{oz}}{12 \; \cancel{oz}} = \dfrac{1 \bullet \cancel{4}}{3 \bullet \cancel{4}} = \dfrac{1}{3}$

9.3 Polygons

Vocabulary

1. A **regular** polygon has sides that are all the same length and angles that all have the same measure.

3. A **hexagon** is a polygon with six sides.

5. An eight-sided polygon is an **octagon**.

7. A triangle with three sides of equal length is called an **equilateral** triangle.

9. The longest side of a right triangle is the **hypotenuse**.

11. A **parallelogram** with a right angle is a rectangle.

13. A **rhombus** is a parallelogram with four sides of equal length.

15. The legs of an **isosceles** trapezoid have the same length.

Concepts

17. 4 sides, quadrilateral, 4 vertices

19. 3 sides, triangle, 3 vertices

21. 5 sides, pentagon, 5 vertices

23. 6 sides, hexagon, 6 vertices

25. scalene triangle

27. right triangle

29. equilateral triangle

31. isosceles triangle

33. square, rhombus, rectangle

35. rhombus

37. rectangle

39. trapezoid

Notation

41. The symbol △ means **triangle**.

Practice

43. $m(\angle C) = 90°$

45. $m(\angle C) = 45°$

47. $m(\angle C) = 90.7°$

49. $m(\angle 1) = 30°$

51. $m(\angle 2) = 60°$

53. $S = (6-2)180$
$S = (4)180 = 720$
$S = 720°$

55. $S = (10-2)180$
$S = (8)180 = 1,440$
$S = 1,440°$

57. $900 = (n-2)180$
$900 = 180n - 360$
$900 + 360 = 180n - 360 + 360$
$1,260 = 180n$
$\dfrac{1,260}{180} = \dfrac{180n}{180}$
$7 = n$ or $n = 7$

7 sides

59. $2,160 = (n-2)180$
$2,160 = 180n - 360$
$2,160 + 360 = 180n - 360 + 360$
$2,520 = 180n$
$\dfrac{2,520}{180} = \dfrac{180n}{180}$
$14 = n$ or $n = 14$

14 sides

Applications

61. Answers will vary.

63. Answers will vary.

65.   POLYGONS IN NATURE

    a. pentagon   b. octagon

    c. triangle   d. pentagon

67.   CHEMISTRY

3 hexagons and 1 pentagon are used.

69.   EASELS

The side view of the easel makes an isosceles triangle.

Writing

71.   A square has 4 angles that are all $90°$ and the opposite sides of a square are the same length, thus a square is a type of rectangle.

Review

73. $x = 0.20 \bullet 110$

    $x = 22$

75. $x \bullet 200 = 80$

    $200x = 80$

    $\dfrac{200x}{200} = \dfrac{80}{200}$

    $x = 0.4$

    $x = 40\%$

77. $0.85 \div 2(0.25) = 0.425(0.25)$

                 $= 0.10625$

9.4 Properties of Triangles

Vocabulary

1. **Congruent** triangles are the same size and the same shape.

3. If two triangles are **similar**, they have the same shape.

Concepts

5. true

7. false

9. true

11. Yes, the triangles are congruent.

13. Yes, the triangles are similar.

15. a and b represent the length of the legs and c represents the length of the hypotenuse.

17. To find c, we must find a number that, when squared, is 25. Since c represents a positive number, we need only find the positive **square root** of 25 to get c.

$$c^2 = 25$$
$$c = \sqrt{25}$$
$$c = 5$$

Notation

19. The symbol $\cong$ is read as "**is congruent to**."

Practice

21. $\overline{AC}$ corresponds to $\overline{DF}$

$\overline{DE}$ corresponds to $\overline{AB}$

$\overline{BC}$ corresponds to $\overline{EF}$

$\angle A$ corresponds to $\angle D$

$\angle E$ corresponds to $\angle B$

$\angle F$ corresponds to $\angle C$

23. yes, by SSS

25. not necessarily congruent

27. yes, by SSS

29. yes, by SAS

31. $x = 6$ mm

33. $x = 50°$

35. yes they are similar

37.
$$c^2 = (3)^2 + (4)^2$$
$$c^2 = 9 + 16$$
$$c^2 = 25$$
$$c = \sqrt{25}$$
$$c = 5$$

39.
$$(17)^2 = (15)^2 + b^2$$
$$289 = 225 + b^2$$
$$289 - 225 = 225 + b^2 - 225$$
$$64 = b^2$$
$$\sqrt{64} = b$$
$$8 = b \text{ or } b = 8$$

41.
$$(9)^2 = (5)^2 + b^2$$
$$81 = 25 + b^2$$
$$81 - 25 = 25 + b^2 - 25$$
$$56 = b^2$$
$$\sqrt{56} = b \text{ or } b = \sqrt{56}$$

43.
$$(17)^2 = (8)^2 + (15)^2$$
$$289 = 64 + 225$$
$$289 = 289$$
yes

45.
$$(26)^2 = (7)^2 + (24)^2$$
$$676 = 49 + 576$$
$$676 \neq 625$$
no

Applications

47. HEIGHT OF A TREE

h = height of the tree

$$\frac{6}{h} = \frac{4}{24}$$

$$4h = 144$$

$$\frac{4h}{4} = \frac{144}{4}$$

$$h = 36$$

The height of the tree is 36 ft.

49. WIDTH OF A RIVER

w = width of the river

$$\frac{20}{w} = \frac{25}{74}$$

$$25w = 1,480$$

$$\frac{25w}{25} = \frac{1,480}{25}$$

$$w = 59.2$$

The width of the river is 59.2 ft.

51. FLIGHT PATHS

x = altitude lost over 5 miles

$$\frac{1,200 \text{ ft}}{1.5 \text{ miles}} = \frac{x \text{ ft}}{5 \text{ miles}}$$

$$1.5x = 6,000$$

$$\frac{1.5x}{1.5} = \frac{6,000}{1.5}$$

$$x = 4,000$$

The plane will descend 4,000 ft as it travels 5 miles.

53. ADJUSTING LADDERS

$$(20)^2 = a^2 + (16)^2$$

$$400 = a^2 + 256$$

$$400 - 256 = a^2 + 256 - 256$$

$$144 = a^2$$

$$\sqrt{144} = a$$

$$12 = a \text{ or } a = 12$$

The base of the ladder is 12 ft from the wall.

55. PICTURE FRAMES

$$c^2 = (20)^2 + (15)^2$$

$$c^2 = 400 + 225$$

$$c^2 = 625$$

$$c = \sqrt{625}$$

$$c = 25$$

The measurement on the yardstick should read 25 in.

57. BASEBALL

$$c^2 = (90)^2 + (90)^2$$

$$c^2 = 8,100 + 8,100$$

$$c^2 = 16,200$$

$$c = \sqrt{16,200}$$

$$c \approx 127.3$$

The distance from home plate to second base is about 127.3 ft.

Writing

59. The Pythagorean theorem states that the square of the hypotenuse of a right triangle is equal to the sum of the squares of both legs of the right triangle.

Review

61. $\dfrac{0.95 \bullet 3.89}{2.997} \approx \dfrac{1 \bullet 4}{3} = \dfrac{4}{3} = 1\dfrac{1}{3}$

63. $32\% \bullet 60 \approx 0.30 \bullet 60 = 20$

65. $49.5\% \bullet 18.1 \approx 0.50 \bullet 18 = 9$

9.5 Perimeters and Areas of Polygons

Vocabulary

1. The distance around a polygon is called the **perimeter**.

3. The measure of the surface enclosed by a polygon is called its **area**.

5. The area of a polygon is measured in **square** units.

Concepts

For problems 7 through 13, your answers may vary from the given solutions.

7.

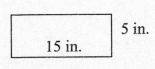

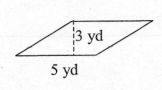

9.

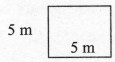

11.

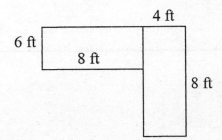

13.

Notation

15. The formula for the perimeter of a square is $P = 4s$.

17. The symbol 1 in^2 means one **square** **inch**.

19. The formula for the area of a square is $A = s^2$.

21. The formula $A = \dfrac{1}{2}bh$ gives the area of a **triangle**.

Practice

23. $P = 8 + 8 + 8 + 8$
$P = 32$ in.

25.   The missing side on the right side of the figure is 2 m long.

$P = 6 + 10 + 6 + 4 + 2 + 2 + 2 + 4$
$P = 36$ m

27. $P = 6 + 7 + 10 + 8 + 6$
$P = 37$ cm

29. $P = 21 + 32 + 32$
$P = 85$ cm

31.   Since it is and equilateral triangle, all sides are the same length.

$s = $ length of a side
$s + s + s = 85$
$3s = 85$
$\dfrac{3s}{3} = \dfrac{85}{3}$
$s = 28.\overline{3}$ ft $= 28\dfrac{1}{3}$ ft

33. $A = s^2$
$A = (4)^2$
$A = 16$ cm^2

35. $A = bh$
$A = (15)(4)$
$A = 60$ cm^2

37. $A = \dfrac{1}{2}bh$
$A = \dfrac{1}{2}(10)(5)$
$A = 25$ in.^{2}

39. $A = \dfrac{1}{2}h(b_1 + b_2)$
$A = \dfrac{1}{2}(13)(9 + 17)$
$A = \dfrac{1}{2}(13)(26)$
$A = 169$ mm^2

41. The picture can be broken into a square and a triangle. The square has sides of length 8 m, and the triangle has a base of 8 m and a height of 4 m.

A = total area, A_s = area of square, A_t = area of triangle

$A = A_s + A_t$

$A = s^2 + \dfrac{1}{2}bh$

$A = (8)^2 + \dfrac{1}{2}(8)(4)$

$A = 64 + \dfrac{1}{2}(8)(4)$

$A = 64 + 16$

$A = 80 \text{ m}^2$

43. The picture is a square with sides 10 yd that has a triangle with a base of 10 yd and a height of 5 yd removed from it.

A = total area, A_s = area of square, A_t = area of triangle

$A = A_s - A_t$

$A = s^2 - \dfrac{1}{2}bh$

$A = (10)^2 - \dfrac{1}{2}(10)(5)$

$A = 100 - \dfrac{1}{2}(10)(5)$

$A = 100 - 25$

$A = 75 \text{ yd}^2$

45. The picture is a rectangle with length of 14 m and a width of 6 m that has a square with sides of 3 m removed from it.

A = total area, A_s = area of square, A_t = area of rectangle

$A = A_r - A_s$

$A = lw - s^2$

$A = (14)(6) - (3)^2$

$A = (14)(6) - 9$

$A = 84 - 9$

$A = 75 \text{ m}^2$

47. One square foot measures 12 inches on each side.

$A = s^2$

$A = (12)^2$

$A = 144 \text{ in.}^2$

Applications

49.   FENCING A YARD

P = perimeter of yard, C = total cost

$P = 2(110) + 2(85)$

$P = 220 + 170$

$P = 390$ ft

$C = 390 \bullet 12.50$

$C = 4,875$

It will cost \$4,875 for the fencing.

51.   PLANTING A SCREEN

n = number of trees she will need

P = perimeter she is planting around

$P = 70 + 70 + 100$

$P = 240$ ft

$n = \dfrac{P}{3} + 1$

$n = \dfrac{240}{3} + 1$

$n = 80 + 1 = 81$

She will need 81 trees. Note we start planting right at the edge, so the additional tree is needed.

53.   BUYING A FLOOR

A square yard is 3 ft by 3 ft.

$A = (3)^2$

$A = 9 \text{ ft}^2$

So, $1 \text{ yd}^2 = 9 \text{ ft}^2$.

C = cost per square foor for linoluem

$C = \dfrac{34.95}{9} \approx 3.883$

$C = \$3.88 \text{ per ft}^2$

Thus, the linoleum is more expensive.

55.   CARPETING A ROOM

24 ft = 8 yd and 15 ft = 5 yd

A = area of room

$A = (8)(5)$

$A = 40 \text{ yd}^2$

C = cost of the carpet

$C = 40 \bullet 30$

$C = 1,200$

The cost of the carpet is \$1,200.

57. TILING A FLOOR

A = area of room

$A = (14)(20)$

$A = 280 \text{ ft}^2$

C = cost of the tile

$C = 280 \bullet 1.29$

$C = 361.20$

The cost of the tile is $361.20.

59. MAKING A SAIL

12 ft = 4 yd and 24 ft = 8 yd

A = area of sail

$A = \frac{1}{2}(4)(8)$

$A = 32 \text{ yd}^2$

C = cost of the nylon

$C = 16 \bullet 12$

$C = 192$

The nylon will cost $192.

61. GEOGRAPHY

The trapezoid has bases that are 205 mi and 505 mi, and the height is 315 mi.

A = area of Nevada

$A = \frac{1}{2}(315)(205 + 505)$

$A = \frac{1}{2}(315)(710)$

$A = 111,825 \text{ mi}^2$

The approximate area is 111,825 mi^2.

63. CARPENTRY

The barn is made up of 2 long walls that are 40 ft long and 12 ft high and 2 short walls that are 20 ft long and 12 ft high.

A = total wall area, A_l = long wall area

A_s = short wall area

$A = 2A_l + 2A_s$

$A = 2lw + 2lw$

$A = 2(48)(12) + 2(20)(12)$

$A = 1,152 + 480$

$A = 1,632 \text{ ft}^2$ to be drywalled

D = the area of a sheet of drywall

$D = (4)(8)$

$D = 32 \text{ ft}^2$

n = sheets of drywall needed

$n = \frac{1,621}{32} = 51$

51 sheets of drywall are needed.

65.   DRIVING SAFETY

For spot 1,

$l \approx 20$ ft, and $w \approx 10$ ft

$A_1 = $ approx. area of spot 1

$A_1 = (20)(10)$

$A_1 = 200$ ft^2

For spot 2,

$b_1 \approx 20$ ft, $b_2 \approx 16$ ft,

and $h \approx 10$ ft

$A_2 = $ approx. area of spot 2

$A_2 = \dfrac{1}{2}(10)(20+16)$

$A_2 = \dfrac{1}{2}(10)(36)$

$A_2 = 180$ ft^2

For spot 3,

$b \approx 28$ ft, and $h \approx 28$ ft

$A_3 = $ approx. area of spot 3

$A_3 = \dfrac{1}{2}(28)(28)$

$A_3 = 392$ ft^2

Writing

67.   Perimeter is the distance around a polygon and area is the measure of the amount of surface that a polygon encloses.

Review

69.
$$\dfrac{3}{4} + \dfrac{2}{3} = \dfrac{3 \bullet 3}{4 \bullet 3} + \dfrac{2 \bullet 4}{3 \bullet 4}$$
$$= \dfrac{9}{12} + \dfrac{8}{12}$$
$$= \dfrac{17}{12} = 1\dfrac{5}{12}$$

71.
$$3\dfrac{3}{4} = 3\dfrac{3 \bullet 3}{4 \bullet 3} = 3\dfrac{9}{12}$$
$$+2\dfrac{1}{3} = +2\dfrac{1 \bullet 4}{3 \bullet 4} = +2\dfrac{4}{12}$$
$$5\dfrac{13}{12} = 6\dfrac{1}{12}$$

73.
$$7\dfrac{1}{2} \div 5\dfrac{2}{5} = \dfrac{15}{2} \div \dfrac{27}{5}$$
$$= \dfrac{15}{2} \bullet \dfrac{5}{27}$$
$$= \dfrac{5 \bullet \cancel{3}}{2} \bullet \dfrac{5}{9 \bullet \cancel{3}}$$
$$= \dfrac{25}{18} = 1\dfrac{7}{18}$$

9.6 Circles

Vocabulary

1. A segment drawn from the center of a circle to a point on the circle is called a **radius**.

3. A **diameter** is a chord that passes through the center of a circle.

5. An arc that is shorter than a semicircle is called a **minor** arc.

7. The distance around a circle is called its **circumference**.

Concepts

9. $\overline{OA}$, $\overline{OC}$, and $\overline{OB}$

11. $\overline{DA}$, $\overline{DC}$, and $\overline{AC}$

13. $\overset{\frown}{ABC}$ and $\overset{\frown}{ADC}$

15.   The diameter is double the radius.

17a.   1 in.

17b.   2 in.

17c. $C = 2\pi r$

$C = 2\pi(1 \text{ in.})$

$C \approx 6.28 \text{ in.}$

17d. $A = \pi r^2$

$A = \pi(1 \text{ in.})^2$

$A \approx 3.14 \text{ in.}^2$

19.   square 6

Notation

21. The symbol $\overset{\frown}{AB}$ is read as **arc AB**.

23. The formula for the circumference of a circle is $\underline{C = \pi d}$ or $\underline{C = 2\pi r}$.

25. If C is the circumference of a circle and D is the diameter, then $\dfrac{C}{D} = \pi$.

27. $\pi(8) = 8\pi$

Practice

29. $C = \pi d$

$C = \pi (12 \text{ in.})$

$C \approx 37.699 \text{ in.}$

$C \approx 37.70 \text{ in.}$

31. $C = \pi d$

$36\pi = \pi d$

$\dfrac{36\pi}{\pi} = \dfrac{\pi d}{\pi}$

$36 \text{ m} = d$ or $d = 36 \text{ m}$

33.   The object has 2 8 ft sides with 2 semicircles that have a diameter of 3 ft on the ends.

P = perimeter of object

C = circumference of the circle =

$P = 8 + 8 + \pi(3)$

$P = 8 + 8 + 3\pi$

$P \approx 8 + 8 + 9.42477$

$P \approx 25.424$

$P \approx 25.42 \text{ ft}$

35.   The object is a rectangle with sides 8 m and 6 m with a semicircle cut into one end with a diameter of 6 m.

P = perimeter of object

C = circumference of the circle

$P = 8 + 6 + 8 + \dfrac{1}{2}\pi(6)$

$P = 8 + 6 + 8 + 3\pi$

$P \approx 8 + 6 + 8 + 9.424$

$P \approx 31.424$

$P \approx 31.42 \text{ m}$

37. $A = \pi r^2$

$A = \pi(3)^2$

$A = 9\pi$

$A \approx 28.27 \text{ in.}^2$

$A \approx 28.3 \text{ in.}^2$

39. The figure is a rectangle with a length of 10 in. and a width of 6 in. that has a semicircle on each end with a diameter of 6 in. The two semicircles give one complete circle with a radius of 3 in.

A = total area of figure,
A_r = area of the rectangle,
A_c = area of the circle
$A = A_r + A_c$
$A = lw + \pi r^2$
$A = (10)(6) + \pi (3)^2$
$A = (10)(6) + 9\pi$
$A \approx 60 + 28.274$
$A \approx 88.274 \approx 88.3 \text{ in.}^2$

41. The figure is a triangle with a base of 12 cm and a height of 12 cm that has a semicircle on one end with a diameter of 12 cm. The semicircle will be half the normal area of a circle with a radius of 6 cm.

A = total area of figure,
A_t = area of the triangle,
A_s = area of the semicircle
$A = A_t + A_s$
$A = \dfrac{1}{2}bh + \dfrac{1}{2}\pi r^2$
$A = \dfrac{1}{2}(12)(12) + \dfrac{1}{2}\pi (6)^2$
$A = \dfrac{1}{2}(12)(12) + \dfrac{1}{2}\pi (36)$
$A = \dfrac{1}{2}(12)(12) + 18\pi$
$A \approx 72 + 56.548$
$A \approx 128.548 \approx 128.5 \text{ cm}^2$

43. The region is a rectangle with a length of 4 in. and a width of 10 in. that has a circle with a diameter of 4 in. removed. The diameter of 4 in. makes the radius 2 in.

A = total area of shaded region,

A_r = area of the rectangle,

A_c = area of the circle

$A = A_r - A_c$

$A = lw - \pi r^2$

$A = (4)(10) - \pi (2)^2$

$A = (4)(10) - 4\pi$

$A \approx 40 - 12.566$

$A \approx 27.434 \approx 27.4 \text{ in.}^2$

45. The region is a parallelogram with a base of 13 in. and a height of 9 in. that has a circle with a radius of 4 in. removed.

A = total area of shaded region,

A_p = area of the parallelogram,

A_c = area of the circle

$A = A_p - A_c$

$A = bh - \pi r^2$

$A = (13)(9) - \pi (4)^2$

$A \approx (13)(9) - 16\pi$

$A \approx 117 - 50.265$

$A \approx 66.734 \approx 66.7 \text{ in.}^2$

Applications

47. AREA OF A LAKE

diameter of 2 mi = radius of 1 mi

$A = \pi r^2$

$A = \pi (1)^2$

$A = 1\pi$

$A \approx 3.14159 \approx 3.14 \text{ mi}^2$

The area is about 3.14 mi^2.

49. GIANT SEQUOIAS

$C = \pi d$

$102.6 = \pi d$

$\dfrac{102.6}{\pi} = \dfrac{\pi d}{\pi}$

$\dfrac{102.6}{\pi} = d$

$d \approx 32.658 \approx 32.66 \text{ ft}$

The diameter is about 32.66 ft.

51. JOGGING

$C = \pi d$

$C = \pi(0.25)$

$C \approx 0.785 \text{ mi}$

Each lap is about 0.785 mi.

L = laps needed to run for 10 mi

$L = \dfrac{10}{0.785}$

$L \approx 12.73$

She needs to run about 12.73 laps.

53. BANDING THE EARTH

To understand the problem it helps to see what is being asked.

Allow r_1 to be the radius of the earth, and allow

r_2 to be the larger radius when the extra 10 ft are

added to the original band. The distance that the

new band is above the Earth's surface is given

by $r_2 - r_1$. This expression is what will be solved

for in the problem.

C_1 = circumference of the band on the Earth,
C_2 = circumference of the band with 10 extra feet
It can be summized that

$C_2 = C_1 + 10$; $C_1 = 2\pi r_1$ and $C_2 = 2\pi r_2$, substitution gives

$\qquad 2\pi r_2 = 2\pi r_1 + 10$

$2\pi r_2 - 2\pi r_1 = 2\pi r_1 + 10 - 2\pi r_1$

$2\pi r_2 - 2\pi r_1 = 10$

$2\pi(r_2 - r_1) = 10$

$\dfrac{2\pi(r_2 - r_1)}{2\pi} = \dfrac{10}{2\pi}$

$\qquad r_2 - r_1 = \dfrac{5 \cdot \cancel{2}}{\cancel{2} \cdot \pi} = \dfrac{5}{\pi}$

$\qquad r_2 - r_1 \approx 1.59 \text{ ft}$

55.   ARCHERY

diameter of 1 ft = radius of 0.5 ft

diameter of 4 ft = radius of 2 ft

A_T = area of target

$A_T = \pi (2)^2$

$A_T = 4\pi$

$A_T \approx 12.566 \approx 12.57 \ \text{ft}^2$

A_B = area of bullseye

$A_B = \pi (0.5)^2$

$A_B = 0.25\pi$

$A_B \approx 0.785 \approx 0.79 \ \text{ft}^2$

What percent of 12.57 is 0.79?

$x \bullet 12.57 = 0.79$

$$\frac{12.57x}{12.57} = \frac{0.79}{12.57}$$

$x \approx 0.0628$

$x \approx 6.28\%$

The bull's eye is about 6.28% of the whole target.

Writing

57.   The circumference of a circle is the distance around the outside of a circle.

59. π is a number that is the ratio of a circles circumference to diameter. Every circle has the exact same ration. π is approximated for calculations to about 3.14159.

61.   If a car has a small turning radius it means that the car can make sharp turns, or if you turned the car in a circle, the circle would be smaller. The smaller the cars turning radius, the smaller the circle will be.

Review

63. $\dfrac{9}{10} = 0.9 = 90\%$

65. UNIT COSTS

   $\$1.29 = 129$ cents

   $\dfrac{129\cancel{c}}{24 \text{ oz}} = 5.375\cancel{c}$ per oz

67. A pentagon has 5 sides.

9.7 Surface Area and Volume

Vocabulary

1.   The space contained within a geometric solid is called its **volume**.

3.   A **cube** is a rectangular solid with all sides of equal length.

5.   The **surface** area of a rectangular solid is the sum of the areas of its faces.

7.   A **cylinder** is a hollow figure like a drinking straw.

9.   A **cone** looks like a witch's pointed hat.

Concepts

11.   A rectangular solid. $V = lwh$

13.   A sphere. $V = \dfrac{4}{3}\pi r^3$

15.   A cone. $V = \dfrac{1}{3}Bh$ or $V = \dfrac{1}{3}\pi r^2 h$

17.   Surface area of rectangular solid. $SA = 2lw + 2lh + 2hw$

19.   1 yd = 3 ft

$$1\,\text{yd}^3 = (1\text{yd})(1\text{yd})(1\text{yd})$$
$$= (3\ \text{ft})(3\ \text{ft})(3\ \text{ft})$$
$$= 27\ \text{ft}^3$$

21.   1 m = 10 dm

$$1\ \text{m}^3 = (1\ \text{m})(1\ \text{m})(1\ \text{m})$$
$$= (10\ \text{dm})(10\ \text{dm})(10\ \text{dm})$$
$$= 1{,}000\ \text{dm}^3$$

23a.   volume   23b.   area   23c.   volume

23d.   surface area   23e.   perimeter   23f.   surface area

25a.   72 in.^{3} 25b.   18 in.^{2} 25c.   24 in.^{2}

Notation

27.   The notation 1 in.^{3} is read as **1 cubic inch**.

Practice

29. $V = lwh$

$V = (3)(4)(5)$

$V = 60 \text{ cm}^3$

31. The area of the base must be found

$B = \dfrac{1}{2}bh$

$B = \dfrac{1}{2}(3)(4)$

$B = 6 \text{ m}^2$

$V = Bh$

$V = (6)(8)$

$V = 48 \text{ m}^3$

33. $V = \dfrac{4}{3}\pi r^3$

$V = \dfrac{4}{3}\pi (9)^3$

$V = \dfrac{4}{3}\pi (729)$

$V \approx 3{,}053.628 \text{ in.}^3$

$V \approx 3{,}053.63 \text{ in.}^3$

35. The area of the base must be found

$B = \pi r^2$

$B = \pi (6)^2$

$B = 36\pi$

$V = Bh$

$V = 36\pi (12)$

$V \approx 1{,}357.168 \text{ m}^3$

$V \approx 1{,}357.17 \text{ m}^3$

37. diameter 10 cm = radius 5 cm

$V = \dfrac{1}{3}\pi r^2 h$

$V = \dfrac{1}{3}\pi (5)^2 (12)$

$V = \dfrac{1}{3}\pi (25)(12)$

$V = 100\pi$

$V \approx 314.159 \text{ cm}^3$

$V \approx 314.16 \text{ cm}^3$

39. The area of the base must be found

$B = s^2$

$B = (10)^2$

$B = 100 \text{ m}^2$

$V = \dfrac{1}{3}Bh$

$V = \dfrac{1}{3}(100)(12)$

$V = 400 \text{ m}^3$

41. $SA = 2lw + 2lh + 2hw$

$SA = 2(3 \bullet 4) + 2(3 \bullet 5) + 2(5 \bullet 4)$

$SA = 2(12) + 2(15) + 2(20)$

$SA = 24 + 30 + 40$

$SA = 94 \text{ cm}^2$

43. $SA = 4\pi r^2$

$SA = 4\pi (10)^2$

$SA = 4\pi (100)$

$SA = 400\pi$

$SA \approx 1,256.637 \text{ in.}^2$

$SA \approx 1,256.64 \text{ in.}^2$

45.   The figure is a pyramid with height of 3 cm and a square base with sides of 8cm set atop a cube with sides of 8 cm. The area of the base of the pyramid is 64 cm^2.

V = total volume of the figure,

V_p = volume of the pyramid,

V_c = volume of the cube

$V = V_p + V_c$

$V = \dfrac{1}{3} Bh + lwh$

$V = \dfrac{1}{3}(64)(3) + (8)(8)(8)$

$V = 64 + 512$

$V = 576 \text{ cm}^3$

47.   The figure is two cones with heights of 10 in. and bases with radii of 4 in.

V = total volume of the figure,

V_c = volume of a cone

$V = 2V_c$

$V = 2\left(\dfrac{1}{3}\pi r^2 h\right)$

$V = \dfrac{2}{3}\pi (4)^2 (10)$

$V = \dfrac{2}{3}\pi (16)(10)$

$V = \dfrac{320}{3}\pi$

$V \approx 335.103 \text{ in.}^3$

$V \approx 335.10 \text{ in.}^3$

Applications

49.   SUGAR CUBES

$V = lwh$

$V = \left(\dfrac{1}{2}\right)\left(\dfrac{1}{2}\right)\left(\dfrac{1}{2}\right) = \dfrac{1}{8} \text{ in.}^3$

$V = 0.125 \text{ in.}^3$

The volume is 0.125 in.^{3}.

51. WATER HEATERS

All measurements must be converted to feet before calculations.

$$27 \text{ in.} = 27 \text{ in.} \bullet \frac{1 \text{ ft}}{12 \text{ in.}} = \frac{9}{4} \text{ ft}$$

$$17 \text{ in.} = 17 \text{ in.} \bullet \frac{1 \text{ ft}}{12 \text{ in.}} = \frac{17}{12} \text{ ft}$$

$$8 \text{ in.} = 8 \text{ in.} \bullet \frac{1 \text{ ft}}{12 \text{ in.}} = \frac{2}{3} \text{ ft}$$

$$V = lwh$$

$$V = \left(\frac{9}{4}\right)\left(\frac{17}{12}\right)\left(\frac{2}{3}\right)$$

$$V = \frac{17}{8} \text{ ft}^3 = 2\frac{1}{8} \text{ ft}^3$$

$$V = 2.125 \text{ ft}^3$$

Over 200 gallons of hot water from 2.125 cubic feet of space...

53. OIL TANKS

diameter 6 ft = radius 3 ft

$$V = \pi r^2 h$$

$$V = \pi (3)^2 (7)$$

$$V = \pi (9)(7)$$

$$V = 63\pi$$

$$V \approx 197.920 \text{ ft}^3$$

$$V \approx 197.92 \text{ ft}^3$$

The volume is about 197.92 ft^3.

55. HOT-AIR BALLOONS

diameter 40 ft = radius 20 ft

$$V = \frac{4}{3}\pi r^3$$

$$V = \frac{4}{3}\pi (20)^3$$

$$V = \frac{4}{3}\pi (8,000)$$

$$V = \frac{32,000}{3}\pi$$

$$V \approx 33,510.321 \text{ ft}^3$$

$$V \approx 33,510.32 \text{ ft}^3$$

The volume is about $33,510.32 \text{ ft}^3$.

57. ENGINES

CR = compression ration

$$CR = \frac{30.4 \text{ in.}^3}{3.8 \text{ in.}^3}$$

$$CR = 8$$

The compression ratio is 8 to 1.

Writing

59.   The volume of an object is a measure of the amount of space it encloses.

61.   Area is measured in square units while volume is measured in cubic units.

Review

63.
$$-5(5-2)^2 + 3 = -5(3)^2 + 3$$
$$= -5(9) + 3$$
$$= -45 + 3$$
$$= -42$$

65.
$$\frac{x+7}{-4} = \frac{1}{4}$$
$$4(x+7) = -4$$
$$4x + 28 = -4$$
$$4x + 28 - 28 = -4 - 28$$
$$4x = -32$$
$$\frac{4x}{4} = \frac{-32}{4}$$
$$x = -8$$

67.
$$\frac{3 \text{ in.}}{15 \text{ in.}} = \frac{1 \cdot 3}{5 \cdot 3} = \frac{1}{5}$$

69. $2.4 \text{ m} = 2{,}400 \text{ mm}$

Key Concept Formulas

1. $d = rt$

3. $P = 2l + 2w$

5. $A = \dfrac{1}{2}bh$

        $A = \dfrac{1}{2}(700)(600)$

        $A = 210{,}000 \text{ ft}^2$

7. $p = c + m$

        $p = 45.50 + 35$

        $p = \$80.50$

9. $d = 16t^2$

        $d = 16(3)^2$

        $d = 16(9)$

        $d = 144 \text{ ft}$

11.

Type of account	Principal	Annual rate earned	Time invested	Interest earned
Savings	$5,000	5%	3 yr	$750
Passbook	$2,250	2%	1 yr	$45
Trust fund	$10,000	6.25%	10 yr	$6,250

Chapter Nine Review

Section 9.1 Some Basic Definitions

1.   points C and D, line CD, and plane GHI 2. $m\left(\overline{AB}\right) = 5$ units

3. $\angle 1$, $\angle B$, $\angle ABC$, and $\angle CBA$ 4. $m\left(\angle 1\right) = 48°$

5. $\angle 1$ and $\angle 2$ are acute,   6.   obtuse angle

$\angle ABD$ and $\angle CBD$ are right angles,

$\angle CBE$ is obtuse,

and $\angle ABC$ is a straight angle.

7.   right angle   8.   straight angle

9.   acute angle   10. $x = 50 - 35$

$x = 15$

11. $y = 180 - 30$ 12. $m\angle 1 = 65°$

$y = 150$ $m\angle 2 = (180 - 65)° = 115°$

13.   The complement would be   14.   The supplement would be

$90 - 50 = 40°$. $180 - 140 = 40°$.

15.   No. They do add to give $180°$, but there are more than 2 angles, so they are not

considered supplementary angles.

Section 9.2 Parallel and Perpendicular Lines

16.   part a is parallel   17. $\angle 4$ and $\angle 6$, $\angle 3$ and $\angle 5$

18. $\angle 4$ and $\angle 8$, $\angle 3$ and $\angle 7$, $\angle 1$ and $\angle 5$,   19. $\angle 1$ and $\angle 3$, $\angle 2$ and $\angle 4$, $\angle 5$ and $\angle 7$,

$\angle 2$ and $\angle 6$ $\angle 6$ and $\angle 8$

20. $m(\angle 1) = 70°, m(\angle 2) = 110°,$
    $m(\angle 3) = 70°, m(\angle 4) = 110°,$
    $m(\angle 5) = 70°, m(\angle 6) = 110°,$
    $m(\angle 7) = 70°$

21. $m(\angle 1) = 60°, m(\angle 2) = 120°,$
    $m(\angle 3) = 130°, m(\angle 4) = 50°$

22. $2x - 30 = x + 10$
    $2x - 30 - x = x + 10 - x$
    $x - 30 = 10$
    $x - 30 + 30 = 10 + 30$
    $x = 40$

23. $4x - 10 + 3x + 50 = 180$
    $7x + 40 = 180$
    $7x + 40 - 40 = 180 - 40$
    $7x = 140$
    $\dfrac{7x}{7} = \dfrac{140}{7}$
    $x = 20$

Section 9.3 Polygons

24. octagon

25. pentagon

26. triangle

27 hexagon

28. quadrilateral
    (it does not show the right angle marks
    so we cannot assume it is a rectangle)

29. 3 vertices

30. 4 vertices

31. 8 vertices

32. 6 vertices

33. isosceles

34. scalene

35. equilateral

36. right

37. yes, it is isosceles

38. no, it is not isosceles

39. $x + 70 + 20 = 180$
$x + 90 = 180$
$x + 90 - 90 = 180 - 90$
$x = 90$

40. $x + 70 + 60 = 180$
$x + 130 = 180$
$x + 130 - 130 = 180 - 130$
$x = 50$

41. $x =$ vertex angle
$x + 65 + 65 = 180$
$x + 130 = 180$
$x + 130 - 130 = 180 - 130$
$x = 50°$

42. You can conclude that the triangle must be an equilateral triangle.

43. trapezoid

44. square

45. parallelogram

46. rectangle

47. rhombus

48. rectangle

49. $m\left(\overline{BD}\right) = 15$ cm

50. $m\left(\angle 1\right) = 40°$

51. $m\left(\angle 2\right) = 100°$

52. true

53. false

54. true

55. true

56. $m\left(\angle B\right) = 65°$

57. $m\left(\angle C\right) = 115°$

58. $S = (4-2)180$

$S = (2)180 = 360$

$S = 360°$

59. $S = (6-2)180$

$S = (4)180 = 360$

$S = 720°$

Section 9.4 Properties of Triangles

60. a. $\angle A$ corresponds to $\angle D$.

b. $\angle B$ corresponds to $\angle E$.

c. $\angle C$ corresponds to $\angle F$.

d. $\overline{AC}$ corresponds to $\overline{DF}$.

e. $\overline{AB}$ corresponds to $\overline{DE}$.

f. $\overline{BC}$ corresponds to $\overline{EF}$.

61. congruent by SSS

62. congruent by SAS

63. congruent by ASA

64. not necessarily congruent, but they are similar triangles

65. yes

66. yes

67. $x =$ height of the tree

$$\frac{7}{x} = \frac{2}{6}$$

$$2x = 42$$

$$\frac{2x}{2} = \frac{42}{2}$$

$$x = 21$$

The tree is 21 ft tall.

68. $c^2 = (5)^2 + (12)^2$

$c^2 = 25 + 144$

$c^2 = 169$

$c = \sqrt{169}$

$c = 13$

69. $(17)^2 = (8)^2 + b^2$

$289 = 64 + b^2$

$289 - 64 = 64 + b^2 - 64$

$225 = b^2$

$\sqrt{225} = b$

$15 = b$ or $b = 15$

70.

$$(52)^2 = a^2 + (41.5)^2$$

$$2,704 = a^2 + 1,722.25$$

$$2,704 - 1,722.25 = a^2 + 1,722.25 - 1,722.25$$

$$981.75 = a^2$$

$$\sqrt{981.75} = a$$

$$a \approx 31.33 \text{ in.}$$

$$a \approx 31.3 \text{ in.}$$

Section 9.5 Perimeters and Areas of Polygons

71. $P = 4s$

$P = 4(18)$

$P = 72 \text{ in.}$

72. $P = 2l + 2w$

$P = 2(3) + 2(1.5)$

$P = 6 + 3$

$P = 9 \text{ m}$

73. $P = 6 + 8 + 4 + 4 + 8$

$P = 30 \text{ m}$

74. From the picture, the short missing side on the right is 4 m long and the top is 10 m long.

$P = 8 + 6 + 4 + 4 + 4 + 10$

$P = 36 \text{ m}$

75. $A = s^2$

$A = (3.1)^2$

$A = 9.61 \text{ cm}^2$

76. $A = lw$

$A = (150)(50)$

$A = 7,500 \text{ ft}^2$

77. $A = bh$

$A = (30)(15)$

$A = 450 \text{ ft}^2$

78. $A = \frac{1}{2}bh$

$A = \frac{1}{2}(40)(10)$

$A = 200 \text{ in.}^2$

79.
$$A = \frac{1}{2}h(b_1 + b_2)$$
$$A = \frac{1}{2}(8)(12 + 18)$$
$$A = \frac{1}{2}(8)(30)$$
$$A = 120 \text{ cm}^2$$

80. The object can be seen as 2 rectangles side by side, with the left rectangle having a length of 12 ft and a height of 14 ft, and the right rectangle being an 8 ft by 8 ft square.

A = total area, A_s = area of square,
A_r = area of rectangle
$$A = A_s + A_r$$
$$A = s^2 + lw$$
$$A = (8)^2 + (12)(14)$$
$$A = 64 + (12)(14)$$
$$A = 64 + 168$$
$$A = 232 \text{ ft}^2$$

81. The object can be seen as a triangle with a base of 12 ft and a height of 4 ft set atop a trapezoid with bases of 12 ft and 20 ft and a height of 8 ft.

A = total area,
A_1 = area of trapezoid,
A_2 = area of triangle
$$A = A_1 + A_2$$
$$A = \frac{1}{2}h(b_1 + b_2) + \frac{1}{2}bh$$
$$A = \frac{1}{2}(8)(12 + 20) + \frac{1}{2}(12)(4)$$
$$A = \frac{1}{2}(8)(32) + \frac{1}{2}(12)(4)$$
$$A = 128 + 24$$
$$A = 152 \text{ ft}^2$$

82. The object can be seen as a triangle with a base of 15 m and a height of 4 m being removed from a parallelogram with a base of 15 m and a height of 10 m.

A = total area,
A_p = area of parallelogram,
A_t = area of triangle
$$A = A_p - A_t$$
$$A = bh - \frac{1}{2}bh$$
$$A = (15)(10) - \frac{1}{2}(15)(4)$$
$$A = 150 - 30$$
$$A = 120 \text{ m}^2$$

83. One square yard measures 3 feet on each side.

    $A = s^2$

    $A = (3)^2$

    $A = 9 \text{ ft}^2$

84. One square foot measures 12 inches on each side.

    $A = s^2$

    $A = (12)^2$

    $A = 144 \text{ in.}^2$

Section 9.6 Circles

85. $\overline{CD}, \overline{AB}$

86. $\overline{AB}$

87. $\overline{OA}, \overline{OC}, \overline{OD}, \overline{OB}$

88. O

89. $C = \pi d$

     $C = \pi(21)$

     $C = 21\pi$

     $C \approx 65.97 \text{ cm}$

     $C \approx 66.0 \text{ cm}$

90. The figure consists of two lengths of 10 cm each and 2 half circles on each end with a diameter of 8 cm. The sum of the 2 half circles is the same as one circle.

    $P = $ perimeter of object

    $C = $ circumference of the circle

    $P = 10 + 10 + \pi(8)$

    $P = 10 + 10 + 8\pi$

    $P \approx 10 + 10 + 25.132$

    $P \approx 45.132$

    $P \approx 45.1 \text{ cm}$

91. diameter 18 in. = radius 9 in.

     $A = \pi r^2$

     $A = \pi(9)^2$

     $A = 81\pi$

     $A \approx 254.46 \text{ in.}^2$

     $A \approx 254.5 \text{ in.}^2$

92.   The figure consists of a rectangle with a length of 10 cm and a height of 8 cm that is between 2 half circles that have a diameter of 8 cm. The diameter of 8 cm makes the radii of the circles 4 cm. The sum of the areas for 2 half circles is the same as the area of one circle.

A = total area of figure,

A_r = area of the rectangle,

A_c = area of the circle

$A = A_r + A_c$

$A = lw + \pi r^2$

$A = (10)(8) + \pi (4)^2$

$A = (10)(8) + 16\pi$

$A \approx 80 + 50.265$

$A \approx 130.265 \approx 130.3 \text{ cm}^2$

Section 9.7 Surface Area and Volume

93. $V = lwh$

$V = (5)(5)(5)$

$V = 125 \text{ cm}^3$

94. $V = lwh$

$V = (10)(6)(8)$

$V = 480 \text{ m}^3$

95.   The area of the base must be found

$B = \dfrac{1}{2}bh$

$B = \dfrac{1}{2}(10)(6)$

$B = 30 \text{ in.}^2$

$V = Bh$

$V = (30)(20)$

$V = 600 \text{ in.}^3$

96.   A hemisphere has half the volume of a sphere.

$V = \dfrac{1}{2}\left(\dfrac{4}{3}\pi r^3\right)$

$V = \dfrac{2}{3}\pi r^3$

$V = \dfrac{2}{3}\pi (12)^3$

$V = \dfrac{2}{3}\pi (1,728)$

$V \approx 3,619.114 \text{ in.}^3$

$V \approx 3,619 \text{ in.}^3$

97. The figure is a cylinder that has a base
with a radius of 5 ft and a height of 16
ft with a hemisphere on top of it with
a radius of 5 ft.

V = total volume of the figure,
V_h = volume of the hemisphere,
V_c = volume of the cylinder

$$V = V_h + V_c$$

$$V = \frac{1}{2}\left(\frac{4}{3}\pi r^3\right) + \pi r^2 h$$

$$V = \frac{2}{3}\pi r^3 + \pi r^2 h$$

$$V = \frac{2}{3}\pi(5)^3 + \pi(5)^2(16)$$

$$V = \frac{2}{3}\pi(125) + \pi(25)(16)$$

$$V = \frac{250}{3}\pi + 400\pi$$

$$V \approx 261.799 + 1,256.637$$

$$V \approx 1,518.436 \text{ ft}^3$$

$$V \approx 1,518 \text{ ft}^3$$

98. diameter 10 in. = radius 5 in.

$$V = \frac{1}{3}\pi r^2 h$$

$$V = \frac{1}{3}\pi(5)^2(30)$$

$$V = \frac{1}{3}\pi(25)(30)$$

$$V = 250\pi$$

$$V \approx 785.398 \text{ in.}^3$$

$$V \approx 785 \text{ in.}^3$$

99. The area of the base must be found

$$B = \frac{1}{2}bh$$

$$B = \frac{1}{2}(500)(433)$$

$$B = 108,250 \text{ ft}^2$$

$$V = \frac{1}{3}Bh$$

$$V = \frac{1}{3}(108,250)(250)$$

$$V \approx 9,020,833.333 \text{ ft}^3$$

$$V \approx 9,020,833 \text{ ft}^3$$

100. The figure is a cylinder that has a base with a radius of 15 ft and a height of 40 ft with a hemisphere on top of it with a radius of 15 ft.

V = total volume of the figure,

V_h = volume of the hemisphere,

V_c = volume of the cylinder

$V = V_h + V_c$

$V = \dfrac{1}{2}\left(\dfrac{4}{3}\pi r^3\right) + \pi r^2 h$

$V = \dfrac{2}{3}\pi r^3 + \pi r^2 h$

$V = \dfrac{2}{3}\pi (15)^3 + \pi (15)^2 (40)$

$V = \dfrac{2}{3}\pi (3,375) + \pi (225)(40)$

$V = 2,250\pi + 9,000\pi$

$V = 11,250\pi$

$V \approx 35,342.917 \text{ ft}^3$

$V \approx 35,343 \text{ ft}^3$

101. $1 \text{ ft}^3 = (1 \text{ ft})(1 \text{ ft})(1 \text{ ft})$

$= (12 \text{ in.})(12 \text{ in.})(12 \text{ in.})$

$= 1,728 \text{ in.}^3$

102. $1 \text{ yd}^3 = (1 \text{ yd})(1 \text{ yd})(1 \text{ yd})$

$= (3 \text{ ft})(3 \text{ ft})(3 \text{ ft})$

$= 27 \text{ ft}^3$

$2(27 \text{ ft}^3) = 54 \text{ ft}^3$

103. $SA = 2lw + 2lh + 2hw$

$SA = 2(3.1 \bullet 2.3) + 2(3.1 \bullet 4.4) + 2(4.4 \bullet 2.3)$

$SA = 2(7.13) + 2(13.64) + 2(10.12)$

$SA = 14.26 + 27.28 + 20.24$

$SA = 61.78 \text{ ft}^2$

$SA \approx 61.8 \text{ ft}^2$

104. $SA = 4\pi r^2$

$SA = 4\pi (5)^2$

$SA = 4\pi (25)$

$SA = 100\pi$

$SA \approx 314.159 \text{ in.}^2$

$SA \approx 314.2 \text{ in.}^2$

Chapter Nine Test

1. $\text{m}\left(\overline{AB}\right) = 4$ units

2. *B* is the vertex.

3. true

4. false, $\left(90° \text{ is a right angle}\right)$

5. false, $\left(180° \text{ is a straight angle}\right)$

6. true

7. $x = 67 - 17$
   $x = 50$

8. $y = 180 - 40$
   $y = 140$

9. $5y - 20 = 3y + 4$
   $5y - 20 - 3y = 3y + 4 - 3y$
   $2y - 20 = 4$
   $2y - 20 + 20 = 4 + 20$
   $2y = 24$
   $\dfrac{2y}{2} = \dfrac{24}{2}$
   $y = 12$

10. CALLIGRAPHY
    $x = 90 - 45$
    $x = 45$

11. $x = $ complement of angle
    $x = 90 - 67$
    $x = 23°$

12. $x = $ supplement of angle
    $x = 180 - 117$
    $x = 63°$

13. $\text{m}\left(\angle 1\right) = 70°$

14. $\text{m}\left(\angle 2\right) = 110°$

15. $\text{m}\left(\angle 3\right) = 70°$

16. $2x + 30 + 70 = 180$
    $2x + 100 = 180$
    $2x + 100 - 100 = 180 - 100$
    $2x = 80$
    $\dfrac{2x}{2} = \dfrac{80}{2}$
    $x = 40$

17.

Polygon	Number of sides
Triangle	3
Quadrilateral	4
Hexagon	6
Pentagon	5
Octagon	8

18.

Property	Kind of triangle
All sides of equal length	equilateral triangle
No sides of equal length	scalene triangle
Two sides of equal length	isosceles triangle

19. $m(\angle A) = 57°$

20. $m(\angle C) = (180 - 57 - 57)° = 66°$

21. x = measure of third angle
$x = 180 - 65 - 85$
$x = 30°$

22. $S = (10 - 2)180$
$S = (8)180 = 1,440$
$S = 1,440°$

23. $m(\overline{AB}) = m(\overline{DC})$,
$m(\overline{AD}) = m(\overline{BC})$,
$m(\overline{AC}) = m(\overline{DB})$

24. $2x + 2(50) = 360$
$2x + 100 = 360$
$2x + 100 - 100 = 360 - 100$
$2x = 260$
$\dfrac{2x}{2} = \dfrac{260}{2}$
$x = 130°$

25. $m(\overline{AB}) = 8$ in.

26. $m(\angle E) = 50°$

27. $\dfrac{6}{4} = \dfrac{9}{x}$
$6x = 36$
$\dfrac{6x}{6} = \dfrac{36}{6}$
$x = 6$

28. $\dfrac{6}{4} = \dfrac{y}{8}$
$4y = 48$
$\dfrac{4y}{4} = \dfrac{48}{4}$
$y = 12$

29.
$$c^2 = (90)^2 + (90)^2$$
$$c^2 = 8{,}100 + 8{,}100$$
$$c^2 = 16{,}200$$
$$c = \sqrt{16{,}200}$$
$$c \approx 127.3 \text{ ft}$$

30.
$$A = \frac{1}{2}bh$$
$$A = \frac{1}{2}(44.5)(17.6)$$
$$A = 391.6 \text{ cm}^2$$

31.
$$A = \frac{1}{2}h(b_1 + b_2)$$
$$A = \frac{1}{2}(6)(12.2 + 15.7)$$
$$A = \frac{1}{2}(6)(27.9)$$
$$A = 83.7 \text{ ft}^2$$

32.   THE OLYMPICS

The steel rod will have to be the circumference of 5 circles with a diameter of 6 ft each.
$$C = 5(\pi d)$$
$$C = 5(\pi \bullet 6)$$
$$C = 30\pi$$
$$C \approx 94.247 \text{ ft}$$
$$C \approx 94.2 \text{ ft}$$

33.   diameter 6 ft = radius 3 ft
$$A = \pi r^2$$
$$A = \pi (3)^2$$
$$A = 9\pi$$
$$A \approx 28.27 \text{ ft}^2$$
$$A \approx 28.3 \text{ ft}^2$$

34.
$$V = lwh$$
$$V = (4.3)(5.7)(6.5)$$
$$V = 159.315 \text{ m}^3$$
$$V \approx 159.3 \text{ m}^3$$

35.   diameter 8 m = radius 4 m
$$V = \frac{4}{3}\pi r^3$$
$$V = \frac{4}{3}\pi (4)^3$$
$$V = \frac{4}{3}\pi (64)$$
$$V \approx 268.082 \text{ ft}^3$$
$$V \approx 268.1 \text{ ft}^3$$

36.   The area of the base must be found   37.   Answers will vary

$B = lw$

$B = (5)(4)$

$B = 20 \text{ ft}^2$

$V = \dfrac{1}{3}Bh$

$V = \dfrac{1}{3}(20)(10)$

$V \approx 66.667 \text{ ft}^3$

$V \approx 66.7 \text{ ft}^3$

38.

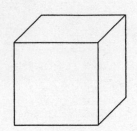

The surface area is six time the area of one face of the cube.

Chapter 1-9 Cumulative Review Exercises

1.   AMUSEMENT PARKS

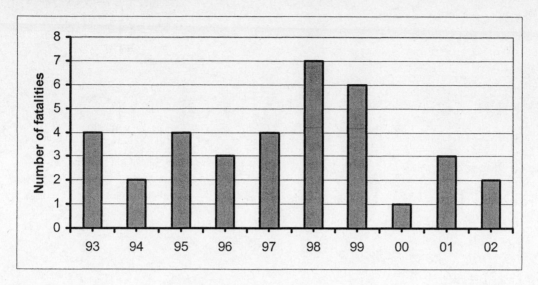

2.   USED CARS

The selling price was $8,995.

3.
$$\begin{array}{r} 35,021 \\ -23,999 \\ \hline 11,022 \end{array}$$

4.
$$41\overline{)1353} = 33$$
$$\begin{array}{r} 33 \\ \underline{123} \\ 123 \\ \underline{123} \\ 0 \end{array}$$

5.   2,110,000

6.
$$220$$

$$\boxed{2} \quad 110$$

$$\boxed{2} \quad 55$$

$$\boxed{11} \quad \boxed{5}$$

$$2^2 \bullet 5 \bullet 11$$

7.   1, 2, 3, 4, 6, 8, 12, 24

8. $\{\ldots -3, -2, -1, 0, 1, 2, 3, \ldots\}$

9.
$$-10(-2) - 2^3 + 1 = -10(-2) - 8 + 1$$
$$= 20 - 8 + 1$$
$$= 12 + 1$$
$$= 13$$

10.
$$5 - 3\left[4^2 - (1 + 5 \bullet 2)\right]$$
$$= 5 - 3\left[4^2 - (1 + 10)\right]$$
$$= 5 - 3\left[4^2 - 11\right]$$
$$= 5 - 3[16 - 11]$$
$$= 5 - 3(5)$$
$$= 5 - 15$$
$$= 5 + (-15) = -10$$

11.
$$\left|-6 - (-3)\right| = \left|-6 + 3\right|$$
$$= \left|-3\right|$$
$$= 3$$

12.
$$\frac{2(2) + 3(-3)}{-4 - (-3)} = \frac{4 + (-9)}{-4 + 3} = \frac{-5}{-1} = 5$$

13. An equation will always have an equals sign while an expression will not.

14.
$$4x - 2(3x - 4) - 5(2x)$$
$$= 4x - 6x + 8 - 10x$$
$$= -12x + 8$$

15. $3(p+15)+4(11-p)=0$

$3p+45+44-4p=0$

$-p+89=0$

$-p+89-89=0-89$

$-p=-89$

$\dfrac{-p}{-1}=\dfrac{-89}{-1}$

$p=89$

Check:

$3(89+15)+4(11-89)=0$

$3(104)+4(-78)=0$

$312+(-312)=0$

$0=0$

16. $5t-7=7t+13$

$5t-7-5t=7t+13-5t$

$-7=2t+13$

$-7-13=2t+13-13$

$-20=2t$

$\dfrac{-20}{2}=\dfrac{2t}{2}$

$-10=t$ or $t=-10$

Check:

$5(-10)-7=7(-10)+13$

$-50-7=-70+13$

$-57=-57$

17. $-x+2=13$

$-x+2-2=13-2$

$-x=11$

$\dfrac{-x}{-1}=\dfrac{11}{-1}$

$x=-11$

Check:

$-(-11)+2=13$

$11+2=13$

$13=13$

18. $4+\dfrac{x}{5}-6=-1$

$\dfrac{x}{5}-2=-1$

$\dfrac{x}{5}-2+2=-1+2$

$\dfrac{x}{5}=1$

$5\left(\dfrac{x}{5}\right)=5(1)$

$x=5$

Check:

$4+\dfrac{5}{5}-6=-1$

$4+1-6=-1$

$5-6=-1$

$-1=-1$

19. SNAILS

$\dfrac{13\text{ in.}}{2\text{ min}}=6.5\text{ in per min}$

20. SHOPPING

$50x¢$

21. $x-5$

22.   LUMBER

$$\text{b.f.} = \frac{2 \cdot 4 \cdot 10}{12}$$

$$\text{b.f.} = \frac{80}{12} = \frac{20 \cdot \cancel{4}}{3 \cdot \cancel{4}} = \frac{20}{3} \text{ b.f.}$$

$$\text{b.f.} = 6\frac{2}{3}$$

The piece is $6\frac{2}{3}$ b.f.

23. $\dfrac{35a^2}{28a} = \dfrac{5 \cdot \cancel{7} \cdot \cancel{a} \cdot a}{4 \cdot \cancel{7} \cdot \cancel{a}} = \dfrac{5a}{4}$

24.
$$
\begin{array}{lll}
45\dfrac{2}{3} = & 45\dfrac{2 \cdot 5}{3 \cdot 5} = & 45\dfrac{10}{15} \\[2mm]
+96\dfrac{4}{5} = & +96\dfrac{4 \cdot 3}{5 \cdot 3} = & +96\dfrac{12}{15} \\[1mm]
\hline
& & 141\dfrac{22}{15}
\end{array}
$$

$$141\frac{22}{15} = 141 + 1\frac{7}{15} = 142\frac{7}{15}$$

25. $\dfrac{x}{4} - \dfrac{3}{5} = \dfrac{x \cdot 5}{4 \cdot 5} - \dfrac{3 \cdot 4}{5 \cdot 4} = \dfrac{5x}{20} - \dfrac{12}{20}$

$$= \frac{5x - 12}{20}$$

26.   BAKING

$$
\begin{array}{lll}
17\dfrac{1}{2} = & 17\dfrac{1 \cdot 2}{2 \cdot 2} = & 17\dfrac{2}{4} \\[2mm]
-3\dfrac{3}{4} = & -3\dfrac{3}{4} = & -3\dfrac{3}{4} \\[1mm]
\hline
\end{array}
$$

$$
\begin{array}{ll}
16\dfrac{2}{4} + \dfrac{4}{4} = & 16\dfrac{6}{4} \\[2mm]
-3\dfrac{3}{4} & = -3\dfrac{3}{4} \\[1mm]
\hline
& = 13\dfrac{3}{4} \text{ cups}
\end{array}
$$

There are $13\frac{3}{4}$ cups left in the bag.

27.

$$-\frac{6}{25}\left(2\frac{7}{24}\right) = -\frac{6}{25}\left(\frac{55}{24}\right)$$

$$= -\frac{\cancel{6}}{5\cdot\cancel{5}}\left(\frac{11\cdot\cancel{5}}{4\cdot\cancel{6}}\right)$$

$$= -\frac{11}{20}$$

28.

$$\frac{15}{8q^4} \div \frac{45}{8q^3} = \frac{15}{8q^2} \bullet \frac{8q^3}{45}$$

$$= \frac{\cancel{15}}{\cancel{8}\bullet\cancel{q}\bullet\cancel{q}\bullet\cancel{q}\bullet q} \bullet \frac{\cancel{8}\bullet\cancel{q}\bullet\cancel{q}\bullet\cancel{q}}{3\bullet\cancel{15}}$$

$$= \frac{1}{3q}$$

29. PET MEDICINES

There is $\dfrac{3}{4}$ oz in the cup.

$d = $ size of a single dose

$$d = \frac{3}{4} \div 8$$

$$d = \frac{3}{4} \bullet \frac{1}{8}$$

$$d = \frac{3}{32}$$

Each dose is $\dfrac{3}{32}$ oz.

30.

$$\frac{x}{2} - \frac{1}{9} = \frac{1}{3}$$

$$18\left(\frac{x}{2} - \frac{1}{9}\right) = 18\left(\frac{1}{3}\right)$$

$$9x - 2 = 6$$

$$9x - 2 + 2 = 6 + 2$$

$$9x = 8$$

$$\frac{9x}{9} = \frac{8}{9}$$

$$x = \frac{8}{9}$$

31.

$$\frac{2}{3}q - 1 = -6$$

$$3\left(\frac{2}{3}q - 1\right) = 3(-6)$$

$$2q - 3 = -18$$

$$2q - 3 + 3 = -18 + 3$$

$$2q = -15$$

$$\frac{2q}{q} = \frac{-15}{2}$$

$$q = -\frac{15}{2} = -7\frac{1}{2}$$

32.

$$\frac{3}{4} + \left(-\frac{1}{3}\right)^2\left(\frac{5}{4}\right) = \frac{3}{4} + \left(\frac{1}{9}\right)\left(\frac{5}{4}\right)$$

$$= \frac{3}{4} + \frac{5}{36} = \frac{3\bullet 9}{4\bullet 9} + \frac{5}{36} = \frac{27}{36} + \frac{5}{36}$$

$$= \frac{32}{36} = \frac{8\bullet\cancel{4}}{9\bullet\cancel{4}} = \frac{8}{9}$$

33.
$$\frac{7-\dfrac{2}{3}}{4\dfrac{5}{6}} = \frac{7-\dfrac{2}{3}}{\dfrac{29}{6}} = \frac{6\left(7-\dfrac{2}{3}\right)}{6\left(\dfrac{29}{6}\right)} = \frac{42-4}{29}$$

$$= \frac{38}{29} = 1\frac{9}{29}$$

34. GRAVITY

w = weight of rock on Earth

$$\frac{1}{6}w = 3$$

$$6\left(\frac{1}{6}w\right) = 6(3)$$

$$w = 18$$

The weight of the rock on Earth is 18 oz.

35a. GLOBAL WARMING

1998 was the largest rise and it was about $0.6°$ F.

35b. 1986 was the largest decline and it was about $-0.4°$ F.

36.

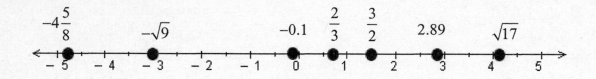

37. 3.1416

38. $154.34 > 154.33999$

39.
$$\begin{array}{r} 3.4 \\ 106.78 \\ 35 \\ +\ \ 0.008 \\ \hline 145.188 \end{array}$$

40. $-5.5(-3.1) = 17.05$

41. $(89.9708)(1,000) = 89,970.8$

42.

$$\frac{0.0742}{1.4} = \frac{0.742}{14} = 14\overline{)0.742} = 0.053$$

$$\begin{array}{r} 0.053 \\ \underline{0} \\ 74 \\ \underline{70} \\ 42 \\ \underline{42} \\ 0 \end{array}$$

43.

$$-8.8 + (-7.3 - 9.5) = -8.8 + (-16.8)$$
$$= -25.6$$

44.

$$\frac{7}{8}(9.7 + 15.8) = 0.875(25.5)$$
$$= 22.3125$$

45.

$$\frac{2}{15} = 0.1\overline{3}$$

46.

$$\frac{(-1.3)^2 + 6.7}{-0.9} = \frac{1.69 + 6.7}{-0.9} = \frac{8.39}{-0.9}$$
$$= -9.3\overline{2}$$
$$\approx -9.32$$

47. DECORATIONS

x = number of balloons she can buy
$$0.05x + 15.15 = 20$$
$$0.05x + 15.15 - 15.15 = 20 - 15.15$$
$$0.05x = 4.85$$
$$\frac{0.05x}{0.05} = \frac{4.85}{0.05}$$
$$x = 97$$

She can buy 97 balloons.

48.

$$1.7y + 1.24 = -1.4y - 0.62$$
$$1.7y + 1.24 + 1.4y = -1.4y - 0.62 + 1.4y$$
$$3.1y + 1.24 = -0.62$$
$$3.1y + 1.24 - 1.24 = -0.62 - 1.24$$
$$3.1y = -1.86$$
$$\frac{3.1y}{3.1} = \frac{-1.86}{3.1}$$
$$y = -0.6$$

49.   INTERNET

t = total distance

$t = 0.26 + 0.19 + 0.15 + 230.11 + 0.21 + 0.81 + 1.59 + 15.46 + 0.16$

$t = 248.94$

The total distance of the trip is 248.94 miles.

50. $\sqrt{64} = 8$ because $8^2 = 64$.

51    $\begin{aligned} 2\sqrt{121} - 3\sqrt{64} &= 2(11) - 3(8) \\ &= 22 - 24 \\ &= -2 \end{aligned}$

52. $\sqrt{\dfrac{49}{81}} = \dfrac{7}{9}$

53.   TABLE TENNIS

$\text{mean} = \dfrac{0.85 + 0.85 + 0.87 + 0.86 + 0.88 + 0.84 + 0.88 + 0.85}{8}$

$\text{mean} = \dfrac{6.88}{8} = 0.86 \text{ oz}$

$\text{median} = \dfrac{0.85 + 0.86}{2} = \dfrac{1.71}{2} = 0.855 \text{ oz}$

$\text{mode} = 0.85 \text{ oz}$

$\text{range} = 0.88 - 0.84 = 0.04 \text{ oz}$

54.

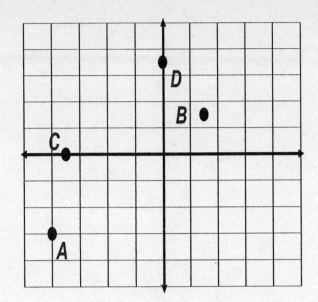

55. $(0, 0)$

56. $3(-2) - 1 = -8$

$$-6 - 7 = -8$$

$$-7 \neq -8$$

no

57.

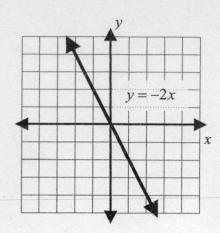

58.

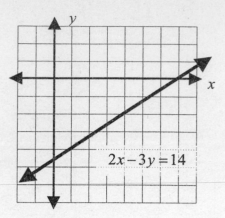

59.

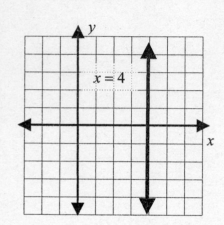

60.

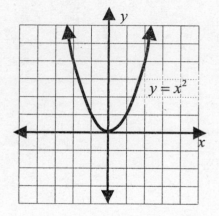

61.
$$-3(-2)^2 - 2(-2) = -3(4) - 2(-2)$$
$$= -12 - (-4)$$
$$= -12 + 4$$
$$= -8$$

62a.   −3 is the base,
2 is the exponent,
$$(-3)^2 = (-3)(-3) = 9$$

62b.   3 is the base,
2 is the exponent,
$$-3^2 = -(3)(3) = -9$$

63. $s^4 \cdot s^5 = s^{4+5} = s^9$

64. $\left(a^5\right)^7 = a^{5 \cdot 7} = a^{35}$

65. $-3h^9(-5h) = (-3 \bullet [-5])(h^9 \bullet h)$

             $= 15h^{9+1}$

             $= 15h^{10}$

66. $(2b^3c^6)^3 = (2)^3(b^3)^3(c^6)^3$

            $= 8b^{3\bullet3}c^{6\bullet3}$

            $= 8b^9c^{18}$

67. $(y^5)^2(y^4)^3 = (y^{5\bullet2})(y^{4\bullet3})$

                $= y^{10} \bullet y^{12}$

                $= y^{10+12}$

                $= y^{22}$

68. $x^m \bullet x^n = x^{m+n}$

69. trinomial with a degree of 2

70. $(5x^2 - 2x + 4) - (3x^2 - 5)$

    $= (5x^2 - 2x + 4) + (-3x^2 + 5)$

    $= (5x^2 + [-3x^2]) - 2x + (4 + 5)$

    $= 2x^2 - 2x + 9$

71. $-3p(2p^2 + 3p - 4)$

    $= -3p(2p^2) + (-3p)(3p) - (-3p)(4)$

    $= -6p^3 + (-9p^2) - (-12p)$

    $= -6p^3 - 9p^2 + 12p$

72. $(3x + 5)(2x - 1)$

    $= (3x + 5)2x - (3x + 5)1$

    $= 2x(3x + 5) - 1(3x + 5)$

    $= 6x^2 + 10x - 3x - 5$

    $= 6x^2 + 7x - 5$

73. $(2y - 7)^2 = (2y - 7)(2y - 7)$

    $= (2y - 7)2y - (2y - 7)7$

    $= 2y(2y - 7) - 7(2y - 7)$

    $= 4y^2 - 14y - 14y + 49$

    $= 4y^2 - 28y + 49$

74. 93% is shaded, 7% is not shaded

75. $x = 0.15 \bullet 450$

    $x = 67.5$

76. $24.6 = 0.205 \bullet x$

    $\dfrac{24.6}{0.205} = \dfrac{0.205x}{0.205}$

      $120 = x$ or $x = 120$

77.

Percent	Decimal	Fraction
57%	0.57	$\dfrac{57}{100}$
0.1%	0.000	$\dfrac{1}{1,000}$
$33\dfrac{1}{3}\%$	$0.\overline{3}$	$\dfrac{1}{3}$

78.   STUDENT GOVERNMENT

Note: For problem 76 percentages indicate the number of tick marks used by each slice of pie. In other words, 45% means that the 45% slice of pie is 45 tick marks wide.

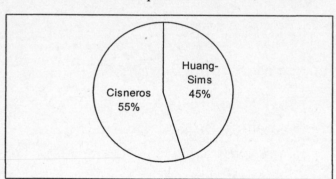

$$\text{Cisneros} = \frac{308}{560} = 0.55 = 55\%$$

$$\text{Huang-Sims} = \frac{252}{560} = 0.45 = 45\%$$

79.   SHOPPING

Since 27% was taken off, that leaves 73% of the price that is the sale price.

73% of what number is 54.75?

$$0.73x = 54.75$$

$$\frac{0.73x}{0.73} = \frac{54.75}{0.73}$$

$$x = 75$$

The regular price is $75.

80.   SALES TAXES

What is $6\dfrac{1}{4}\%$ of 18,550?

$$x = 0.625 \bullet 18,550$$

$$x = 1,1.59.375$$

$$x = \$1,159.38$$

The tax would be $1,159.38.

81. **COLLECTIBLES**

The amount of increase was $625.

What percent of 125 is 625?

$$x \bullet 125 = 625$$
$$125x = 625$$
$$\frac{125x}{125} = \frac{625}{125}$$
$$x = 5$$
$$x = 500\%$$

82. **PAYING OFF A LOAN**

$$2 \text{ months} = \frac{2}{12} = \frac{1 \bullet \cancel{2}}{6 \bullet \cancel{2}} = \frac{1}{6} \text{ year}$$

$$I = 1,500 \bullet 0.09 \bullet \frac{1}{6}$$

$$I = 22.5$$

total repayment $= 1,500 + 22.50$

total repayment $= \$1,522.50$

He would repay $1,522.20 total.

83. **SAVING FOR RETIREMENT**

$$A = 5,000\left(1 + \frac{0.08}{12}\right)^{12 \bullet 50}$$

$$A = 5,000(1 + 0.00666666667)^{600}$$

$$A = 5,000(1.00666666667)^{600}$$

$$A = 5,000(53.8782902)$$

$$A \approx 269,390.916$$

$$A = 269,390.92$$

The account will be worth

$ 269,390.92.

84.

$$\frac{3 \cancel{\text{ cm}}}{7 \cancel{\text{ cm}}} = \frac{3}{7}$$

85.

$$\frac{13 \text{ weeks}}{1 \text{ year}} = \frac{13 \cancel{\text{ weeks}}}{52 \cancel{\text{ weeks}}} = \frac{1 \bullet \cancel{13}}{4 \bullet \cancel{13}} = \frac{1}{4}$$

86. **COMPARISON SHOPPING**

$$\frac{\$24}{400 \text{ in}^2} \approx \$0.06 \text{ per in}^2$$

$$\frac{\$42}{600 \text{ in}^2} \approx \$0.07 \text{ per in}^2$$

The 400 in^2 board is a better buy.

87.

$$\frac{5-x}{14} = \frac{13}{28}$$

$$28(5-x) = 182$$

$$140 - 28x = 182$$

$$140 - 28x - 140 = 182 - 140$$

$$-28x = 42$$

$$\frac{-28x}{-28} = \frac{42}{-28}$$

$$x = -\frac{42}{28}$$

$$x = -\frac{3 \cdot \cancel{14}}{2 \cdot \cancel{14}} = -\frac{3}{2}$$

88.   INSURANCE CLAIMS

$x =$ number of policies

$$\frac{3 \text{ complaints}}{1{,}000 \text{ policies}} = \frac{375 \text{ complaints}}{x \text{ policies}}$$

$$3x = 375{,}000$$

$$\frac{3x}{3} = \frac{375{,}000}{3}$$

$$x = 125{,}000$$

The company had 125,000 policies.

89.   SCALE DRAWINGS

$x =$ length of the house

$$\frac{\frac{1}{4} \text{ in}}{3 \text{ ft}} = \frac{\frac{25}{4} \text{ in}}{x \text{ ft}}$$

$$\frac{0.25 \text{ in}}{3 \text{ ft}} = \frac{6.25 \text{ in}}{x \text{ ft}}$$

$$0.25x = 18.75$$

$$\frac{0.25x}{0.25} = \frac{18.75}{0.25}$$

$$x = 75$$

The house is 75 ft long.

90.

$$168 \text{ in.} = 168 \cancel{\text{ in.}} \cdot \frac{1 \text{ ft}}{12 \cancel{\text{ in.}}}$$

$$= \frac{168}{12} \text{ ft}$$

$$= 14 \text{ ft}$$

91.

$$15 \text{ yd} = 15 \cancel{\text{ yd}} \cdot \frac{36 \text{ in.}}{1 \cancel{\text{ yd}}}$$

$$= 15 \cdot 36 \text{ in.}$$

$$= 540 \text{ in.}$$

92.

$$212 \text{ oz} = 212 \cancel{\text{ oz}} \cdot \frac{1 \text{ lb}}{16 \cancel{\text{ oz}}}$$

$$= \frac{212}{16} \text{ lb}$$

$$= 13.25 \text{ lb}$$

93.
$$30 \text{ gal} = 30 \text{ gal} \bullet \frac{4 \text{ qt}}{1 \text{ gal}}$$
$$= 30 \bullet 4 \text{ qt}$$
$$= 120 \text{ qt}$$

94.
$$25 \text{ cups} = 25 \text{ cups} \bullet \frac{8 \text{ fl oz}}{1 \text{ cup}}$$
$$= 25 \bullet 8 \text{ fl oz}$$
$$= 200 \text{ fl oz}$$

95.
$$738 \text{ min} = 738 \text{ min} \bullet \frac{1 \text{ hr}}{60 \text{ min}}$$
$$= \frac{738}{60} \text{ hr}$$
$$= 12.3 \text{ hr}$$

96. $654 \text{ mg} = 65.4 \text{ cg}$

97. $500 \text{ mL} = 0.5 \text{ L}$

98. $5,980 \text{ dm} = 58.9 \text{ dam}$

99.
$$F = \frac{9}{5}C + 32, \text{ when } C = 75$$
$$F = \frac{9}{5}(75) + 32$$
$$F = 135 + 32$$
$$F = 167° \text{ F}$$

100.   THE AMAZON
$240,000 \text{ m} = 240 \text{ km}$

101.   TENNIS

between 5,700 cg and 5,800 cg

102a.   OCEAN LINER
$$13 \text{ ft} = 13(0.3048 \text{ m})$$
$$= 3.9624 \text{ m}$$
$$\approx 4 \text{ m}$$

It got about 4 m to the gallon.

102b.
$$3,000,000 \text{ gal} = 3,000,000(3.785 \text{ L})$$
$$= 11,355,000 \text{ L}$$

It held 11,355,000 liters of fuel.

103.   COOKING
$$10 \text{ lb} = 10(0.454 \text{ kg})$$
$$= 4.54 \text{ kg}$$

It weighs about 4,54 kg.

104.   A right angle is 90°

105. An acute angle is more than $0°$ but less than $90°$.

106. x = the supplement of $105°$

$x = 180 - 105$

$x = 75°$

107. x = the compliment of $75°$

$x = 90 - 75$

$x = 15°$

108. $m(\angle 1) = 50°$

109. $m(\angle 2) = 130°$

110. $m(\angle 3) = 50°$

111. $m(\angle 4) = 50°$

112. $m(\angle 1) = 75°$

113. $m(\angle C) = 30°$

114. $m(\angle 2) = 105°$

115. $m(\angle 3) = 105°$

116. JAVELIN TROW

$x = 180 - 90 - 44$

$x = 46$

$y = 180 - 46$

$y = 134$

117. $S = (5 - 2)180$

$S = (3)180 = 540$

$S = 540°$

118. $c^2 = (12)^2 + (5)^2$

$c^2 = 144 + 25$

$c^2 = 169$

$c = \sqrt{169}$

$c = 13 \text{ m}$

119. $P = 2l + 2w$

$P = 2(9) + 2(12)$

$P = 18 + 24$

$P = 42$ m

$A = lw$

$A = (9)(12)$

$A = 108$ m^2

120. $A = \dfrac{1}{2}bh$

$A = \dfrac{1}{2}(14)(18)$

$A = 126$ ft^2

121. $A = \dfrac{1}{2}h(b_1 + b_2)$

$A = \dfrac{1}{2}(7)(12+14)$

$A = \dfrac{1}{2}(7)(26)$

$A = 91$ in.^{2}

122. diameter 14 cm = radius 7 cm

$C = \pi d$

$C = \pi(14)$

$C \approx 43.982$ cm

$C \approx 43.98$ cm

$A = \pi r^2$

$A = \pi(7)^2$

$A = 49\pi$

$A \approx 153.938$ cm^2

$A \approx 153.94$ cm^2

123. The figure is a rectangle with a length of 20.2 yd and a width of 19.2 yd that has two semicircles removed on each end with a diameter of 19.2yd. The two semicircles give one complete circle with a radius of 9.6 yd.

A = total area of figure,

A_r = area of the rectangle,

A_c = area of the circle

$A = A_r - A_c$

$A = lw - \pi r^2$

$A = (20.2)(19.2) - \pi(9.6)^2$

$A = (20.2)(19.2) - 92.16\pi$

$A \approx 387.84 - 289.529$

$A \approx 98.311 \approx 98.31$ yd^2

124. $V = lwh$

$V = (5)(6)(7)$

$V = 210$ m^3

125.   diameter 10 in. = radius 5 in.

$V = \dfrac{4}{3}\pi r^3$

$V = \dfrac{4}{3}\pi (5)^3$

$V = \dfrac{4}{3}\pi (125)$

$V \approx 523.598$ in.^{3}

$V \approx 523.60$ in.^{3}

126.   diameter 8 m = radius 4 m

$V = \dfrac{1}{3}\pi r^2 h$

$V = \dfrac{1}{3}\pi (4)^2 (9)$

$V = \dfrac{1}{3}\pi (16)(9)$

$V = 48\pi$

$V \approx 150.796$ m^3

$V \approx 150.80$ m^3

127.   diameter 6 in. = diameter 0.5 ft

diameter 0.5 ft = radius 0.25 ft

$V = \pi r^2 h$

$V = \pi (0.25)^2 (20)$

$V = \pi (.0625)(20)$

$V = 1.25\pi$

$V \approx 3.926$ ft^3

$V \approx 3.93$ ft^3

128. $SA = 2lw + 2lh + 2hw$

$SA = 2(15 \bullet 24) + 2(15 \bullet 18) + 2(24 \bullet 18)$

$SA = 2(360) + 2(270) + 2(432)$

$SA = 720 + 540 + 864$

$SA = 2{,}124$ in.^{2}

Appendix I Inductive and Deductive Reasoning

Vocabulary

1. **Inductive** reasoning draws general conclusions from specific observations.

2. **Deductive** reasoning moves from the general case to the specific.

Concepts

3. circular

4. decreasing

5. alternating

6. increasing

7. alternating

8. circular

9. ROOM SCHEDULING
    10 am

10a. QUESTIONAIRES
     11 students

10b. 18 students

10c. 21 students

Practice

11. 1, 5, 9, 13, 17, . . .

    increased by 4 each time

12. 15, 12, 9, 6, 3, . . .

    decreased by 3 each time

13. $-3, -5, -8, -12, -17, \ldots$

    decreased by 2, then 3, then 4 . . .

14. 5, 9, 14, 20, 27, . . .

    increased by 4, then 5, then 6 . . .

15. $-7, 9, -6, 8, -5, 7, -4, 6, \ldots$

    alternating increase by one then
    decrease by one

16. 2, 5, 3, 6, 4, 7, 5, 8, . . .

    alternating numbers both increasing by
    one

17.   9, 5, 7, 3, 5, 1, 3, . . .

alternating numbers both reducing by 2

18.   1.3, 1.6, 1.4, 1.7, 1.5, 1.8, 1.6, . . .

alternating numbers both increasing by 0.1

19. $-2, -3, -5, -6, -8, -9, -11, \ldots$

alternating numbers being decreased by one then by 2.

20.   8, 11, 9, 12, 10, 13, 11, . . .

alternating numbers both increasing by one.

21.   6, 8, 9, 7, 9, 10, 8, 10, 11, 9, . . .

groups of 3 numbers sets all being increased by one.

22.   10, 8, 7, 11, 9, 8, 12, 10, 9, 13, . . .

groups of 3 numbers sets all being increased by one.

23.

24.

25.

26.

27.   A, c, E, g, I, . . .

2 letters away each time and alternating caps and lower case

28.   R, SS, TTT, UUUU, . . .

one letter away each time but increasing pattern by one with each term

29.   d, h, g, k, j, n, m, . . .

alternating letters are becoming 3 letters "bigger" each time

30.   B, N, C, N, D, N, . . .

alternating letters are "increasing" by one with N alternating between them

31. Maria must be the teacher since we know that the baker is Luis and Paula and John are married leaving Maria as unmarried.

32. If the tiger were put in cage 4, then the monkey would have to be in cage 1, and the lion would have to be in cage 2 since it cannot be next to the tiger, leaving the zebra to be in cage 3.

33. If the vehicles are parked in 4 spaces side by side that are numbered 1 to 4 from left to right, the Buick is parked in the space 1 (the left space). For the Ford to be between the Mercedes and the Dodge, the Mercedes must be in space 4 (the right space) since it cannot be next to the Buick.

34. The order of finish would be diver A won, diver B came in second, diver C third, and diver D came in fourth.

35. From top to bottom the flag colors would have to be green (on the top), blue, yellow, then red (on the bottom).

36. From the given information is can shown that Andres is the painter and barber, Barry is the musician and bootlegger, and Carl is the gardener and chauffeur.

37.   JURY DUTY

There are 18,935 respondents who have not served on either a criminal or civil jury.

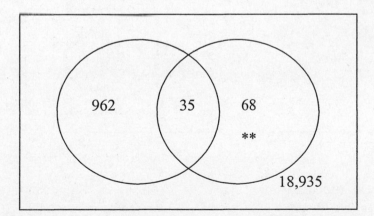

38.   POLLS

46 people voted "neither" on the poll.

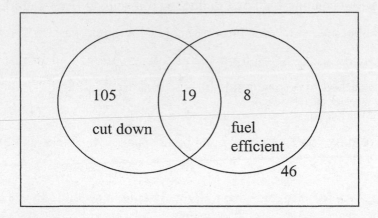

39.   THE SOLAR SYSTEM

There are no planets that are nither rocky nor have moons.

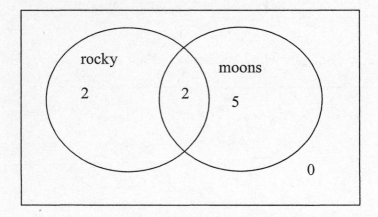

40.   Answers will vary. 100 cars were tested for which features they have. 20 have fuel injection, 30 have turbo-chargers, and 10 have both fuel injection and turbo-chargers. How many cars did not have fuel injection or turbo-chargers?

Writing

41. Deductive reasoning moves from the general case to the specific, whenever we apply a general principle to a particular instance, deductive reasoning is being used.

42. Answers will vary.

43. When a general conclusion is drawn from specific observations, inductive reasoning is being used.

44. Answers will vary.

Appendix 1 Review Inductive and Deductive Reasoning

1.   12, 8, 11, 7, 10, 6, . . .

      alternately decreasing by one each

      time

2.   5, 9, 17, 33, 65, . . .

      each term is increasing by an increasing

      power of 2, that is

      $$5 + 2^2 = 9, \ 9 + 2^3 = 17, \ 17 + 2^4 = 33, \ \text{etc...}$$

3.

4.

5.   c, b, a, f, e, d, i, h

      each group of 3 letters were decreasing by one letter each time

6.   If you picture the stalls from left to right, the cow is in the left stall, the horse next to it,

      the pig next to the horse, and the sheep is in the last stall to the right.

7.   Jim and Sandra eat lunch with the math teacher, so neither can be the math teacher.

      Since Jim is married to the math teacher, the only other female teacher is Mary, so she

      must be the math teacher.

8.   BUYING A CAR

      43 cars have automatic transmissions and no CD players.

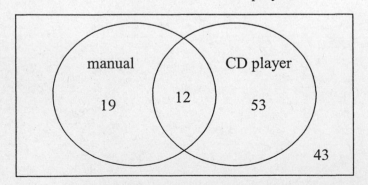